OUR WANDERING CONTINENTS

OUR
WANDERING CONTINENTS

AN HYPOTHESIS OF CONTINENTAL DRIFTING

BY

ALEX. L. Du TOIT, D.Sc., F.G.S.

WITH 48 DIAGRAMS

"Africa forms the Key"

HAFNER PUBLISHING COMPANY
NEW YORK

L

FIRST PUBLISHED   1937
REPRINTED   1957

PRINTED IN GREAT BRITAIN BY
BRADFORD AND DICKENS, LONDON, W.C.1

TO

THE MEMORY

OF

ALFRED WEGENER

FOR HIS

DISTINGUISHED SERVICES

IN CONNECTION WITH THE

GEOLOGICAL INTERPRETATION

OF

OUR EARTH

PREFACE

RECENTLY the writer was invited to prepare an account of the late palæozoic Fold Systems in the Southern Hemisphere, but had to decline for the vital reason that no intelligible or convincing interpretation of those structures seemed possible that failed to recognise Continental Sliding. It appeared, too, that current opinion rather worked on the lines " that things are *there*, because they *are* there," and, while always ready to describe, was seldom prepared to predict. Meanwhile the quantity of undigested facts and deductions was piling up, though, as Kober has protested, " Our time is pressing for Synthesis."

So many critics have merely condemned the ideas of Taylor and Wegener without seriously attempting to sift the vast body of relevant and highly significant data and discover whether current interpretation can actually be so well and truly founded as supposed.

This attempt by the writer to explain the elaborate architecture of the Globe has necessitated a critical revision of Geological Principles as well as a review of the geology of the whole of the Earth. To compress the fascinating surface history of our planet within reasonable limits has been difficult, a good deal of simplification has been inevitable, but a certain amount of repetition has been unavoidable, since the major issues have had to be treated from various angles. Taylor, Argand and Staub have each over-emphasised the tectonic aspect, while concentrating upon the Northern Hemisphere, wherefore the stratigraphical viewpoint will be given its rightful place and fuller attention paid to the Southern Hemisphere, for which the evidence is, as it happens, clearer and less equivocal.

So numerous then become the congruities, so remarkable the so-called " coincidences " and so close the agreement between prediction and observation, that, whether the explanation here offered for the *causes* of such postulated Drift be valid or not, the author feels that a great and fundamental truth is embodied in this revolutionary Hypothesis.

With so vast a field it has not been possible to cite all original sources of information or always to give due credit to those who by fact or idea have contributed to this many-sided problem.

The infinitely complex puzzle of Earth Evolution cannot obviously be solved all at once, and the admittedly imperfect scheme outlined here is put forward rather as a basis for further and more detailed researches. In doing so, one may well revive Waterschoot van der Gracht's plea for a little more tolerance of spirit from critics, more particularly in regard to points of a minor or dubious character.

In conclusion the author wishes sincerely to thank all those who have kindly contributed information or who have permitted the reproduction of original diagrams, among whom must be mentioned Messrs F. B. Taylor, Fort Wayne (Fig. 1) ; H. B. Baker, Detroit (Figs. 2 and 3) ; A. Holmes, Durham (Figs. 5, 19, 20 and 47) ; the editors of the *Geological Magazine*, Cambridge (Fig. 46) ; Methuen & Co., London, and Fr. Vieweg und Sohn, Braunschweig (Fig. 4) ; and Gebrüder Borntraeger, Berlin (Figs. 39, 41 and 42), on behalf of Dr. J. S. Lee, Professor W. Köppen and the late Professor A. Wegener.

ALEX. L. DU TOIT.

P.O. Box 4565,
   JOHANNESBURG,
      June 1st, 1937.

CONTENTS

ILLUSTRATIONS

CHAPTER I

CURRENT THEORIES VERSUS "CONTINENTAL DRIFT"

Introduction. Orthodoxy as opposed to Drift. Merits of the Hypothesis. Criticisms of the Hypothesis. Validity of Current Postulates. Inherent Weaknesses of Current Theories.

INTRODUCTION

LOOKING back dispassionately into the history of Geology it is interesting to observe how deeply conservatism appears to have become entrenched. Particular theories have come to be so widely accepted, that any doubts regarding their validity are apt to be overlooked. Indeed there is some danger lest the science become stereotyped through too close an adherence to accepted beliefs, and in no instance is this more striking than in the general attitude displayed towards the *Hypothesis of Continental Drift or Sliding*—otherwise known as the *Displacement Hypothesis*.

Under those terms must be included all hypotheses postulating more than a limited amount of horizontal movement of the land-masses and not merely Taylor's or Wegener's conceptions, which are but two out of a number that have been expressed or that could indeed be formulated. There is furthermore nothing contained in those writings to suggest that such kind of drift would have been restricted to any particular period and would not have operated from an early stage in the earth's history, only such remote episodes are too obscure to be elucidated just yet.

It is regrettable that too many persons are acquainted merely with the writings of Alfred Wegener, or perhaps with summaries thereof, and may have become antagonised by the various incongruities and inaccuracies which the work of that author unquestionably displays. Only a minority have carefully followed more recent pronouncements, and more especially the notable contributions in this field by R. Staub.

The resulting pointed and spirited discussions have nevertheless been all to the good, though at the same time a certain amount of unfair criticism and not a little hostility have been shown towards these new ideas of earth structure. It is seem-

ingly not yet appreciated that, as in the case of every new hypothesis, the complete and perfect picture cannot straightway be
presented.   Unfortunately, there is as yet no proper consensus
of thought regarding Continental Drift, for the opinions so far
expressed have been partial and also diverse in their outlook,
though a closer agreement will doubtless be reached in time.

Waterschoot van der Gracht has rightly emphasised the need
for concentrating upon the broader issues, in which the co-
operation of specialists would be of great assistance, leaving the
details to be filled in later.

He has also put in a plea for more tolerance, though the
writer's viewpoint is that, the most effective defence being in
attack, a good deal is, on the contrary, to be gained from an ex-
posure of some of the weaknesses and fallacies of current or
orthodox Geology, by which is meant that collective body of
opinion which does not admit of extensive horizontal displacements of the continents.   Such deficiencies are by no means
insignificant, as will briefly be set forth below.   Only a few of
the critics have as yet been prepared to concede with readiness
the point that current views might require some appreciable
revision in the light of the new thesis, even though the latter be
ultimately rejected.

ORTHODOXY AS OPPOSED TO DRIFT

When dealing with diastrophism and the nature of the earth's
interior, one is apt to overlook the fact that what has to be termed
" current theory " is no general doctrine, but embraces the most
diverse of views.   As individually interpreted, geosynclines and
rift-valleys are ascribed alternatively to tension or to compression; fold-ranges to shrinkage of the Earth, to isostatic adjustment or to plutonic intrusion; some regard the crust as weak,
others as having surprising strength; some picture the sub-
crust as fluid, others as plastic or solid; some view the land-
masses as relatively fixed, others admit appreciable intra- and
inter-continental movement; some postulate wide land-bridges,
others narrow ones, and so on.   Indeed on every vital problem
in geophysics there are, and indeed must be, fundamental differences of viewpoint.

In spite of these basic differences such opinion is almost
unanimous in affirming that the mechanism demanded for
Continental Drift is of too colossal a character, that the lateral
movements of continental blocks, which are demonstrated by
the fold-belts of the earth, are on a limited scale only, that those
blocks have in the main retained throughout geological time
their positions relative to one another as well as to the Earth's
axis, and that the close relationships revealed by their terrestial

life can quite readily be explained by land-bridges built up and again destroyed in various times and in various places. All this is strongly doubted by the protagonists of displacement.

While in the future some closer approximation in viewpoint will become possible, it must frankly be recognised that the principles advocated by the supporters of Continental Drift form generally the antithesis of those currently held. *The differences between the two doctrines are indeed fundamental and the acceptance of the one must largely exclude the other.* Indeed, under the new hypothesis certain geological concepts come to acquire a new significance amounting in a few cases to a complete inversion of principles, and the inquirer will find it necessary to re-orient his ideas. For the first time he will get glimpses—albeit imperfect as yet—of a pulsating restless earth, all parts of which are in greater or less degree of movement in respect to the axis of rotation, having been so, moreover, throughout geological time. He will have to leave behind him—perhaps reluctantly—the dumbfounding spectacle of the present continental masses, firmly anchored to a plastic foundation yet remaining fixed in space ; set thousands of kilometres apart, it may be, yet behaving in almost identical fashion from epoch to epoch and stage to stage like soldiers at drill ; widely stretched in some quarters at various times and astoundingly compressed in others, yet retaining their general shapes, positions and orientations ; remote from one another throughout history, yet showing in their fossil remains common or allied forms of terrestial life ; possessed during certain epochs of climates that may have ranged from glacial to torrid or pluvial to arid, though contrary to meteorological principles when their existing geographical positions are considered—to mention but a few such paradoxes !

MERITS OF THE HYPOTHESIS

Instead he will be introduced to the concept of an earth in which the periodic, though variable, softening of the sub-crust through radioactive heating enables the skin to creep differentially over the core with consequent wrinkling ; one continental block in its travel develops, if unopposed, fold-chains along its leading edge, or else impinges upon a second one with mutual folding along the fringe of impact, to be replaced or paralleled by a new trough as the assailing mass is arrested or the assaulted one is propelled forward ; masses either become welded together, to share thereafter in the same general geological and biological environment, or are ruptured with diverging histories and life ; differential creeping changes their position relative to the poles and brings about sweeping changes of climate ; land-bridges—

that thorn in the side of orthodoxy—do not need to be invented so freely ; while the past and present distribution of land, sea and terrestial and marine life finds comprehensive and intelligible explanation.

An outstanding consequence of the Hypothesis is the orderly and interrelated nature of all associated phenomena. The drifting away of a part of a continent automatically severs its connection with the remainder and with its life, removes it generally to a different latitude and climate, causes it ultimately to come to rest through the crumpling up of its leading margin, induces tilting of the rearward part with possibly far-reaching effects in the way of warping and rifting, institutes a new system of drainages and gives rise to eruption and intrusion of igneous matter which tends towards an intermediate composition at the leading edge and an alkalic one at the rear. Areas now widely separated stand revealed as fragments of greater masses and their observed similarities find their explanation therein and have not to be gratuitously interpreted as due to an extraordinary set of coincidences—geographical, stratigraphical, tectonic, biological and climatic.

Current ideas, on the contrary, regard the continental masses as having existed over great periods as more or less *independent units* save when temporarily connected, and consequently any close relation between their respective histories would not strictly be expectable, though such is flatly contradicted by the evidence. The explanations currently given for the grander features of the globe bristle with difficulties, primarily because of this tendency to regard each mass as a distinct entity. Instead of such units behaving more or less arbitrarily they become under our hypothesis parts of a living whole, each influenced by and reacting upon its neighbours in a definite and orderly fashion throughout geological time. Indeed it is modestly suggested that the *Displacement Hypothesis represents the "Holistic" outlook in Geology.* Furthermore, unlike current views of Earth structure, this illuminating *hypothesis can be tested on the basis of prediction.* Several remarkable deductions in the case of South America and South Africa have been thus verified by field work.

The foregoing advantages being so overwhelmingly in favour of the "New Geology", the critic may, and with justification, ask why the Hypothesis has apparently found so few whole-hearted supporters. The answer is, first, that it cuts at the basis of customary geological interpretations and is hence not particularly welcome, and, secondly, that no forces have so far been, nor according to its opponents can be, invoked competent to move the continents about as supposed.

The first objection is largely a psychological one, and has

to be overcome by the marshalling of the relevant data, their skilful analysis and presentation and the closer study of the evidence, wherever such is in apparent conflict with preconceived ideas. Incidentally this would involve the rewriting of our numerous text-books, not only of Geology, but Palæogeography, Palæoclimatology and Geophysics.

The second is admittedly a serious one, but, as Rastall has stressed, is no more weighty than the lack of an adequate recognised cause for past ice-ages, a puzzle that has exercised the ablest minds for about a century. This difficulty, which the Author optimistically fancies is by no means insoluble, ought not, however, to be allowed to prejudice the issue and arrest the progress of the new ideas in the face of the host of geological facts, certain of them apparently interpretable in no other way, that collectively point to some scheme of wholesale continental drifting.

While whole-hearted supporters are still few, quite a number of geologists, biologists and physicists have candidly admitted the ability of the Hypothesis to explain, or to assist in explaining, certain puzzling problems of tectonics, ice-ages, animal and plant migration and so on, and would be prepared to subscribe to it, if more convincing evidence for its probability could be adduced. To those who properly take leave to doubt those revolutionary ideas, a careful examination of the vast number of congruences put on record in these pages is earnestly requested. Then the so-called "coincidences" will be found to mount up so enormously, as to raise the Displacement Hypothesis outside of the region of mere speculation. In the face of geological opinion of today the writer has been forced by the sheer weight of such evidence into accepting some form of Continental Drift to explain the "face of the Earth"; indeed a world without some form of crustal drifting would appear to him as unreal as one lacking in biological evolution. This must form the sole excuse for the rather positive phrasing made at various points in the text instead of the repeated use of such qualifying expressions as "if the hypothesis be correct," which is mere pedantry.

Regarding the mechanism demanded it is not intended to say much, though the most promising lines of attack will be pointed out. To the geophysicist pre-eminently must be left the search for and evaluation of an adequate force for such crustal movement as is indicated by the full weight of geological data and inference.

CRITICISMS OF THE HYPOTHESIS

Opposition has been directed perhaps more against Wegener's opinions than the general principle of drift. Every attempt will be made in this work to meet such objections when well-founded, but unfortunately many of the criticisms, coming even from persons of eminence, are at times only partly justified, trivial or even fatuous.

Not a few critics seem to be under the impression that no real alternatives to their own particular interpretations are possible, and it is difficult to combat such opinions when the fundamental viewpoints are so utterly different. Others have placed undue weight on negative evidence—a dangerous policy in view of the serious limitations to geological and intellectual knowledge. Thus, the former union of South America and Africa has seriously been questioned on the ground that no Permo-Triassic reptiles of Karroo type have been found in Brazil. Actually they have! In pointing out the difficulties introduced by Isostasy, Schuchert voices the pathetic appeal " Geologists must find a way to sink such land-bridges." Why? Save to extricate orthodoxy from an impasse! For the words " sink such land-bridges " we might with at least equal fairness substitute " move the continents."

Criticism of the *non posse* type has been common, indeed almost every conceivable argument has been put forward to prove that the continents just could not move. We are seriously told that a block could not drift because a kind of vacuum would then be developed in the rear, having at the least an equal and opposite effect! We are solemnly advised that the ocean floor could not be pushed up into folds because if the former were resistant the continent could not advance, and if non-resistant the floor could neither acquire nor retain folding! These same persons nevertheless insist that marginal ranges are due to the thrust of the ocean against the land, which makes one think of an urchin trying to push back a large policeman by the knees. Some indeed cannot visualise how fragments could possibly break off in the rear of a moving block and remain behind as islands!

Others too have not hesitated to back up with mathematical formulæ their rather positive conclusions, but such need not be too seriously considered. In doing so they are less circumspect than the meteorologists, who, not unmindful of unhappy pronouncements in the past regarding the reality of the tropopause, have wisely refrained from questioning the physical possibility of past glaciations. The particular equations used—which commonly have to be simplified for mathematical calculation—are based upon postulates regarding the properties of the crust, which those alive to the potentialities of other alternatives will not always be prepared to concede. They usually treat of the various shells of the Earth as approximately uniform in radius and in composition, which is not only a big assumption, but improbable. Furthermore they either fail to take into account or else dismiss with scant consideration the vital principle—radioactivity—through which the Earth receives its periodical rejuvenation. Finally, no concession is made for possibly unknown factors, as disclosed for example in the newly formulated paramorphic principle enunciated in Chapter XI.

Under such circumstances we have no option but to conclude that *geological evidence almost entirely must decide the probability of this Hypothesis.*

VALIDITY OF CURRENT POSTULATES

Geology has not wholly recovered from the cramping influence of the Uniformitarian doctrine of Lyell. Having passed through various diastrophic " revolutions," the present Earth is at one particular stage in this, the latest, cycle, yet seldom is it recognised that the conditions today within and below the crust may not be constant but experiencing related cyclic changes also. During various past epochs such things as internal temperature, geothermic gradient, depth of zones of discontinuity, degree of isostatic adjustment, state of magmatic activity, etc., might have been somewhat different to those of the present, just as is admittedly the case with climate as well as the rates of erosion and sedimentation.

We have not only to keep the above in mind when dealing with the past, but to view with some suspicion quite a number of the so-called principles that are currently implied or accepted :

(*a*) The usual picture of the earth as being built up of almost concentric shells (decreasing regularly in density upwards) receives scant support from the study of earthquake waves, which reveals considerable variation in depth of the chief surfaces of discontinuity beneath different regions. Future work should merely magnify such departure from the ideal, just as newer soundings but serve to accentuate the unevenness of the ocean floor. Considering the prodigious effects of certain diastrophic episodes, such irregularities within the crust are only to be expected, and mathematical analyses postulating a regular spheroid should be looked at askance.

(*b*) Admitting (*a*) it would follow that the amount of radioactive matter in the crust might vary widely not only between the land and water areas but between the several continents and oceans themselves. Indeed at any instant various large sections could well have reached different stages in their geothermal cycle. Joly's well-known hypothesis is therefore in need of some revision in this respect.

(*c*) Quite a number of geologists, for instance Schuchert and Willis, still subscribe to the archaic doctrine of the Permanence of Ocean Basins, and that in the face of the geological maps of the globe and nearly all palæogeographical restorations. Were it even partially true we should know relatively little about the marine strata and their invertebrate remains. From the general absence of abyssal types of deposit from among the stratified rocks many persons have concluded that the oceans were dominantly shallow, during the Palæozoic at least, and that only subsequently did deepening take place. If that be correct, then many of the so-called epi-continental seas could not have been very different in depth and other characters from the normal " oceans " of the time. Those numerous land-bridges postulated by the orthodox would speak eloquently against the permanence of deep basins like those of today.

Others again have adduced reasons for believing that the oceans have throughout time been augmented by " juvenile waters " from volcanic sources. If such be admitted, the problem of linking and unlinking during the past would be much simplified. The argument that the seas which trespassed repeatedly upon the lands were merely of an epi-continental character is of small moment in our inquiry, for, whether they were 1000 or only 10 m. deep, they would still have constituted barriers to terrestial migration.

It is generally accepted that the dominant features of the present oceans were determined by the world-wide diastrophism of the Tertiary. The picture of their evolution is, however, vastly different under the Hypothesis from that generally drawn.

(d) It will be conceded that, with a general distribution of land and sea during the past not vastly different from that of the present, the climatic girdles should have been rather similar to those of today, though their mutual boundaries might have fluctuated appreciably.

There is, on the contrary, a wealth of evidence for surprisingly great regional variations in climate that periodically culminated in glacial periods not only in high but in medium and low latitudes, for warm temperatures within the polar circles and for intense aridity over immense territories. Endeavouring to explain such radical changes under accepted meteorological principles one becomes involved straightway in a maze of difficulties. Over and over again great elevation has been invoked to account for such refrigerations regardless of the fact that the glacial deposits in question are in many places associated or even intercalated with marine horizons. The terrestial deposits of the Triassic throughout the world are so largely of a steppe-like or arid type that an extraordinary climatic distribution must be presumed for their formation during that particular epoch.

Until the main outlines and positions of the land-masses of the past can be settled with some degree of probability, it would seem hopeless to build up, as has been done by so many geologists, elaborate maps showing hypothetical oceanic circulations, anticyclonic centres and prevailing winds. The moment any considerable horizontal movement of the continents is admitted, most of the hitherto recognised standards must go by the board, for then the Earth's poles can no longer be regarded as having remained fixed relatively to the lands.

INHERENT WEAKNESSES OF CURRENT THEORIES

The following curious facts are very imperfectly or not at all accounted for :

(1) That each of the continental blocks shows youthful marginal foldings along but a section of its borders, and an enlargement in that direction despite the resulting compression, making the structure thereof characteristically *asymmetrical* ;

(2) That such folds run in the New World on the western side, in the Old World on the eastern side, of those masses and in between trend nearly equatorially ;

(3) The elevated character of those portions of the remaining coast-line known or suspected to be due to faulting ;

(4) The general absence between the opposed shores in the remnants of the ancient southern continent of Gondwana of marine strata ranging in age from Devonian to Triassic ;

(5) The close similarities displayed by the various palæozoic fold-belts on opposite sides of the Atlantic despite the great distances separating such " loose-ends," as Bucher terms them. Thus the North Atlantic shows four widely sundered tracts of intense (Taconian) overthrusting, which are obviously relics of but a single tectonic zone ;

(6) The astounding resemblances between the stratigraphies and past life of the widely separated land-masses ;

(7) The extraordinary distribution of the Carboniferous glacials in the Southern Hemisphere (including India) that are today situated in the anticyclonic belt (with highest snow-line), two of them having the ice-movement directed *away from the equator* ;

(8) The rift-valley systems of Africa and other lands and their unique physiography ;

(9) The enormous extent of terrestial deposits formed between the Carboniferous and the Triassic ; and

(10) The discovery of peculiar faunas in isolated and remote positions, as for example certain marine Permian mollusca in Brazil characteristic of New South Wales.

These are but a few of the numerous problems ordinarily only explainable in a rather nebulous fashion, which find their ready solution under our Hypothesis.

Impressive is the fact that, in every case into which palæo-geography enters, *current theory demands something on a far greater scale than is needed under our viewpoint* :

(1) For the Tertiary fold-girdle the total distance becomes one-fourth longer (p. 34) ;

(2) For the palæozoic fold-belts of the Northern and Southern Atlantic the distance is increased by one-half and is trebled respectively ;

(3) For the numerous land-connections of the Past land-bridges thousands of kilometres in length have repeatedly to be invoked and sunk. Failing such narrow links, the only alternative thereto is the colossal continents pictured by De Lapparent, Frech, Haug, etc.,—truly relics of the past !

(4) For the Tertiary basalts of Great Britain, Faroes, Iceland and Greenland the area embracing them is many times greater (p. 224) ; and

(5) For the Carboniferous glaciation the total area ice-capped must be far larger, since in four countries the glacial centre is indicated as having lain outside the existing lands (pp. 72-6).

A serious objection arises from the concept of the *general fixity of the continents,* or more properly of their more stable parts or " shields." The farther apart they lay, the less the chance of their evolutionary history having been able to follow identical lines. Now the outstanding feature of geological analysis is the wonderfully similar histories—stratigraphical, tectonic, climatic, biological and eruptive—of particular pairs of the land-masses at various periods, which is finely displayed between Greenland and Scandinavia, South America and Southern Africa, East Africa and India, and South Africa and Australia. That they should have reacted so similarly while at so great a distance apart one from the other is improbable, accepting orthodox views, whereas under the Displacement Hypothesis such is not only reasonable but *inevitable.* To use a homely analogy one could readily picture a number of bathers standing in shallow water holding hands and plunging up and down in unison, but scarcely if they were out of sight of one another.

To sum up, it will be seen that current theories rest upon foundations which are far from secure, that they fail to explain with conviction many of the larger features of the globe, and that they demand diastrophic changes on a far greater scale. It is freely admitted that our Hypothesis of former continental *rapprochement* is faced by not a few difficulties, though, it would seem, not any greater than those striking at the basis of orthodox theories. The trenchant criticisms of the Displacement Hypothesis are hence a long way from being justified.

CHAPTER II

HISTORICAL

Introduction. Taylor: Baker: Wegener: du Toit: Wing Easton: Evans: Argand: Daly: Krige: Smit Sibinga: Joly: van der Gracht: Staub: Holmes: Bailey: Rastall: Bogolepow: Borchert: Maack: Nissen: Watts: Sörgel: Penck: Lake: Keith: Washington: Termier: Krenkel: Chamberlin: Schuchert: Willis: Leme: Douglas: Kreichgauer: Kober: Gutenberg. Conclusion.

INTRODUCTION

SEARCH for the germ of the vital concept of Continental Drift takes us back for three centuries at least, but, as it happens, the evolution of such ideas is more a matter of interest to the historian than to the scientist.

Regarding those who have expressed such opinions in the past, it will suffice to mention Francis Bacon (1620), Buffon (c. 1780), Young (c. 1810), R. Owen (1857), A. Snider (1859), H. Wettstein (1880), O. Fisher (1882), C. B. Warring (1887) and W. H. Pickering (1907). A crustal creep towards the equator forms the keynote of the hypothesis of D. Kreichgauer (1902), while shifting of the polar axis has been favoured by quite a number of persons, though that view is generally discredited today.

It can be stated that the first definite and convincing presentation came from F. B. Taylor in 1908 (published in 1910), followed by H. B. Baker in 1911 and by A. Wegener in 1912. Taylor's able synthesis does not seem to have produced an undue stir, mainly because its heterodox character was then, as perhaps now, not fully realised. Wegener's opinions on the contrary constituted a more direct and flagrant challenge to recognised principles and his views have consequently been the more vigorously combated.

Below will be reviewed, first the ideas of those favouring some form of Continental Drift, and thereafter the more important of the criticisms made by its opponents, the many minor points raised by them being dealt within, or else being covered by, the text in subsequent chapters. A large number of prominent writers have had occasion to discuss the subject from one or more aspects, and it has admittedly not been possible to study everything that has been written either for or against, but only the more notable contributions, so that this work makes no pretence at being exhaustive ; indeed such a review would doubtless prove rather tedious. Because of their instructive and at times relevant character the alternative hypotheses of Kreichgauer, Joly, Kober and Gutenberg are briefly surveyed, while reference will be made in the sequel to other theories of earth evolution. In a summary of this kind it is naturally difficult to give due credit to all those who have in some way or another contributed to this many-sided hypothesis, which, it must be admitted, is still in the process of being roughed-out.

F. B. TAYLOR in his " Bearing of the Tertiary Mountain Belt on the Origin of the Earth's Plan " (1910) was the first to present a clear picture of continental drift by showing that those particular foldings indicated definitely that the land-masses must have moved bodily for considerable distances away from the poles and towards the equatorial zone. Such meant that North America had drawn away from Greenland, that South America and Africa had parted from the mid-Atlantic Ridge, and that Asia had flowed outwards into the bordering ocean. He was indeed able to use Asia to interpret the structural evolution of Europe. See Fig. 1.[1] The most probable cause of such crustal drift was held to lie in flattening of the earth toward the poles by *tidal action*. In his " Greater Asia and Isostasy " (1926) he developed those ideas further, showing that the crustal sheet of Southern Asia had moved as a unit outwards, and suggested that the upthrusting of its expanding island festoons had dragged up some of the deeper-seated matter in front, thereby creating oceanic " foredeeps " in advance of the migrating arc. His contribution " Sliding Continents and Tidal and Rotational Forces " (1928) stressed his previous views and particularly the radial drift of the masses from the poles. Noting the abrupt commencement of the Tertiary Orogeny, he hints that the capture of the Moon out of space during the Cretaceous might have increased tidal action and so caused the continents to start sliding.

Throughout Taylor ignores stratigraphical and palæontological relationships and any tectonic analogies older than Cretaceous, while, save in the case of North America and Greenland, he has made no attempt to reassemble the jig-saw puzzle of the continental fragments, as was so spectacularly done by Wegener. Furthermore his graphic picture of the Tertiary mountain deformation rather obscures that of the magnitude of the postulated continental displacements.

[1] This is essentially the same as his Fig. 7 of 1910.

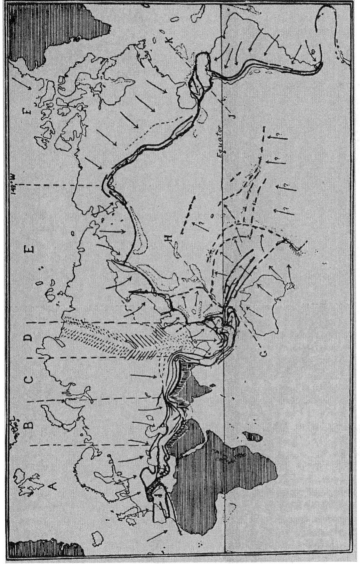

FIG. 1.—Map of Tertiary Movements of Continents and Mountain Trend Lines—after F. B. Taylor (1928). Heavy black lines, mountain ranges; faint dotted lines, foredeeps and irregular deep basins in the ocean floor; vertical shading, foredeeps on land filled with sediments. The arrows show the general direction of crustal movement in all of the continents.

It must, nevertheless, be recognised that Taylor in able fashion advocated Continental Drift by 1908,[1] and that his several contributions on the subject will be of lasting value.

HOWARD B. BAKER in 1911 presented—illustrated by means of the " displacement globe "—" The Origin of Continental Forms," the several sections of which account were published

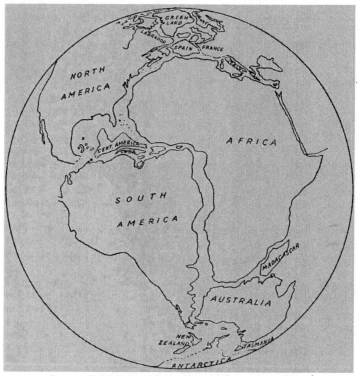

FIG. 2.—The " Replacement Globe " of H. B. Baker (1911)

between 1912 and 1914. In 1932 appeared *The Atlantic Rift and Its Meaning*, a mimeographed volume of limited edition, which for that reason is reviewed at some length. Like Wegener he postulates a single continent or pangæa, which split, though not centrally, from Alaska across the Arctic and down the full length of the Atlantic to the Antarctic, the unequal parts drifting off in opposite directions towards the Pacific region with subsidiary fracture and the rotation of certain portions. His reassembly of the fragments is pictured in Fig. 2, though regret-

[1] His paper was read on Dec. 29th, 1908.

tably no restoration of the Eastern Hemisphere is given. Taylor's grouping is followed for the Northern Hemisphere, but a tighter fit is obtained by closing up the disjunctive basin of the Mediterranean with the rotation of Italy, Spain and Newfoundland, thereby bringing North America next to North-West Africa and not far from South America. Experiment with a terrestial globe will, however, show that appreciable distortion is introduced in attempting to secure so close a fit.

The Southern Hemisphere is less satisfactorily treated, Australia being jammed between South Africa, Antarctica and Patagonia in the face of strong structural and other differences, while New Zealand is pushed even further to the west. The position of India is not indicated in the drawing, and scarcely anything is said about this important land.

Although the author enlarges upon the numerous geological parallels with skill, he gives insufficient attention to stratigraphical and tectonic similarities, and places too much weight upon the resemblances between opposed coast-lines, which nevertheless is not out of accord with his particular hypothesis. For the same reason he is not much concerned with the implications of past climates. On the other hand, the biological evidence for former connections between the various lands is presented at considerable length, many fresh and striking instances being quoted, to which the reader is referred.

Unlike any other writer, he views the movement away from the " Atlantic rift " of the angular fragments and their convergence upon the Pacific not as a progressive and slow act, but as *a simultaneous and rapid flight* occupying quite a brief period during the late Miocene or early Pliocene. The explanation proffered, though ingeniously worked out, is bizarre. The orbital eccentricities of the Earth and Venus are supposed to have varied enough to have brought those planets for a short while within such a range, that the consequent tidal distortion in the Earth caused a thick layer of crustal matter to be stripped off from the Pacific region. *Such went to form the Moon!* The consequent loss of most of the oceanic waters of the globe was then compensated for by water captured in turn by the Earth from the breaking up of the hypothetical fifth planet of the Solar System, now represented by the Asteroids !

It is inconceivable that such an astounding and stupendous catastrophe during the later Tertiary would not have left an indelible impress upon the geological record over most of the globe, not the least of which would have been a wholesale destruction of life. As against this could be cited many instances of unbroken terrestial and marine Tertiary successions, while other serious consequences arising out of this novel hypothesis will readily suggest themselves.

This work is throughout characterised by a refreshing
originality of outlook and presentation; even if the astronomical explanation prove unacceptable, it provides a wealth
of argument in favour of continental drift, which is all the
more weighty in that it has been reached from an approach
quite different from those of other supporters of that hypothesis.

In the mimeographed " Structural Features crossing the
North Atlantic " (1936) he brings out the close agreement in
the palæozoic folding on both sides of the ocean, Fig. 3, and

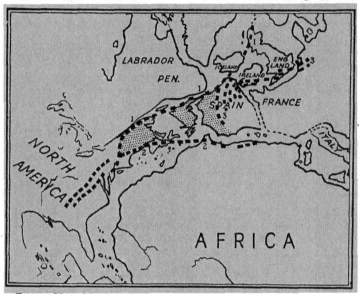

Fig. 3.—Plan of the three Transatlantic orogenies—after H. B. Baker (1936)

declares certain astounding tectonic correspondences between
Newfoundland and Iberia, which would suggest the relatively
recent separation of those masses, regarding which certain
criticisms are made in Chapter VIII.

A. WEGENER, Professor of Meteorology and Geophysics in
the University of Graz, developed his ideas independently of
Taylor and published them as " Die Entstehung der Kontinente"
in 1912. His book, *Die Entstehung der Kontinente und Ozeane*
appeared in three editions, in 1915, 1920 and 1922, the last
being translated into English as *The Origin of Continents and
Oceans* by J. G. A. Skerl in 1924.

The fourth German edition (1928) not only contains important new matter, but has been rewritten to meet various

criticisms, though probably few English-speaking persons are aware of this revision that marks so great an advance over his previous writings.   His tragic death towards the end of 1930, while engaged in exploring Greenland, has undoubtedly been an immense blow to Science.

Wegener rejects the permanence of ocean basins and shrinkage of the Earth as the prime cause of folding. He starts off with a single land-mass or "pangæa" that proceeded to break up during the Mesozoic, the portions thereof drifting generally westwards, though at varying rates. This is indicated diagrammatically in Fig. 4, which by now must be well known to all. Despite the high viscosity of the sub-crustal matter, it is considered competent to yield to small forces if they are applied over a sufficient length of time, the blocks of sial drifting forward through or upon the sima as icebergs do over the calm or current-affected ocean.

A mass of evidence—geological, tectonic, palæontological, palæoclimatic and geophysical—is analysed to show that the continents have drawn apart, in certain cases for huge distances, thereby giving the disjunctive basins of the Atlantic and Indian oceans. Astronomical observations are quoted to support the claim that certain spots are today experiencing changes in their geographical positions, which, although small, are measurable. In his restoration of the Lands, Wegener brings their present shore-lines almost together, but in his fourth edition he recognises the weight of opinion against such a procedure and accepts instead a matching at about the edges of the continental shelves.

The westerly drift postulated by him readily accounts for the marginal foldings along the western sides of the Americas, but the explanation advanced, that the island-arcs of the Western Pacific are fold-margins left behind during the westward drift of Asia and Australia, is not convincing.   Little attention is given to the Alpine fold-belt between the stretch across from the East to the West Indies, which plays so vital a part in the hypotheses of Taylor and others. For that very reason he is unable to provide a land-bridge between North America and Europe and between South America and Africa during the Tertiary and has to postulate an absurdly late date for the main movement under which those continents became separated. On the other hand, he rightly insists that the numerous land-bridges that have been conjured up to span the oceans in various times and places are not demanded under his hypothesis. His reassembly of the lands reveals furthermore, in simple though striking fashion, zonal distributions for not only the life but the climates of the past, such as have proved difficult or impossible to explain under current theories.

C

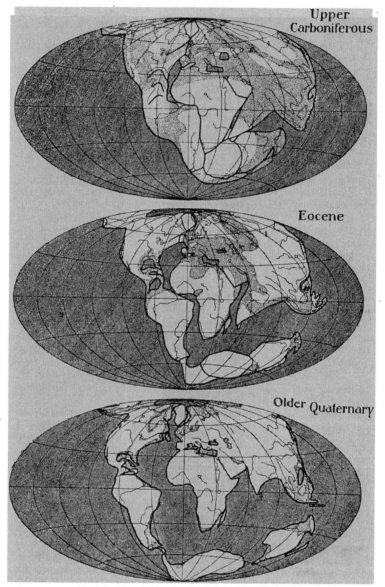

FIG. 4.—Reconstructions of the map of the world for three periods according to the Displacement Theory—after A. Wegener. Lined, ocean ; dotted, shallow seas : latitude and longitude arbitrary. (With the permission of Methuen & Co. and Fr. Vieweg & Sohn.)

The apparent changes in the positions of the poles at various epochs are interpreted by Wegener as due to actual shifting of the polar axis, the feasibility of which has been stoutly denied by most physicists. Such incongruity, which incidentally has lost Wegener considerable support, is easily met by other hypotheses, which picture the axis of rotation as fixed while the crust slides around over the denser core.

Reviewing the various possible forces such as centrifugal action, tidal action, precession and differences in crustal density areally, Wegener has had to admit that all these would seem too small to account for the gigantic displacements of the kind envisaged by him, but inclines in his last edition (p. 184) towards convection currents in the sub-crust, such as have been associated with the name of Ampferer and in more recent times with Joly's hypothesis of radioactivity.

While the credit for having first formulated a working hypothesis of Continental Drift must be awarded to Taylor, it was Wegener who presented the first comprehensive picture of the Earth from that aspect, drawing freely upon the other Sciences as well as Geology, one that profoundly stirred geological thought and rescued it from the cramping influences of conservatism and tradition.

To the memory of that great pioneer this modest contribution by the writer is accordingly dedicated.

A. L. DU TOIT in " Land Connections between the other Continents and South Africa in the Past " (1921) deals popularly with the Gondwana portion of the land-masses, its unity from the Palæozoic as shown geologically and biologically, its weakening along the periphery through marine sedimentation, its centrifugal fragmentation during the Mesozoic, the throwing-up of outer ring-chains in the Tertiary and the breaking of the latter, when tension produced the Atlantic and Indian oceans. For the first time is set forth an explanation for its varying biological relationships, which reveal first a unity, next a disagreement—through dispersion of the fragments—then a reunion during the Tertiary—through the ring-folds—and lastly an isolation — as those links became further stretched and fractured.

In " A Geological Comparison of South America with South Africa " (1927) is set forth the host of correspondences between the sides of the South Atlantic in the shape of the Devonian system, Carboniferous glacials, Permo-Triassic strata, etc., and the significance stressed of phasal variations away from the respective coasts. It is pointed out that the mesozoic foldings of the Cape and Argentina meet at right angles the older structures that trend parallel to the two Atlantic shores, and, as

Holmes has since phrased it, *the crossing begun in the one continent is completed in the other.* Such phenomena are explicable only on the assumption of drift. In the reconstruction, unlike that by Wegener, the edges of the two continents are not brought into contact, but are regarded as having been set originally not less than some hundreds of kilometres apart, the space in between having been for the most part land since lost (Fig. 13). The collective evidence points to the Falkland Islands as having formerly been situated between the Cape and Argentina, their stratigraphy and structure being almost identical with those of the Cape.

In " Some Reflections upon a Geological Comparison of South Africa with South America " (1929) the close resemblances are again stressed, which extend even to the diamondiferous occurrences and manganese deposits. The significance of " Fault-line Coasts " finds its first mention here.

N. WING EASTON in " Het Ontstaan van den Maleischen Archipel, bezien in het licht van Wegener's Hypothesen " (1921) points out certain basic differences between the portions of that region situated to the east and to the west of central Celebes. The tectonic synthesis by Suess incidentally comes in for review and some criticism.

He pictures the Archipelago as having originally formed part of the *great southern continent*, the western portion having been derived from that part of Antarctica situated to the west of Wilkes Land—contiguous to India, Madagascar and South Africa—and the eastern from that to the east of Wilkes Land with Australia set next South Victoria Land. New Caledonia, New Guinea, Halmaheira and the eastern side of Celebes were united to one another and to Australia, but advanced much further to the north than the Sumatra-Java-Ceram arc.

He pointedly remarks that under such a scheme the islands should in their northerly drift have passed through the southern dry girdle, and that evidences of such former aridity have actually been preserved among the volcanic products, for example, in Sumatra, Java and Central Celebes. As it happens, such an environment would be in full accord with the positions of the East Indies in regard to the poles deduced by Köppen, Wegener and Davidson Black for the Pliocene and early Pleistocene. Nevertheless, the Archipelago could all the same have belonged to, and formed part of, Asia—as currently supposed—and not to Australia, as suggested by Wing Easton.

J. W. EVANS in a note entitled " The Wegener Hypothesis of Continental Drift " (1923) clearly pointed out the radially outward movement of the land-masses from Africa, and followed this up by writing the Preface to the English translation of Wegener's book. In two outstanding presidential addresses (25 ; 26) he brought together a considerable body of data upon crustal tension and compression which has proved invaluable in this problem.

E. ARGAND'S monumental, though diffuse, essay "La Tectonique de l'Asie" (1924) is essentially a study of massed crustal folding with frankly tectonic outlook, but outdoing Suess. The keynote is the "plasticity of Asia" during the Alpine Orogeny and its generally radial and outward flow from the Angara "shield" towards the Ural geosyncline and the Indo-Pacific region. Unlike Grabau (23-4 ; 28) he gives no clear-cut pictures of the land-masses and geosynclinal seas of the past as disclosed by the stratigraphical evidence. His scheme views the Hercynian (Altaide) elements as presenting far less obstruction to the later Alpine diastrophism than other workers in Asia are prepared to admit. Accepting Wegener's general picture of Gondwana, Argand proceeds to follow Taylor in visualising the Alpine-Himalayan plication as due to the squeeze between that continent and Eurasia, and illustrates the bilateral symmetry of the zone of conflict by serial sections of which the most frequently reproduced are those drawn through the Alps and Tibet respectively (his Figs. 19 and 13).

In the first-mentioned region Africa moving from the south overrides Europe in curving piled-up chains, which through their momentum continue their northward drift and so tear themselves away from the arrested mass of Africa, thus producing the rift-basin of the Mediterranean. While worked out cleverly, this interpretation has come in for some harsh criticism, though, to be candid, those who have so objected can only *describe*, but not *explain* in convincing terms, the tectonics of that highly complicated terrain.

In the second region Asia overwhelms the depressed "prow" of the Indian block, which travelling from the south-west forces itself beneath its opponent, wherefore the stupendous Himalayan ranges and the Tibetan plateau.

Though apparently erring in ascribing an undue degree of "plasticity" to the continental blocks and assuming movements more precise and detailed than the data would seem to warrant, Argand's views have profoundly influenced geological thought and have greatly furthered the hypothesis of Continental Drift. Wisely Argand has made no attempt to explain the cause of such displacements, which he merely accepts as the unescapeable deduction from the wealth of geological evidence available to the unfettered mind.

R. A. DALY'S " Downsliding Hypothesis " (1923) or " Land-slide Hypothesis " (1926) views the older continental masses as moving under the influence of gravity by reason of an initial tilting produced through the combined actions of erosion, con-traction of the Earth and changes in its speed of rotation, the crust tending to fold or fracture along the weaker oceanic borders. Sliding is supposed to take place upon a deep-seated stratum of basaltic glass. There are, however, some serious difficulties connected with the detailed application of this theory, particularly on the colossal scale postulated, though the general idea would seem correct. Daly's views [1] on the physical basis of Drift are worthy of careful study.

L. J. KRIGE'S address, " On Mountain Building and Con-tinental Sliding " (1926), develops some encouraging ideas rather like those of Daly, though here again it may be doubted whether the mechanism invoked would be commensurate with the scale of the deduced horizontal movement especially since the effect of the " free side " of the continent is ignored. The vital importance of Isostasy, however, finds adequate recogni-tion.

In " Magmatic Cycles, Continental Drift and Ice Ages " (1930) he investigates the rôle of the peridotite shell and shows that at long intervals its radioactive content causes it to melt and produce a revolution communicating heat to the overlying basic shell enough to cause melting in that layer, especially beneath the continents, thereby putting the latter into tension and so inducing folding, fragmentation and drifting.

G. L. SMIT SIBINGA in " Wegener's Theorie en het Ontstaan van den oostelijken O. I. Archipel " (1927) sets forth weighty evidence that decidedly opposes the contraction hypothesis for this region. His detailed analysis furthermore discloses a *Double Molucca Arc*, which follows more or less the boundary between land (to N.W.) and sea (to S.E.) during the Mesozoic. By straightening-out in imagination those curved structures he obtains a definite picture of the responsible crustal movements through which such S-shaped forms have been produced. The outer arc is accordingly ascribed to the collision of the Australian with the Asiatic mass, and the Banda Sea to the consequent pressing-in of the Banda loop.

The superiority of the displacement hypothesis as an in-terpreter of the tectonics of this region finds full expression in this illuminating paper.

J. JOLY in " The Surface History of the Earth " (1925) pictures

[1] Daly (33), 251-65.

the crust as resting in approximate isostatic equilibrium upon a sub-crust composed of basalt, today in the solid state, though periodically melted through the accumulation of radioactive heat, thereupon enabling one of the orogenic " revolutions " of geological history to take place.

While Joly does not admit Drift in the sense of Taylor and Wegener, he points out that a general westerly creeping of the crust over its core must come about because of tidal action. It can, however, be remarked that the uneven distribution of radioactive substances in the rocks, more especially those forming the lands as compared with those flooring the oceans, would render any regular slipping of the crust in the highest degree improbable, hence *differential crustal creep* must be the logical consequence of his theory. Joly's hypothesis with all its physical implications would indeed appear to embody the vital principle of Continental Drift.

W. A. J. M. v. WATERSCHOOT VAN DER GRACHT. Paramount in the Symposium on the *Theory of Continental Drift* (1928) is the restrained contribution by this writer, who in his able rejoinder makes short work of most of the objections by the other contributors to that instructive publication. Although accepting Wegener's scheme of a general westerly drift of the lands, he emphasises, like Taylor, the equatorward movement as shown by the Alpine foldings. Differential drifting of the crust over its core is regarded as a more likely explanation of past climates than the actual shifting of the polar axis. He points out how simply, yet effectively, the climatic zones of the various epochs could be accounted for under Köppen and Wegener's scheme of reassembling the continents. His explicit distinction between " strength " and " rigidity " disposes of the academic arguments that such a drifting mass could not wrinkle-up the ocean floor in advance of the block. While recognising the minuteness of the forces available to move such masses, so far as was known, he sees a possible cause in Joly's " thermal revolution cycle " with its periodic loosening of the bonds between the crust and sub-crust.

In two later papers (1931 ; 1933) van der Gracht compares Europe and North America in detail and stresses the remarkable correspondences not only of the late Palæozoic strata, sedimentary facies and faunas, but of the orogenic episodes during which they were folded, the evidence of which would collectively favour in unquestioned fashion the closer geographical union of those two countries in the past. He points out that the thinness of the crust as compared with the enormous areas implicated renders quite probable some distortion within the blocks themselves, which would doubtless extend to the con-

tinental shelves as well, thereby preventing a perfect fit being
obtained between the existing opposed coasts. The opening of
the Atlantic is regarded as having begun in the Cretaceous
and continued down to the present. With most of these views
the writer finds himself in accord.

R. STAUB. 1928 saw the appearance of that outstanding
work, which has unfortunately not attracted the attention it
deserves, "Der Bewegungsmechanismus der Erde," in which,
curiously, no credit is given to Taylor, Joly or Holmes, though
Suess, Argand and Kober are often cited with approval, and
sometimes Wegener.

Its keynote is the repeated interaction of two great northern
and southern land-masses—Laurasia and Gondwana—that culminated in their fracture, the overriding of Europe by Africa
and of India by Asia, and the crowding of the various parts
towards the Pacific, which is viewed as a rigified depression
resulting from the tearing away of the sial to form the Moon
during the remote past. Staub presents a clear-cut and detailed
picture of the behaviour of these two primary masses, the
problem being treated essentially and with skill from the tectonic
aspect, though little stratigraphical evidence is submitted.

He works out, though not so graphically or satisfactorily as
Köppen and Wegener, the climatic girdles for past epochs,
determines the relative positions of the poles and deduces therefrom that the main orogenic belts ran approximately parallel to
the equator of the time, which provides the clue to their origin,
namely, *centrifugal action* due to the Earth's rotation. Such
centrifugal influence has caused the two units to approach one
another, compressing the equatorial geosyncline in between,
while *sub-crustal streaming* has drawn them apart again.
Through such alternating "polar flight" and "polar drift"
Laurasia and Gondwana have successively impinged upon or
else parted from one another. The regularity throughout
geological history postulated by him (his Fig. 43) is, however,
exaggerated. Like Wegener, a general westerly drift is also
conceded. He not only emphasises the connection of the
"Pacific" suite of igneous rocks with orogenic zones and of
the "Atlantic" suite with regions of epeirogeny, but regards
the peculiar "greenstones" or "ophiolites" as marking an
early phase within the geosyncline and as representing the
primitive magma of the sub-crust.

From the continual use of the word "drift," the apparently
considerable movements suggested from or towards the poles,
the various diagrams on which large arrows would indicate a
bodily transfer of the continental blocks—which in certain cases
was even radial—and the recognition of torsion and furthermore

of a westerly creep of the lands as a whole, the reader is bound to acquire a vivid impression of wholesale continental fracturing and dispersal like that visualised by Taylor and Wegener. Such is, however, not the case for enormous sections of the Atlantic and Pacific are designated " broken-down, stretched and torn-off parts of Laurasia and Gondwana," which sank beneath the waves during the Alpine orogeny. His large coloured map also shows many submerged horsts bounded by faults. One is therefore taken back to Neumayr's concept of huge former continents and incidentally of a colossal Gondwana.

Staub is, however, vague regarding the degree of such stretching, the date of such a cataclysm, how such vast blocks could have foundered in defiance of isostasy or why the related stupendous transfer of oceanic waters should not be firmly impressed on the geological record. Furthermore, his postulated " drift " and " flight," while correct for the Tertiary, becomes less and less truly " polewards " as one goes back in time, as will appear from his own admission of different polar positions in the past and from a study of his diagrams, particularly his Fig. 42. Insufficient attention has been given to the ever-varying position of the continents in regard to the polar axis.

The impression gathered by the writer is that Staub, while recognising and making most successful use of the principle of drift, has hesitated to take the bold course of admitting that the one and only conclusion from his particular scheme is large-scale drifting in the sense of Taylor and Wegener. His hypothesis suffers therefore from several of the great disabilities set forth in Chapter I. He points out that his hypothesis differs radically from that of Wegener. The guiding principle thereof is an eternal conflict of forces—between that due to the Earth's rotation and that due to the poleward sub-crustal streaming—which, as the one or the other obtains the preponderance, causes alternate movement of the continental blocks from the poles to the equator and from the equator to the poles.

Staub's synthesis is both original and forceful and the most important, detailed and scientific contribution to the subject of Earth Evolution that has yet been made. With much of his masterly interpretation in respect of the movements of the lands the writer finds himself in very close agreement.

A. HOLMES has contributed mainly on the theoretical side. His concise " A Review of the Continental Drift Hypothesis " (1929) shows graphically (his Fig. 2) the disjunctive nature of the Atlantic as revealed by the dislocation in both north and south of the three great intersecting trans-oceanic fold-systems. The Tertiary plications are held to support the views of Suess, Taylor and Argand as indicating movements of the masses to which they are marginal, but not quite in the sense of Wegener. The rôle of the Ural geosyncline is for the first time emphasised and also the significance of past climates and of mineral occurrences.

Here, too, but more thoroughly in " Radioactivity and Earth Movements " (1928–29), he ably and convincingly develops the idea that *convection currents in the sub-stratum* are responsible for crustal flow and can explain the evolution of geosynclines, mountain-building, island arcs, rift-valleys, distribution of volcanos and earthquakes, and other allied phenomena. The part played by the deep-seated shells and the influence of *radioactivity* are skilfully worked out.

One of Holmes' stimulating diagrams is reproduced here, Fig. 5.

E. B. BAILEY in his illuminating address " The Palæozoic Mountain Systems of Europe and America " (1929) demonstrates the wonderful stratigraphical and tectonic parallels between Western Europe and Eastern North America. He notes the way in which the inner Caledonian front, that trends south-westwards from Finmark to South Wales, is crossed by the younger Hercynian front with general westerly direction, and observes right across the Atlantic the outer Caledonian (Taconian) front of Newfoundland, New York and Maryland being crossed obliquely and in analogous fashion by the later Appalachian (Hercynian) front that runs from New England to Arkansas. He is hence led to conclude that the crossing of the two fold-systems, begun in Europe, is completed in America, and he questions whether the phenomena have not to be explained as due to Continental Drift.

This able synthesis should be read in conjunction with van der Gracht's[1] more detailed and masterly comparison between Europe and North America during the late Palæozoic, and the graphic restoration by Baker.[2] In Chapter VIII Bailey's valuable interpretation is extended to Greenland and Spitzbergen with amplification in certain respects and modification in others.

R. H. RASTALL'S article " On Continental Drift and Cognate Subjects " (1929) is essentially a reply to those who have dogmatically asserted that no forces are available to produce drift. In it he trenchantly points out that no reasonable physical explanations have as yet been forthcoming for three accepted geological realities, namely—the Alpine " decken," the sinking of land-bridges and present and past ice-ages.

M. BOGOLEPOW in his stimulating " Die Dehnung der

[1] van der Gracht (31 ; 33).   [2] Baker (36), Fig. 2.

Lithosphäre " (1930) urges the widespread and interrelated character of Continental Drift and *Stretching of the Crust* as evidenced for example by the African Rifts, Indian Ocean, Malay Archipelago and mid-Atlantic Ridge.   Zones of com-

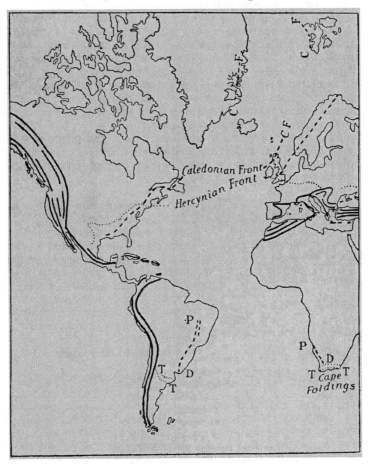

FIG. 5.—Palæozoic Fold-Systems of the opposing lands of the Atlantic—after A. Holmes (1929).

pression and torsion, arcuate and spiral structures, fracture systems and volcanicity, which are all intimately associated, are comprehensively viewed as surface manifestations of world-wide flowage of matter beneath a brittle crust.   He is hence a strong supporter of the Hypothesis.

H. BORCHERT'S " Über den Werdegang der subpazifischen
Schicht und verwandte Probleme " (1932) is noteworthy in that
it presents a concrete picture in word and diagram of the
presumed process of rifting.   Such embodies faulting, rising of
basic magma, growing and sinking of heavy ferro-magnesian
crystals and ascending of differentiated matter into the fractured
crust until the crystallised sial-sima surface has become exposed
to the ocean. . This process is repeated with each stretch, while
basalt is poured out as a " plaster " over the ruptured
" tesselated " ocean floor.   On the contrary, the compression of
the rigid basic and ultrabasic block of the Pacific along its
margins furnishes an eruptive product of andesitic composition
that escapes into the circum-oceanic zone.

Borchert rejects sub-crustal currents as the prime cause of
drift and to a large extent crystallisation *per se*.   He seems to
place more weight upon forces set· up through changes in the
rotation of the Earth such as precession, etc.

R. MAACK'S " Die Gondwanaschichten in Süd-Brasilien und
ihre Beziehungen zur Kaokoformation Südwestafrikas " (1934) is·
particularly instructive since it is based upon personal explora-
tions in both countries, while the conclusions reached are very
close to those advanced by du Toit, although its author has been
handicapped by lack of literature.

The agreement—stratigraphical, tectonic and palæontological—displayed by these opposed lands is found to apply
even in the finer detail and thoroughly to support the hypothesis
of Drift.   Maack's reconstruction is shown in simplified form
in Fig. 14.   The varied new information therein presented
makes this an outstanding contribution to the subject.

H. NISSEN'S fantastic book "The Origin of the Moon" (1934)
ascribes that satellite to the removal of crust over the Pacific
region to the depth of 350 miles through a " titanic cyclone from
the sun," and that, too, during the Permian!   Vast continental
drifting towards the hollow was thereby initiated, thus accounting
for the Hercynian and subsequent orogenies.   Tectonic movements are postulated of such an amazing character that Sweden
becomes the former " roof " of Norway, Scotland of that of
Newfoundland, etc. !   The work nevertheless contains some
useful ideas.

W. W. WATTS in his address " Form, Drift and Rhythm of
the Continents " (1935) takes up the attitude of the impartial
inquirer into general hypotheses in the light of the facts upon
which they have been based rather than that of the protagonist
of Drift, and asks for more data.

The principal contributions adverse to the Hypothesis are the following :

W. Sörgel in " Die Atlantische Spalte " (1916) either disputes or denies the majority of Wegener's statements or deductions, which is not surprising since he disagrees with such views that positive gravity anomalies rule over the oceans, that sial blocks float in the sima, that isostatic compensation is unfavourable to the sinking of land-bridges and so on, though he is correct in questioning the lateness of the separation of North America from Europe postulated by Wegener. His detailed criticism of the correspondences between the two sides of the Atlantic is a marked example of failure to consult original sources of information on the subject. The very relationships which Sörgel so stoutly denies both in word and diagram (his Figs. 3, 3a, 3b) are precisely those borne out by all the available evidence—a testimony to the general correctness of Wegener's intuition.

A. Penck's " Wegeners Hypothese der kontinentalen Verschiebüngen " (1921) is a denial that isostasy favours a grouping together of the lands and of the oceans, that sial blocks could have moved through the sima, and so on. He doubts Wegener's views on land-bridges, minimises or disputes the resemblances between the Atlantic shores and argues that the probable errors in the longitude observations made in Greenland have been far greater than the differences obtained by calculation. He criticises Wegener's location for the poles during the Permo-Carboniferous, but does not attempt to explain the extraordinary distribution and relationships of the corresponding glacials. Altogether a rather non-constructive paper !

P. Lake's " Wegener's Displacement Theory " (1922) is typical of criticisms directed not so much against the principle of the hypothesis, as the particular presentation by Wegener. He points out various errors of fact which would seriously weaken the argument for continental approximation, but lamentably fails to investigate whether the whole of the available evidence is of an adverse kind. Like so many others, yet with a full picture before him of the amazing distortion of Europe during the Alpine diastrophism, Lake nevertheless seems to imagine that the Americas ought to have remained undistorted during their lengthy drift westwards, during which journey some rotation must inevitably have occurred—a frame of mind as rigid as the continents of current conception !

A. Keith in his " Outline of Appalachian Structure " (1923)

evinces towards the hypotheses advanced by Taylor and Wegener an attitude typical of the orthodox mind. For various *a priori* reasons based essentially upon presumed qualities of the crust and sub-crust, mixed up with a good deal of misunderstanding about the hypotheses themselves, he is convinced of the physical impossibility of sliding. Proceeding next to analyse the current views of mountain building he curiously finds " that each theory has serious drawbacks." Faced by such an impasse he has to revive the old idea of *batholithic intrusion*, currently regarded as of secondary rank in the orogenic scheme, and is compelled like Bailey Willis to seize upon " crystallisation " as the prime cause of such enormous horizontal compressive movements. Elsewhere it will be argued that such molecular behaviour would incidentally contradict the " law of minimum work " and that batholithic intrusion is far more probably a consequence of continental sliding.

Keith's weightiest criticism applies to the small size of the Arctic basin as compared with the apparent southward creep of the masses, that under Taylor's Hypothesis ought to have come out from it during the Tertiary. Study of the folding in Asia nevertheless shows that the amount ascribable to the Tertiary is quite small in the north, though becoming much greater in the south, where, however, it becomes exaggerated due to the presence of the earlier Altaide compression. Furthermore, the large distances today between Scandinavia, Greenland and the American Archipelago would be due only in part to their presumed radial movement from the polar regions. The question will be discussed later.

H. S. WASHINGTON in " Comagmatic Regions and the Wegener Hypothesis" (1923) compares the igneous suites along both sides of the Atlantic, and concludes that the distribution of certain peculiar types is adverse thereto, a view that has been widely quoted. Arguments based on negative evidence are unsafe, and such will apply forcibly in this case because of the lack of knowledge concerning critical areas. Furthermore, the boundaries of petrological provinces need not always, as assumed by Washington, have *run at right angles to* the present shores.

His contention, that the (pre-Cambrian) anorthosites and the palæozoic granites of Norway are practically wanting in Greenland, loses force when it is realised that the great crystalline complex along the eastern side of Greenland is generally admitted to be essentially post-Cambrian, while certain of the intrusive granites therein closely resemble those of Norway and Finland. The alkaline plutonics of southern Norway parallel those of southern Greenland and Quebec. The coasts of western Greenland and Labrador are too little known for useful comparison.

The apparent lack of correspondence in Triassic volcanicity in Europe and North America is dealt with in Chapter VIII.

His assertion that the charnockite suite of West Africa and the soda-granites of Nigeria are not represented in South America is not quite true, since gabbros and norites are now known from British Guiana, while syenites were long ago reported from Ceara (Katzer). Lastly, the number of occurrences of alkaline types along both sides of the South Atlantic is steadily mounting.

Washington's conclusions have manifestly been formed upon very incomplete data.

P. TERMIER'S contribution " The Drifting of the Continents" (1924) reveals the gloomy spirit of its author, who, although recognising and indeed dilating on the merits of Continental Drift, finds himself overwhelmed by apparent contradictions, and, without attempting their solution, capitulates with the cry " The Theory of Wegener is to me a beautiful dream."

E. KRENKEL in his " Geologie Afrikas " rejects drift for reasons that are disposed of in the following pages. His argument that the floor of the Atlantic shows the character of a subsided land surface can be combated, and is scarcely convincing.

R. T. CHAMBERLIN'S " Some of the Objections to Wegener's Theory " (1928) is too full of superficial statements or sweeping generalisations and misunderstandings regarding Wegener's views to necessitate detailed reply. His assertion, for example, that " The framework of the present continents was developed in pre-Cambrian time " may cause surprise to many in Europe and Asia. He has nevertheless rescued from oblivion the saying of an eminent geologist, " If we are to believe Wegener's hypothesis we must forget everything which has been learned in the last seventy years and start all over again." Such is, of course, an exaggeration, though astronomers had to do more or less the same thing when the Principle of Relativity was forced upon them.

C. SCHUCHERT in " The Hypothesis of Continental Drift " (1928) examines at length the views of Taylor and Wegener, and effectively exposes their weak points. Using plasticene on the " replacement globe " he discovers wide misfits between the Old and New Worlds as assembled by Wegener. The resulting protest loses much of its point by the fact that (a) the selected fit is neither the only nor the most suitable one, (b) no allowance is made for the considerable " telescoping " within the various fold-belts—which must in each case total hundreds of kilo-

metres, and (*c*) Tertiary fold-regions, such as the Caribbean,
are taken as integral parts of the land-masses, whereas their
shaping was done *after* most of the alleged Atlantic rifting had
taken place.

The reality of Gondwana is admitted and the stratigraphical
and other correspondences on both sides of the Atlantics are
recognised, but quite narrow land-bridges are deemed competent
to account for them, though Schuchert deplores the apparent
lack of geophysical causes to explain their subsequent sinking.
Incidentally he endorses the much-criticised principle of the
permanency of ocean basins. He concedes Wegener's connection of the " Caledonian " foldings of Europe with those of
Newfoundland, but denies that the Appalachians form the
continuation of the Hercynian system, although the latter
relationship is accepted by probably the majority of geologists
today.

Köppen and Wegener's continental groupings in relation to
the polar axes of past geological periods come in for adverse
comments, but the orthodox and rigid alternative proposed by
Schuchert (his Fig. 20)—one which doubtless makes appeal to
the conservatively minded—is to the writer meteorologically
mysterious.

Schuchert's " Gondwana Land Bridges " (1932) and BAILEY
WILLIS' " Isthmian Links " (1932) are to a great extent complementary papers. Both incidentally provide many and
excellent arguments for narrow, inter-continental land-bridges
during the *Tertiary*, but merely touch the fringe of the real
problem—that of pre-Tertiary Gondwana, its manner of growth,
union and dismemberment. It is clear, for instance, that those
selfsame land-bridges could never have been in existence during
the late Palæozoic and Mesozoic, otherwise the distribution of
the marine faunas must have been vastly different.

The atmospheric and oceanic circulations postulated by
Willis to explain the " Permian " ice-age of Gondwana can be
criticised as being far more precise than the fragmentary data
would warrant, and is based on the questionable assumption
that the position of the polar axis has not shifted appreciably.
The problem is dealt with in Chapter XIII.

A. BETIM. P. LEME, the Brazilian geologist, in voicing his
objections in " État des connaissances géologiques sur le
Brésil " (1929), makes the fundamental error of pressing South
America close up against Africa, thereby spoiling the fit, and
he further exaggerates minor differences (such as by comparing
the respective *maximal* thicknesses of formations), while
ignoring phasal variation. His criticisms are effectively disposed
of by Maack (34).

G. V. DOUGLAS, " On the Theory of Continental Drift "
(1934), will not accept the principle of drift because of his
" positive belief that the dominating factor in earth mechanics
is contraction." Just why those two agencies should be regarded
as antipathetic is not, however, made clear.

Every one of the other theories of Earth Structure has
necessarily some bearing on our hypothesis, but of them all
only three are so outstanding as to deserve special analysis :

P. D. KREICHGAUER'S " Die Äquatorfrage in der Geologie"
(1902), of which a second and revised edition—ably reviewed
by Richarz (28)—appeared in 1926, is a striking and suggestive
work with a sound mathematical basis. The author develops
the view that, whereas the interior shells of the earth are through
the rotation of the latter bounded by ellipsoidal surfaces, the
outermost one or crust tends towards a constant thickness,
such having been essentially determined by the temperature
gradient. Under rotation a depression would develop equatorially, within which crumpling would take place as shrinkage of
the earth's interior proceeded through cooling. This equatorially directed force is, as Richarz has stressed, by no means
inconsiderable.

The resulting plications would be *dominantly equatorial* with
a subordinate set at right angles thereto, i.e. *meridional*. From
the fact that the great geosynclines and their related mountain
belts diverge more and more from those two directions as they
are traced back into the past, a creep of the crust as a whole
over the interior has to be presumed, and therefore an apparent
shift of the poles, a view confirmed by the geological evidence
favouring equivalent movement of the climatic girdles, as thereupon detailed by the author in convincing fashion (see Chapter
XIII).

Kreichgauer recognises distortion of the continents arising
out of their repeated compression and stretching, but does not
admit of Continental Drift. The basic ideas contained in this
revolutionary work are accepted by the writer—with the
modifications demanded by such drift—and more particularly
the corollary of the systematic screwing around of the continental
masses, which would indeed seem to be one of the fundamentals
of Earth Structure.

L. KOBER has in " Die Bau der Erde " (1921) advanced views
even more extraordinary than those put forward by supporters
of Drift. The keynote is that *each continental block* as seen
today consists of a relatively rigid foundation mass or
" *Kratogen* " built up and welded together during the past,

D

completely surrounded by a much younger mobile zone or
" *Orogen* " of marine sedimentation, which was laterally compressed, particularly during the Cretaceo-Tertiary, through
contraction of the Earth. A fold-zone entering the sea is held
to bifurcate, and the branches thereof to proceed medially down
the length of the ocean basin until they come to join other
similarly arranged belts, which in like fashion curve and fork,
thus producing a series of contiguous fold-rings, each including
one of the land-masses—" cartouches " we can call them. Such
submerged sections are made to coincide with the various
oceanic " swells " and their island " peaks," presumably
because many of the latter include, or are believed to be underlain by, continental rocks.

It will be of interest to compare the relative lengths of the
Tertiary fold-girdle under Current views, the Displacement
Hypothesis and Kober's Scheme respectively. Measured on a
globe the visible continental portions, the island arcs and the
submerged extensions across the Atlantic and between Cape
Horn and New Zealand total about 65,000 km. Under our
Hypothesis that figure becomes cut down to about 51,000 km.
The complicated curves depicted by Kober on the contrary run
beneath the ocean for no less than 80,000 km., thus giving a
total of 130,000 km. of foldings—just double that of Current
views, and demanding furthermore twice as much shrinkage !

The squeeze between any two blocks is pictured as having
produced a compound structure made up of two sets of marginal
chains (" *Randketten* "), each backfolded or backthrust upon
its related block, and an intervening " *Zwischengebirge*," translated as " Intermontane space " (Longwell), " Median mass "
(de Böckh) or " Median area ". In contrast to the scheme of
Suess, which involves unilateral symmetry, Kober regards all
major fold-systems as possessing such a threefold structure,
though to give effect thereto he often has to bracket movements
of widely different ages. Actually Kober's scheme would seem
to apply mainly to the Tertiary girdle within the region extending
from the West Indies eastward to the East Indies, a section
viewed under our Hypothesis as resulting from the mutual
interference of *two* separate, marginal fold-systems. Save in
this particular instance the two hypotheses are quite incompatible.

Another important consequence of Kober's scheme is that
the area of a continental block must have *decreased* since the
Palæozoic, which is contrary to the abundant evidence pointing
to their general expansion. The volume will nevertheless be
found to contain many useful observations and ideas.

Kober's " Die Orogentheorie " (1933) develops the theme
even further and can be described as a super-hypothetical work,
which omits stratigraphy and exalts tectonics, though the terms

" Caledonides," " Variscides " and " Alpides " are used in so wide a sense as to do violence to tectonic geology. The diagrammatic generalisations furthermore conceal innumerable real inconsistencies with his hypothesis. Great weight is placed upon existing oceans as geosynclinal basins, and indeed, so permanent do these become, that it is difficult to see how terrestial life could ever have become interchanged; while as for Gondwana, it presumably never existed! No reference is made to Taylor or van der Gracht.

B. GUTENBERG's " Fliesstheorie " (1932) or " Theory of Continental Spreading " (Lake) [1] supposes that the stripping away of much of the Earth's crust to form the Moon left a single primitive continent in the neighbourhood of the South Pole. Under centrifugal action this sial cap became spread out to reach today to far beyond the equator, with weaknesses that developed into the Arctic, Atlantic and Indian oceans, the bed of the Pacific still consisting of sima. Periodic contraction of the Earth produced folding, mostly within the equatorial zone or along the edges of the sial sheet. The slow passage of any part of the sheet from higher to lower latitudes would readily account for certain of the recorded climatic changes of the past.

The important land-distribution, stretching across the equator right to within the Arctic region, invalidates the hypothesis, unless a second sial cap be postulated around the North Pole, whereupon the scheme would merely become a variant of the arrangement by Staub or du Toit with its two parent continents set in the universal ocean, though farther apart to begin with.

In his " Structure of the Earth's Crust and the Spreading of the Continents " (1936) Gutenberg produces valuable seismological evidence to show that stresses can be accumulated even at great depths. His argument, that the presence of continuous sial beneath the Atlantic invalidates Wegener's ideas, is unsound since the floor could equally well today consist, in part at least, of acid and sub-acid differentiates from the lower layers formed during the process of rifting, as Borchert (32) has made evident.

CONCLUSION

Among the many others who have expressed opinions adverse to the Hypothesis stand E. W. Berry, A. Born, Bowie, Bucher, Coleman, Gregory, Hobbs, Jeffreys, Kober, Kuenen, Longwell, Mushketov, Singewald, Stille, von Huene, von Ihering and W. B. Wright, and almost certainly the majority of geophysicists.

[1] This criticism is based on the summary by Lake (33).

As supporters or else sympathetic thereto can particularly be mentioned Brouwer, Kossmatt, Matley, Salomon-Calvi, Holtedahl, Groeber, Windhausen and Wing Easton, and not a few botanists such as Seward and zoologists such as Barnard, de Beaufort, Michaelsen, Ökland and von Ubisch.

As active protagonists are Taylor, Baker, Argand, du Toit, van der Gracht, Holmes, Molengraaff, Longfellow, Rastall, Smit Sibinga and Staub.

It is naturally difficult to formulate from the variety of views that have been expressed by these latter a true consensus of opinion, but the collective ideas of such supporters would perhaps favour some general scheme involving :

(*a*) Two great parent masses throughout the Palæozoic— Laurasia and Gondwana ;

(*b*) A fragmentation that began in the later Mesozoic and is still in progress ;

(*c*) A dispersal of the fragments radially outwards and also equatorwards, with a tendency towards a westerly creep as well ;

(*d*) A drift of the crust relative to the polar axis, thereby bringing about major climatic changes ;

(*e*) Some distortion of the masses during their drift ;

(*f*) A transfer of part of the Pacific waters to fill up the tension-basins making the other oceans ;

(*g*) The recognition of drift as a process that has operated throughout geological time ; and

(*h*) A cause or causes that lie seemingly not outside but within the Earth itself.

CHAPTER III

GEOLOGICAL PRINCIPLES

INTRODUCTION

EVERY " Hypothesis of Continental Displacement " must of necessity have as its basis the known or inferred properties of the crust or lithosphere of this unstable Earth of ours, though unfortunately such forthwith involves the very branch of geological science wherein opinion is most diverse and in many respects least convincing.

That being the case, it becomes imperative that the presentation of the formulated scheme of crustal evolution should be prefaced by a brief statement of the cardinal principles upon which such hypothesis has been reared. This is not the place for discussing the validity of these principles, but, further on, reasons will be given for arriving at certain viewpoints which are to a large extent at variance with those of current geology: It will then be made apparent that most of the arguments used by opponents of " drift " to " prove " the physical impossibility of the latter have no really sound basis.

For a particularly clear and impartial presentation of current and alternative views the reader is referred to Nevin's *Principles of Structural Geology.*

With such a preliminary we shall proceed to a brief review of the *Crust* and the *Sub-crust* and the *Stable* and the *Mobile portions* thereof.

CRUST AND SUB-CRUST

Following Bucher[1] the crust or lithosphere can be defined as " the outermost portion of the earth, which on the whole possesses sufficient strength to offer resistance to deformation

[1] Bucher (33), 39.

37

and to transmit long-continued stresses within certain limits."
It rests upon the yielding sub-crust, sometimes referred to as
the "Asthenosphere" (Barrell). In general too the active areas
of the crust have been constituted rather by the Lands than the
Oceans.

Opinions differ rather widely concerning the depth, density
and composition of the several subdivisions of the crust, but
fortunately such differences as have been expressed are not
actually of vital consequence in our problem. The following
subdivision is tabulated from Chapter XI :

		Layer	Density	Depth in km.	Composition	Zone
CRUST	SIAL	First or Upper	2·0-2·9	0-10	Stratified rocks and ocean.	
			2·6-3·0	10-25	Granite, gneiss, diorite, etc.	Brittle
		Second or Intermediate	3·0-3·6	25-50	Gabbro, amphibolite, eclogite	Paramorphic
	SIMA	Third or Lower	3·0-3·5	50-70	Peridotite ..	Isostatic compensation
SUB-CRUST				70-100		

The material composing the First and Second layers being
rich in *Si*lica and *Al*umina is conveniently referred to as *Sial* ;
that of the Third, being characterised by *Si*lica and *M*agnesia,
is termed the *Sima*.

THE STABLE OR RIGID AREAS

Each of the continents contains one or more extensive areas
composed of rocks that have experienced but slight deformation
since the beginning of the Palæozoic era and which are normally
composed of Archean or pre-Cambrian strata, granites, gneisses,
etc. They have usually been worn down to such a degree that
much or all of their subsequent sedimentary coverings has been
removed, and their normally peneplained surfaces are hence
rather appropriately known as " shields," " platforms " or
" nuclei "—according to Kober, " kratogens "—(Fig. 6).

The impression all too frequently gathered from a study of
the literature is that such shields owe their undeformed character
to their superior rigidity or strength, but, as Bucher and others
before him have repeatedly emphasised, the materials composing

these areas are not as a rule different from those exposed in the adjacent highly deformed zones and have no manifest claim to superior "strength." They have escaped apparently because the previously folded rocks bounding them—which may include direct extensions of the shield itself—have, following recognised mechanical laws, more readily yielded to any renewal of lateral pressure.

It can nevertheless be suggested that the apparently greater rigidity of these ancient masses may in part have been due to properties developed in depth through their repeated and

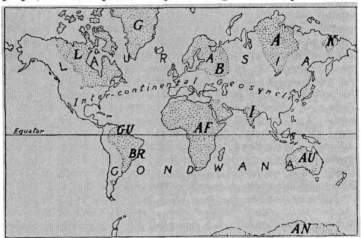

FIG. 6.—Showing the positions of the Continental Shields and of the persistent Geosyncline parting Laurasia and Gondwana: A, Angara; AF, Africa; AN, Antarctica; AU, Australia; B, Baltica; BR, Brazil; G, Greenland; GU, Guiana; I, India; K, Kolyma; L, Laurentia.
(The Mercator's Projection used exaggerates areas in high latitudes).

profound erosion—to cooling and hence thickening of the crust, to general lack of intrusion, to weaker development of joints and shear-planes, etc. The mass would indeed have become comparable to a rusted steel spring. On the other hand, the development of tension in the same or in some other direction would readily bring about subsidence or fracture and thereby initiate a zone of weakness of which subsequent compression could take advantage, while the older fold-belt in its turn became rigified and escaped further deformation.

GONDWANA AND LAURASIA

Of such shields the more important are for the *Southern Hemisphere* that of Brazil—Guiana—Uruguay, Africa—Arabia

—Madagascar, peninsular India, Western and Central Australia, and Antarctica. These can all be viewed as integral portions of the basement of the great southern continent of " Gondwana," encircled by an ocean that trespassed deeply over its margins from time to time though failing to reach its interior until the late Cretaceous.

Those of the *Northern Hemisphere* are Central and Eastern Canada—Greenland (" Laurentia " or " Eria "), Fennoscandia (" Baltica "), North-central Siberia (" Angara "), North-eastern Siberia (" Kolyma "), and possibly South China-Indo-China (" Cathaysia "), the individuality of the last-named being somewhat uncertain. These can similarly be viewed as components of the equally large northern continent of " Laurasia " (Laurentia + Asia), which formed a single land-mass during only a few relatively brief periods, but for the rest of the time consisted of two or more relatively stable masses parted by widely trans-gressive, though impermanent seas.

As A. Holmes has well pointed out, the continuity of sial (upper crust) with sial does not necessarily mean continuous land. The lower levels of the continental platforms have always been more or less flooded by oceanic waters, just as they are today in the Baltic or in northern Canada, and hence there is no need to visualise Laurasia or Gondwana as having been ever free from epi-continental seas, though the latter certainly shifted their position quite considerably from time to time.

Normally these two major units were parted by an important sea known as the " Tethys " from at least the mid-Palæozoic onwards with general east-west trend, its depth varying enorm-ously in space and time, from shallow to abyssal, and its margins fluctuating accordingly. At several periods the shallowing of the Tethys at some point or points in its mid or eastern section put Gondwana into connection with Laurasia for a short while, until in the Tertiary era Africa, India and Eurasia (Europe + Asia) became permanently welded together. On the contrary, North and South America were quite separate right down to the mid-Tertiary, while Indo-China and Australia became united only on a few occasions during the past, and are today not linked together.

While the behaviour of both Laurasia and Gondwana has been wonderfully similar throughout geological history, it should particularly be noted that their responses have quite frequently been slightly out of phase, Gondwana having usually led during the mid-Palæozoic and Laurasia during the late Palæozoic and early Mesozoic.

This conception of a primitive, double land-mass is the logical outcome of Taylor's Hypothesis and is also that of Argand and Staub, and thus differs radically from the single Pangæa pictured by Wegener.

It should be realised, although usually ignored by geo-logists in their presentation of this problem, that the Polar Regions would appear to form the key to the grander structure of the Earth, hence the considerable and repeated attention given in this work to the Arctic and Antarctic.

EPEIROGENY

To G. K. Gilbert we owe the two contrasted terms, *Orogeny*, that is to say, a localised deformation of the crust, and *Epeirogeny*, a broader displacement in the vertical sense, either evenly or with limited warping into swells and basins, with or without simple faulting, most readily manifested in widespread advances or retreats of the oceanic waters over the lands, alternatively expressed as *negative* or *positive movements of the strand line*. Such normally shallow invading seas, frequently highly irregular in plan, can be styled *epi-continental* or *epeiric*, and as a rule are distinct from the more regular, furrow-like and commonly deeper troughs known as " *geosynclines.*" This is well brought out in palæogeographical maps. Such displacements have ever been characteristic of the continental shields, where they find their record in unconformities, overlaps or erosion surfaces.

Speaking generally, the important epeirogenic movements have fallen within the intervals between orogenic periods, and these two processes—epeirogeny and orogeny—are clearly genetically related. The causes of epeirogenic movements are pre-eminently the unloading of the lands through erosion, the weighting of the lower portions of the land surface or sea floor by sediment, and volume changes taking place in the sub-crust connected with radioactive heating, rock melting and igneous crystallisation and, subordinately, the growth or waning of ice-sheets and the eruption of lavas. The isostatic response of the crust ought to be nearly perfect, though naturally slightly delayed.

In addition to such movements, which have affected each of the continents somewhat differently, there are well marked simultaneous changes in sea-level of a world-wide character produced by the rising or sinking of the floor of an ocean-basin or by the locking-up in, or release of water from, ice-caps. To these the name *Eustatic* movements was given by Suess.[1] Such are best exemplified by raised or drowned marine beaches or the concealed platforms of coral islands. On a larger scale they show themselves in certain of the world-wide marine trans-gressions of the past, for instance those of the Middle Devonian, Middle Jurassic or Upper Cretaceous.

The well-defined and rhythmic nature of such eustatic move-

[1] Suess (06), II, 538

ments is not readily understandable under current theories,
chiefly because there is difficulty in explaining, save arbitrarily,
the corresponding deformation of the ocean bottom, though the
theory of thermal cycles of Joly and Holmes clearly points the
way.

A simple explanation of the phenomena is nevertheless to
be found in Continental Drift, since the volumes of both lands
and oceans would have been subject to variation in a definite
manner with each horizontal displacement of a crustal block,
as will be discussed later, so bringing epeirogeny and orogeny
into definite genetic relationship.

ISOSTASY

That the more rigid crust rests as a whole in *Isostatic equilibrium*—to use Dutton's term—upon its weaker sub-crust
cannot be doubted, or that loading and unloading of the surface
tends to be balanced or " compensated " by the sub-crustal
transfer of denser matter. The author has furthermore developed (Chapter XI) a new principle connected with the deeper
part of the crust—termed the Paramorphic Zone—under which
such vertical responses of the surface must become amplified
(p. 231). Its acceptance will almost certainly introduce differences into the results obtained from geophysical calculation.
It will indeed so modify Airy's hypothesis regarding the elevation and maintenance of mountain chains as to remove certain
difficulties connected with their presumed " roots " by which
they must be supported. The considerable thickness and
rigidity of the solid crust render it improbable that the smaller
topographical irregularities would be individually compensated.
The limiting size cannot readily be specified, but such an area
may perhaps range between 50 and 100 km. across.

It must be stressed that one should not imagine the ocean
floors, as is so frequently asserted or implied, as consisting
entirely of dense sima, even in the extreme " deeps " of the
Pacific, for the upper part must contain a certain thickness of
sial, probably materials of intermediate or even acid composition, the presence of which is definitely indicated from seismic
and other evidence.

It is, moreover, currently stated that above the zone of isostatic compensation—or zone of lateral flow in the sima—vertical
crustal columns of unit area possess practically identical masses.
The now-established fact that earthquakes often originate at
depths of several hundred kilometres is on the contrary proof
of internal readjustments taking place well below the zone of
compensation, which renders the above conclusion rather
doubtful. We cannot indeed be sure of finding equal-weighted

columns save within an area of no great size, possibly less than
1000 km. across. The postulated centripetal movement of the
land-masses towards the Pacific basin, if such be correctly
interpreted, would indeed point to a distinct lack of regional
balance in the spheroid.

Another weakness in the current scheme of compensation
lies in picturing the crust not from the dynamic, but the static
viewpoint, regarding it rather as a flexible layer in which the
horizontal components of the various forces tend towards zero.
The whole weight of evidence suggests on the contrary considerable and ever-present tangential forces, which have varied
greatly not only in amount but in sign in different parts of any
area even of limited dimensions; furthermore, such forces may
have persisted, though with varying magnitude, over considerable periods. Our hypothesis will enable us to understand how
this could be. In our opinion *Isostatic Compensation is regarded as a most powerful agent in the dynamic evolution of the
Earth.*

THE MOBILE AREAS

In contrast to the relatively broad and stable " swells " and
" basins ", which have been subject to epeirogenic movements,
are the unstable, narrower, linear or arcuate " welts ", " furrows "
or " troughs " that constitute the *Mobile* or *Orogenic Belts* of
the earth. They disclose all stages in the conversion of the
Geosynclines from relatively wide depressions of sedimentation
to restricted, crumpled-up and subsequently upheaved *Geanticlines* for, as has been truly said, " the geosynclines are the
cradles of the mountains." In such " revolutions " the parent
geosyncline is either reduced, displaced laterally or destroyed.

Such structures, which may be of enormous length—in
certain cases encircling the globe—are currently classified either
as *Intra-geosynclines*—that is to say, originating within the
continent itself—or *Inter-geosynclines*—forming between one
continent and another. Folding has developed within them,
commonly along one, less frequently along both sides, examples
thereof being the Rocky Mountains and the Alps respectively.
It is not, however, essential that the structure should be bounded
by land on both sides; indeed, an even more important type in
our opinion is the *Marginal Geosyncline* forming off-shore,
limited—if such a word can be used—on its outer side by a
submarine rise or moving anticline, here called an " *Advance
Fold* " (Fig. 48). The latter is held to result from the same set
of tectonic forces, and may ultimately rise above sea-level to form
an island chain or even a continuous belt—the " *Border Lands* "
of Schuchert—and incidentally a fresh source of sedimentation.
The main folding has then taken place either within this youth-

ful zone or along the edge of the mainland, examples thereof being the western and eastern sides of the Pacific respectively. Such a structure is considered to have played a highly important part in the evolution of the continents, not only occasionally, but at all times, though particularly during the Tertiary.

A modification thereof is the shallower moat-like depression encircling a whole continent, for example that surrounding Gondwana during the Palæozoic and early Mesozoic. This can appropriately be termed a "*Fossa*." Its progressive deepening over any extended period would manifestly have weakened the bonds holding together the central shield, which might ultimately result in the latter breaking into several portions. Such is held, for instance, to have been the case with Gondwana.

GEOSYNCLINAL SEDIMENTATION

Geosynclines have normally been initiated by *tension*, more rarely by compression, within and below the crust, and thereafter deepened not only by such forces, but, and in no small degree too, by sedimentary loading ; this view was long ago advocated by Dutton. The writer strongly favours this theory of sedimentary control in geosynclinal evolution, and will submit reasons therefor in Chapter XII.

While their floors have usually been depressed far below sea-level, the materials filling the hollows are not necessarily marine, but may include each and all of the other types— terrestial, deltaic, estuarine, paralic (i.e. containing repeated marine intercalations) or volcanic.

During the quieter initial and medial stages sedimentation is commonly normal or "*epeirogenic*," usually with three successive phases indicative of (1) transgression, (2) inundation and (3) regression, comprising the customary arenaceous, argillaceous, calcareous and occasionally deep-water or "bathyal" varieties of sediment. During the later or diastrophic stages sedimentation becomes "*orogenic*" and is marked out by coarse clastic deposits derived from the erosion of the emerging fold-ranges with maximal thickness usually next the latter, where local unconformities and overlaps are commonly to be found. Volcanic outbursts have furthermore been not unusual along one side of the folding.

While the lesser geosynclines have passed through the full orogenic cycle—from trough to fold-range—in perhaps a quarter or half a geological period, the more important ones have by the continued deepening and horizontal shifting of the furrow persisted over several periods at least. Such tendency of the geosyncline to "migrate" forms in a way a measure of its activity and age. Complications would inevitably be introduced by factors varying both in time and space and more particularly through the intermittent nature of the pressures within and outside the belt. The thickness of the strata that have thus accumulated exceeds in many cases 5 km.; and in the more enduring geosynclines or those that have been subsequently rejuvenated may attain to from 5 km. to 10 km. and even to 20 km. in exceptional instances.[1] It is important to note that highly folded sediments are usually several times thicker than the same formation in the undisturbed regions.

STRUCTURAL PLAN

Notable, as emphasised by Suess and almost universally subscribed to, is the general lack of symmetry of the orogenic zone in its later stages, due, primarily, to the unilateral or one-sided nature of the pressures, and, secondarily, to the uneven accumulation of sediments within the structure itself. Kober's generalised scheme of a double folding with bilateral symmetry is not admitted save as a special and restricted case.

Characteristic are the four structural elements, namely— following orthodox nomenclature : the wide and stable *foreland*, the generally narrowing trough of sedimentation or *foredeep*, the crumpling *geanticline* commonly invading the latter and the more variable and unstable *backland* marking the side from which the pressures have come. The upheaval of the folded zone need not synchronise with the compression, but commonly takes place at a relatively late stage.

Under our hypothesis the entire crustal segment is pictured as having moved horizontally, the " foreland " more than the " backland," thereby squeezing together the intervening fore-deep and particularly the farther side of the latter. Generally the foredeep has become pushed or underthrust beneath the fold-section which has tended to overturn backwards upon it, and thereby assist in its depression. Less frequently that rôle has been undertaken by the sunk-land region. A strong fore-deep would generally imply that it is supporting a fair propor-tion of such mountain belt (Fig. 38).

Conversely, as erosion proceeds not only do these chains rise, but the edge of the foredeep as well, and its recently formed deposits become elevated and thereby exposed to the forces of erosion.

So far we have tacitly assumed that the fold system has developed at *right angles* to the direction of the crustal pressures. While such seems to be on the whole true for marginal and for

[1] The exceptional value attained in the Cordilleran geosyncline of North America is doubtless connected with the distinctly high density of the sediments contained therein.

most of the longer and narrower geosynclines, it is far from universal, a point which, as Lee [1] has insisted, is too frequently overlooked. Where the pressures have been directed obliquely to the trough, where two troughs have been involved that meet at an angle, where rotation of a block has occurred, or where strong differential movements have taken place, a complicated system of diverging folds has resulted, sometimes without any obvious relation to the inferred pressures. Unequal advances in different sectors may furthermore lead to diagonal, branching and échelon folds of various patterns, to arcuate and linked fronts and, under more intense compression, to nappes, slides and vortices, a subject that has so lucidly been discussed by Bucher.

Two main classes of major structures can normally be distinguished : the *Alpine* or *Alpinotype* in which the geosynclinal sediments themselves have principally been affected, and the *German* or *Germanotype* in which older folded elements, usually of a less regular or fragmentary character, are also involved that would naturally interfere with the development of the younger foldings, an instance being those of Tibet. A renewal of pressure, though not necessarily in the same direction as before, would result in what is known as " *Posthumous folding.*"

IGNEOUS ACTIVITY

Typical of most orogenies is the intrusion of acid or sub-acid plutonic rocks as stocks and batholithic bodies introduced into the cores of anticlines or with cross-cutting relationships. These are generally considered to represent the acid differentiate either from basic igneous matter in depth or from the fusion of the lower part of down-folded portions of the crust, i.e. of the " roots of the mountains."

Despite severe criticism Harker's broad generalisation of the existence of " *petrographical provinces* " still holds. Zones of strong compression are thus characterised by igneous suites of intermediate composition and dominant calcic content—granodiorites, diorites, dacites, andesites and occasionally peridotites that have ascended from great depths—finely exemplified in the circum-Pacific girdle. Notable too is the way in which its volcanos are everywhere ranged along the inner edge of the fold-belt.

Areas of intense crustal tension—and especially disjunctive basins—are on the contrary marked by suites of alkalic content and variable composition, as along both sides and down the centre of the Atlantic. We may also add a third category, where simple epeirogeny has ruled and where more or less

[1] Lee (29), 362.

" neutral " types have usually been erupted, for example, the heart of the Pacific.

When it is recollected that certain territories experienced tension after strong compression, that localised areas of tension could normally develop in a compressional zone or *vice versa*, and that the sub-crustal magmas have often become differentiated or reworked as it were, an explanation is forthcoming for the many apparent though rather local exceptions to the above scheme which have so troubled petrologists.

THE MAJOR REVOLUTIONS

The so-called " revolutions " that punctuate the round of Geological Time consist of major orogenic cycles on which are superimposed minor rhythms, spaced wider apart in the Palæozoic and closer together in the Tertiary. The larger diastrophisms, termed " Caledonian," " Hercynian," " Alpine," etc., have controlled the destinies of enormous regions—incidentally revealing the mobility of this earth throughout geological history. Their enormous zonal development shows that during such orogenic crises the dynamical state of the earth was very different to that prevailing during the much longer intervals of quiescence. It is suggested further that such crises were essentially brought about by continental sliding, which then alone became possible.

The earlier view, still widely held, is the importance of " direction," foldings having similar courses being referred to the same orogeny even in distantly separated areas. Thus one finds foldings directed N.E.–S.W. in Brazil spoken of as " Caledonian " without any direct proof thereof. Doubt is at once cast thereon by the fact that fold-systems have a tendency not only to follow great or small circles across the globe but to show rapid changes in direction from point to point, especially those of the arcuate type.

The later, and the only reasonable view, arising out of Bertrand's presentation, is the *grouping together of individual foldings of the same age independent of their actual direction*. In recent years, however, the time limits have been drawn far too widely, thereby bringing under the one name movements that were not strictly synchronous. Palæontological considerations show that the crustal responses of Laurasia and Gondwana were very similar, yet at times somewhat out of phase, wherefore the use of such terms as " Caledonian " or " Hercynian " to label certain movements, say in Australia, tend merely to obscure the true position. For that reason local names will have to be introduced for certain of the orogenies of the Southern Hemisphere, until their dating has been more precisely determined.

TABLE OF EPOCHS AND OROGENIC REVOLUTIONS

Eras	Epochs	Principal Orogenies	Approx. Age in Millions of Years
Quaternary	Pleistocene		
Tertiary	Pliocene Miocene Oligocene Eocene	→ →　} ALPINE ; ANDEAN → → } LARAMIDE	20 40
Mesozoic	Cretaceous { U / L	→ } → INTER-CRETACEOUS ; PATAGONIDE → LATE CIMMERIAN ; NEVADIAN	60
	Jurassic		110 150
	Triassic { (Rhætic) / U / L	→ EARLY CIMMERIAN ; PALISADE → PFALZIAN ; GONDWANIDE	170
Palæozoic	Permian { U / L	→ SAALIAN } APPALACHIAN → LATE URAL	200
	Carboniferous { U / L	→ ASTURIAN ; ARBUCKLE } HERCYNIAN- → SUEDETIO ; ALTAIDE WICHITA → BRETONNIC	
	Devonian { U / M / L	→ EARLY URAL } ACADIAN → MID-DEVONIAN	
	Silurian { (Downtonian / U / M / L	→ HIBERNIAN → ARDENNIAN } CALEDONIAN	300
		→ TACONIAN	
	Ordovician		350
	Cambrian		450
Proterozoic	Pre-Cambrian	(Not considered)	
Eozoic	Archæan		

U = Upper; M = Middle; L = Lower.

CRUSTAL CREEP

For the major structures the amount of crustal shortening is indisputably large, hundreds of kilometres in some cases, even if we drastically cut down the extravagant estimates of certain geologists, especially for the Alpine and Himalayan regions. In the case of inter- or intra-geosynclines, the one crustal segment is currently regarded as having moved bodily nearer the other and "telescoped" it, so that farther along the Earth's circumference a certain section or sections ought by rights to have been *correspondingly stretched*.

This dilemma has driven a large number of geologists, among whom are Suess, Jeffreys, Stille, Kober and Bucher, into advocating *shrinkage of the earth* as the prime cause of its superficial compression. In the case of the marginal geosynclines the sinking of the ocean floor is indeed held to have pushed in the edges of the higher-standing continents, an assumption that is rather improbable for dynamical reasons. On the contrary, not a few authorities have doubted on physical grounds whether the amount of contraction to be anticipated could be as great as that to be deduced from field observation, while there are other weighty objections, such as the lengthy pauses between the revolutions, when shrinking would have to be presumed as dormant, as will be discussed in Chapter XVII.

The majority usually try to evade this difficulty, drawing all too frequently cross-sections of the crust that stop short at the ocean, which is conveniently permitted to hide the troublesome remainder of the earth's circumference and obscure its unruly behaviour. The impression gained from such schemes is that the hypothesis of a shrinking earth is merely a relic of the past retained to explain away this outstanding, although not the only important, difficulty in current diastrophic theory.

On the contrary, each and all of these troubles can be successfully met by the Hypothesis of Drift, under which the continental blocks are conceived to have moved not arbitrarily but in definite and orderly fashion towards or away from one another with contraction or expansion of the neighbouring oceans. Taken in conjunction with Radioactivity, it provides a harmonious picture of the unity of orogenic forms, the interrelation of orogeny and epeirogeny, alternating tension and compression in the same geosyncline, advance and posthumous folding, maximal sedimentation, etc., with isostatic adjustment playing a modifying and vital part throughout.

E

OVERTHRUSTING VERSUS UNDERTHRUSTING

The mechanism of the low-angle thrusts, recumbent folds and " nappes," so characteristic of the major fold-systems, has been the subject of strenuous discussion. The majority favour the view that the upper block has been driven forwards and upwards or " overthrust " upon the lower, which has remained passive. A. Heim has maintained that the thrust has always been directed towards the convex or outer side of the tectonic arc. A minority, among whom are notably Hobbs and Lawson, picture the upper block as passive and experiencing little lateral displacement, and the lower one as having been pushed beneath it or " underthrust." Quite a number, however, Mansfield and Nevin, for example, admit that both the masses could well have moved towards one another and that both over- and under-thrusting could have occurred.

Unquestionably the factors involved must not only have been complex, but have varied greatly during the progress of the thrusting. Whether the planes of thrust—or the axes of over-turned folds—would have become tilted in the one direction or the opposite, would appear to have depended largely upon the relative heights or stands of the opposed crustal segments, as Bailey Willis has pointed out, the depths of the accompanying foredeep or ocean trough, the materials involved and so on. Indeed, during subsequent stages the resistance offered by the piling-up slices might become so great as to lead to the develop-ment of thrusting in a contiguous section, quite possibly in the reverse direction, as has happened, for example, in the Alps. The numerous experiments conducted with models have only been helpful up to a point, since the conditions of the sub-crust (and its paramorphic zone, p. 231) have not been reproduced in them or allowances made for erosion of rising chains or filling of the deepening troughs during such deformation. The process in Nature must, furthermore, have occupied much time.

Current geology, it should be noted, is contented to regard the problem as a more or less academic one, and the movements of the two masses as purely *relative* ; as a matter of fact it is actually incompetent to decide the true motion of either body. Under the Displacement Hypothesis, on the contrary, the absolute motions of the blocks must be in accordance with the logical conception of such theory and therefore determinable. In the majority of instances—almost invariably in the case of the marginal fold-systems—the absolute movement of the lower block is visualised as having been *towards* the ocean and down-ward, the upper or displaced mass having *lagged somewhat* during the advance of the continent and so appearing to have moved landwards. In the current sense " *underthrusting* " has occurred. In a minority of instances the absolute movement of the upper mass has been *greater than* the advance of the continent itself, and such displaced mass has been pushed oceanwards and upwards over the materials composing the coastal belt. In the conventional sense it has been " *overthrust.*" Numerous instances will be submitted later where the directions of overturning and overthrusting are in full accord with the above general scheme. This is not surprising, since the moving continental block is regarded as the cause of such deformation with all its consequences.

DRIFTING OF BLOCKS

The horizontal movement of a block can be pictured as becoming possible through a temporary softening or melting of the basement under radioactive heating, although its motion is admittedly not easy to explain unless sub-crustal currents are also postulated. The crumpling-up of the leading edge is easily visualised, but the phenomena of the rifting in the rear are more difficult to interpret.

Shearing can, however, be conceived as taking place over one of the crustal shells—probably well down in the intermediate layer—with the dense sima displaced by the advancing fold-margin passing backwards in a counter-current and ascending behind in the opening rift. The horizontal sliding of the block would by removing an enormous load cause a considerable expansion in the paramorphic zone below, which, coupled with the sima ascending to restore compensation, would limit the depth of the newly formed ocean, while basaltic matter would be poured out over the sea-floor from numerous tension-fractures. A faint tilt of the block in the forward direction would generally result, thereby supporting the sliding hypothesis suggested by Daly. The mechanism of Drift is, however, discussed more fully in Chapter XVII

CRITERIA FOR CONTINENTAL DRIFT

For drift to be established with any degree of probability, it is essential that the following lines of argument apply not only singly, but so far as possible collectively.

1. *Physiographical*

(*a*) General similarity, though not necessarily full parallelism, of opposed coast lines—as, for example, the east of South America and west of Africa. This has, however, to be used with discretion for reasons set forth below (*b*) Fracture patterns along coasts shown by fault-line scarps, fiords and rias—Greenland and Labrador.

(*c*) Elevated peneplains with disturbed or reversed drainages—Greenland and Scandinavia—a criterion hitherto unnoticed (p. 257).

(*d*) Submarine features producible by drifting blocks, namely, swells, ridges and especially island festoons situated upon advance-folds developed by the forward wave of sub-crustal matter—Malayan arcs.

2. *Stratigraphical*

(*a*) Equivalent formations on opposite coasts with due regard to their mode of origin, lithology, facies, attitude, metamorphism, fossils, etc.—Cambro-Ordovician of eastern Canada and Scotland.

(*b*) Similar variation, when traced along opposite shores—Upper Triassic of East Africa and Madagascar.

(*c*) Contrasted phasal variation, when traced at right angles to or away from opposed shores—a most important criterion. Let us consider the case of two equivalent formations, one beginning at or near the first coast at A and extending inland to A^1, the other starting at or near the second at B and stretching away inland towards B^1. If the facial resemblance between A and B is greater than that between A and A^1, or between B and B^1, some drift can be suspected—the Siluro-Devonian sandstone series of the Cape and Argentina (pp. 62 and 110).

(*d*) Overlaps and unconformities on comparable horizons—Downtonian of Spitzbergen and Scandinavia, Keuper of New England and Portugal.

3. *Tectonic*

(*a*) Comparable geosynclinal troughs in each mass having more or less similar trends and histories.

(*b*) Comparable fold-systems passing out to sea at the opposed shores—Atlas and Venezuelan Andes.

(*c*) Crossings of fold- or fault-systems of varying ages—those of Palæozoic in North Atlantic (p. 161).

(*d*) Arcuate and échelon folds indicative of differential flowage in the crust—East Asiatic arcs.

(*e*) Rift-valleys, indicative of continental fracturing—East Africa.

4. *Volcanic*

(*a*) Synchronous intrusion of batholiths in equivalent fold systems—those of the Appalachian-Hercynian orogeny.

(*b*) Plateau basalts and associated dyke swarms—Iceland and Scotland.

(*c*) Petrographical provinces with similar eruptive suites of varying ages—Brazil and western side of Africa.

5. *Palæoclimatic*

(*a*) Strata denoting a special environment, particularly extreme climatic types such as tillite, varved shale, laterite, " evaporite " (salt, gypsum, etc.), æolian sandstone, coal, coral limestone, banded ironstone, etc.—with innumerable examples.

(*b*) Distribution of the above with reference to the climatic girdles of the past as indicating relative polar shift.

(*c*) Glacial deposits with special reference to ice-centres and ice-movement—Gondwana.

The above are criteria possessing great weight.

6. *Palæontological*

(*a*) Terrestial faunal and floral provinces with identical or allied species—coal basins of eastern U.S.A. and Western Europe, diprotodonts of Australia and South America.

(*b*) Relative distribution of primitive and specialised forms —scorpions in South Hemisphere.

(*c*) Distributions of fresh-water fishes—East Indies.

(*d*) Littoral and neritic forms indicating sea shallower than the ocean today parting the opposed shores, and hence denoting a former connecting shore, shelf, ridge, or swell—Oligocene fauna of West Indies and Mediterranean.

(*e*) Unique marine or terrestial forms—*Mesosaurus* in Brazil and South Africa, *Laplatasaurus* in Argentina and Madagascar.

7. *Geodetic*

Repetition of longitude and latitude measurements showing successive differences greater than the probable errors of such observations.

REASSEMBLY OF THE FRAGMENTS

With the help of such criteria we can proceed to replace those blocks in the positions which they inferredly occupied before the initiation of drifting in the Jurassic. During the lengthy period that has since elapsed, diverse influences have operated to modify their outlines for which full allowance must, of course, be made. The problem is hence far more complex than the proverbial jig-saw puzzle of current fancy. Modifications in their shape could have resulted from compression of the block within tectonic belts, internal distortion during drift, crustal warping of margin, erosion of fractured edge, coastal sedimentation, changes in sea-level with displacement of the strand-line, and development of the continental shelf.

The flattening out of the Appalachian foldings, when fitting North America to Europe, would extend America towards the south-east for a distance of fully 320 km.,[1] probably much more, while " ironing out " of the Cape foldings would prolong Africa to the south by at least 150 km. (p. 86). Much greater values have been deduced in the case of the Alps. The " plasticity " of the continents postulated by Argand has rightly come in for strong criticism, though, as pointed out by van der Gracht and others, a certain amount of distortion could be quite expectable. The geological map of any wide region will show the latter seamed by zones of deformation of various ages, sometimes interfering, often much fractured, and dotted with batholithic intrusions ; usually only small portions ("nuclei ") have escaped. In certain cases not merely the outline, but the entire fabric has become a caricature of that of the past. Because of the important principle of similitude emphasised by the late Archibald Barr, the elastic deformation of a continental block under horizontal shear must far exceed current expectations because of its excessive dimensions and non-rigid base. In addition, the internal movement through minute displacements along constituent mineral grains, joints, cleavage-planes, etc., cannot be negligible. The tectonic pattern of the earth is full of instances of crustal distortion obviously due to differential horizontal movement, well marked in the narrower masses such as Alaska, Central America and Japan.

Nevertheless the crust should not be pictured as a plastic mass to be moulded to the whim of the reconstructor : allowances for presumed distortion should only be made where and when the collective evidence becomes overwhelming. The several reconstructions by Wegener, Argand and Baker unquestionably involve not a little distortion in their attempts to secure a close fit, and van der Gracht's comments thereon are worthy of study. Actually we are most concerned with the correct replacement of the original central broken edges or " sutures," the peripheral portions of the reassembled unit being of less importance in the problem because of their repeated fluctuations under epeirogenic movements.

Most persons view the continental shelf as an integral part of the continental block, and criticise Wegener for endeavouring to fit together the masses by their present coast-lines instead of by the submerged margins of the shelves. It should be noted that in his fourth edition he has recognised the desirability of so doing. For Baker's reconstruction, wherein the opposed shores are usually brought almost into contact, that argument

[1] Keith (23), 335.

loses much of its force because of the quite recent date for the separation postulated by him.

The author cannot agree that the shelf-margin has necessarily to be viewed as marking the fracture-edge of the continent. In certain places the shelf has been built out recently and is still being extended, in others it would seem to include far-travelled masses—for instance, the Falkland Islands, which show no structural relationship to near-by Patagonia. It is, on the contrary, thought with van der Gracht that this comparatively youthful and marine-developed feature—the shelf—may well conceal displacements of considerable magnitude and that greater weight may have to be placed on the evidence from phasal variation shown by the strata of the opposed lands.

In the case of South America and Africa the evidence would indeed suggest that the crustal matter forming their respective coasts today were perhaps never nearer together during the past than from 250 to 350 km. (p. 107). In any restoration the ragged intervening space between the present shores would have been land mostly, lost after fracture and during drift by erosion and/or submergence, much of it perhaps going to form the continental shelves. In other parts of the globe such secondary losses might have been very much less, and the blocks would then have to be brought closer together in the reconstruction.

It might incidentally be pointed out that such an inevitable " erosion gap " in the fitting has a high significance when the rock materials, structures, life, etc., of the opposed lands are under comparison. Not a few critics write as though an absolute identity must be found before any drifting could be conceded. One has merely to compare parallel lines taken a few degrees apart drawn across any continent to realise the importance of structural, phasal and biological variation within even those short intervening distances.

Through our limitations of knowledge and ideas any graphic restoration of the lands must perforce be tentative, and the scheme shown diagrammatically in Figs. 7 and 16 must therefore be viewed in that light. While following those of Taylor and Argand, it departs considerably from that of Wegener, and introduces several important modifications in the case of the Southern Hemisphere. For convenience the orientations of Africa and of Eastern Europe are taken as remaining unaltered, and to that roughly meridional tract will be referred all statements of direction as affecting the restored masses. The minimum of allowance is moreover made for compression.

The Falklands are set between Bahia Blanca and the Cape, Madagascar is moved northwards (not south-westwards as is generally done), Antarctica is transferred far to the north, Australia is rotated through some 50 degrees, while New

Guinea, New Caledonia and New Zealand are brought next its eastern side.

The reassembly of North America and Eurasia largely follows conventional lines, though some slight rotation is favoured for Spain and perhaps Newfoundland, and Spitzbergen and some condensation for the East Indies. The main difficulty is in making allowance for the deformation of eastern Siberia and Alaska.

Taking everything into consideration the fit thus obtained is a marvellous one, though even greater agreement could be secured if allowance were to be made in various places for the inevitable distortion of the blocks during their drift, as will be set forth later. It will be noted that in this scheme the seas are now spread out widely between North and South America and between Eastern Asia and Australasia, the Pacific enlarges enormously, the Atlantic and Indian oceans disappear, while the Arctic shrivels to a fraction of its size.

More thorough comparisons between the various masses should permit a much better and closer fitting to be effected in the future.

CHAPTER IV

GONDWANA—SECTION I

Introduction. Palæozoic Grouping of Gondwana. Late Silurian Transgression. Lower to Middle Devonian Transgression. The "Samfrau" Geosyncline. Upper Devonian Regression. Devonian Faunal Provinces. Devono-Carboniferous Orogeny. Lower Carboniferous Geographical Changes. Upper Carboniferous Movements. Carboniferous Glaciations. Glacials. Ice Radiation. Circumpolar Grouping. Post-Glacial Beds.

INTRODUCTION

So far back as 1887 M. Neumayr visualised a continuous continent extending from South America to Africa and sending a branch north-eastwards to India, and in this he was supported by W. T. Blanford in 1890. It was, however, Suess who further synthesised the data and conferred on this mass the name of "*Gondwána-Land*."

It is nevertheless a curious commentary to find even today persons who refuse to admit the former existence of such a continent ; to them Gondwana is as nebulous as the traditional "Atlantis." It will, however, be conceded that no truly comprehensive account of our hypothetical southern mass has ever been penned.

The past decade has witnessed vast accessions to geological and palæontological knowledge concerning the lands of the Southern Hemisphere, and no longer is it possible to doubt the reality of Gondwana. Especially can be cited the notable contributions on South America by Windhausen (31), du Toit (27) and Gerth (32 ; 35) : on Africa by Gregory (21), Fourmarier (28), Krenkel (25 ; 28), Mouta (33) and Veatch (35) : on Madagascar by Besairie (30) : on India by Wadia (26) : on Malaya by Scrivenor (31) : on Australasia by David (14 ; 32), Süssmilch (35), Bryan (25 ; 26) and Benson (23 ; 24)—to mention only some of the more comprehensive studies. The literature being vast, scattered and rather inaccessible, our review has purposely been made somewhat lengthy in order to establish beyond question the former unity of the southern lands and to confirm the general correctness of our continental reconstructions.

PALÆOZOIC GROUPING OF GONDWANA

If the several existing components of Gondwana be assembled to form a larger whole, as in Fig. 7, some striking peculiarities are brought out by the recurrent marine submergences between early Palæozoic and early Mesozoic   These affected the same general stretch of country and were almost entirely *marginal*, even when the negative movement was considerable.   Such becomes intelligible only if those lands composed a unit, when the transgressions would have constituted spillings from the surrounding ocean over more or less the same coastal regions. If, however, each component be viewed as an independent continent, it is hard to understand why marine inundation should repeatedly have affected one, or at most two sides of each mass throughout the enormous periods involved without causing the submergence at some time or another of the other free sides. For example, no marine strata of age between Devonian and Lower Triassic have yet been found on either side of the mid-sections of Africa and Peninsular India or on the southern side of Australia.

Such strongly supports the hypothesis of the former closer union of those masses into the single continent—Gondwana. That in turn involves a more or less stable " shield " subject to epeirogenic movements and surrounded by an unstable trough, orogen or " fossa."   That fossa need not have had the same character everywhere ; it could have been (*a*) the open ocean with floor sloping away from the continental shelf (as is common), (*b*) a " deep " with the farther side made by a " swell " that failed to reach the surface (as off the west coast of South America), (*c*) a sea bounded by an island festoon (as in the east China Sea), or (*d*) one hemmed in by land (as in the Sea of Japan).   The nature of the fossa must obviously have varied in different places and at different times.   Into it would have been shed waste from one, less commonly from two sides, and it would therefore have acted as a " collector-trough," its subsequent history being controlled by the normal Orogenic Cycle.

It is arresting to observe, both in South America and Australia, that, despite the repeated *marginal depression and compression*, there has been since the Devonian an *outward movement* on the whole of the shore-line—i.e. in the westerly and easterly direction respectively, with a *net gain in land area*. The same would seem to apply to the extreme north and south of Africa, though in less degree.   Against that must be set off the losses sustained during the fragmentation of Gondwana, due to the foundering of relatively narrow strips during its break-up or to the erosion of the fractured edges by the sea, an amount which could hardly have exceeded, say, 500 km.

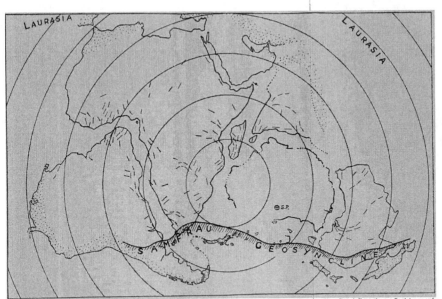

Fig. 7.—Reassembly of Gondwana during the Palæozoic Era. The space between the various portions was then mostly land. Short lines indicate the pre-Cambrian or early Cambrian "grain." Diagonal ruling shows the "Samfrau" Geosyncline of the late Palæozoic. Stippling marks out regions of late Cretaceous and Tertiary compression. (Lambert's Equal Area Polar Projection.)

between one mass and another, and which is believed in general
to have been much less.

LATE SILURIAN TRANSGRESSION

For our purpose it is needless to go back further than the
Silurian, which was one of slow uplift and peneplanation over
most of Gondwana with accumulation upon its surface, over
the north-western and south-western sides at least, of the
products of its weathering and the deposition of arenaceous
sediments within the marginal fossa.

During the Upper Silurian negative movement of the
Continent deepened the main geosyncline of Eastern Australia
and brought in the sea over Tasmania, past Melbourne to
Cobar and thence by way of Charleville to Chillagoe at least,[1]
Dutch New Guinea, Burma, China and onwards. On the
opposite side of Gondwana the Amazon syncline had only
shortly before come into being.

Over large portions of the remaining periphery of the
continent the slow subsidence of the peneplain and the re-
working by water of the waste thereon led to the formation of
pale unfossiliferous grits and sandstones of uniform siliceous
character, often strongly false-bedded. With steady sinking this
type of accumulation became extended inwards across the
faintly inclined coastal region in a covering of considerable
thickness and enormous extent unconformably upon the highly
eroded basement.

Such is represented, for example, in the Muth Quartzites of
India—Spiti to Afghanistan. In Western Africa we find a
great development thereof between Fezzan, Sudan border and
Sierra Leone, a distance of 26 degrees, around the domed-up
region of Ahoggar (Arak, Tassili, Ahnet, Tamanrasset), in
French Sudan from Hombori within the bend of the Niger over
the Bandiagara plateau to the Black Volta and thence westwards into Guinea and northern Sierra Leone and over the
Senegal basin to Bakel and northwards far into Mauritania.
Under the name of the Voltaian System (Kitson) it occupies a
wide area in Gold Coast on the southern side of the very flat
arch made by these beds (Fig. 8). The French geologists
delimiting this formation in the territory of the Upper Niger
have found it to be split over a wide area by a horizon of dolomite.
Lying for the most part horizontal, it becomes openly folded to
the north-west in Senegal. The base rests with strong unconformity upon crystallines and sediments, some so young as
Ordovician, which have been folded along axes trending N.-S.
in the north, but veering to S.S.E. and S.E. in the south.

[1] David (32), 46.

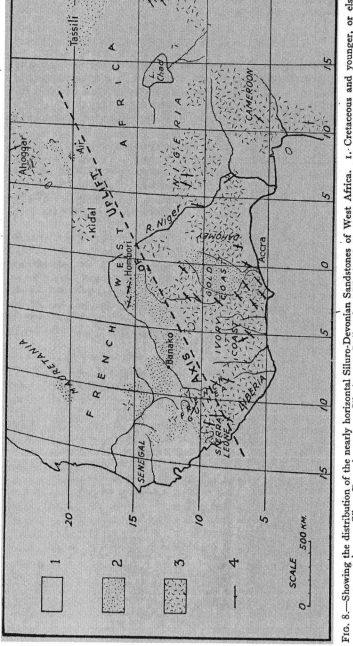

FIG. 8.—Showing the distribution of the nearly horizontal Siluro-Devonian Sandstones of West Africa. 1. Cretaceous and younger, or else uncertain; 2. Siluro-Devonian; 3. pre-Silurian; 4. Strike of pre-Silurian formations. (Compiled from various sources.)

In South America this formation is possibly represented in the horizontal unfossiliferous Roraima sandstones of British and Dutch Guiana and eastern Venezuela. Unquestionable, however, are the arenaceous beds of the lower part of the Amazon syncline, the sandstones at the base of the fossiliferous Devonian in Bolivia (Icla Sandstone); western Argentina—Salta and Jachal; western Brazil—Diamantino; southern Brazil—Paraná basin (Furnas Sandstone); Uruguay (Carmen Sandstone); southern Argentina—Sierras of Buenos Aires; Falklands and Cape (Table Mountain Sandstone). The massive West Coast conglomerates and quartzites of western Tasmania display a remarkable similarity, but are reputed to be older. The lowest beds of the Reefton Devonian in New Zealand (S.I.) are quartzitic also.

From the huge areal extent and presumed manner of accumulation of this arenaceous facies upon a broad subsiding continental margin, its age cannot everywhere be the same. In the northern Sahara and Mauritania it is essentially marine and carries a Lower Devonian fauna; in the southern Sahara (Tassili), French Guinea, Bolivia and western Argentina (Jachal) the basal portion has yielded *Monograptus priodon* indicating an Upper Silurian (Wenlock or Ludlow) age; in the Amazon syncline, where incidentally there is a conformable succession upwards into the Devonian, its base is even slightly older. Elsewhere the formation is unfossiliferous, but is succeeded in many areas by marine Lower, although not Lowest, Devonian. Its age can therefore be taken as essentially *Siluro-Devonian*.

It is suggested that *wind* played an important part in its accumulation by blowing inland quartz-grains pulsated on an extremely flat shore, to which they were returned by rivers. A cold, moist climate is indicated: indirectly by the generally bleached nature of the sands — like those of " podsols " —and lack of plant life, and directly by a fairly widespread glacial zone in the south-western Cape, intra-formational folding proving that the ice came from a source situated out in the present South Atlantic.

In the Cape the formation becomes finer-grained and loses its typical quartz pebbles when traced southwards, i.e. away from its source, while in the Sierras of Buenos Aires the sandstones are less coarse than their northern equivalents in Uruguay and Paraná.

It should be noted that in Gondwana there was on the whole continuity of deposition from late Silurian to Devonian in contrast to Laurasia in which sedimentation normally began with the Devonian, save over limited areas, where the transitional Downtonian series was represented (p. 151)

LOWER TO MIDDLE DEVONIAN TRANSGRESSION

Eustatic depression of the continent,[1] and probably encroach-
ment by the marginal fossa, brought in the ocean over the
identical extensive regions previously mentioned, i.e. over
Algeria, Morocco, Sahara, Mauritania, Gold Coast, practically
all the central territories of South America (with north-west
Argentina as an island), the Falklands and southern Cape.
Madagascar and most of Antarctica lay apparently just north
of its waters.

The sediments in this *Western sector* of Gondwana — in
marked contrast to the Eastern—show wonderful uniformity,
being mainly dark grey, blue or greenish mudstones, micaceous
sandy shales and thin sandstones together with thicker sandstone members, while limestones are normally absent. Shallow-
water conditions are denoted by their lithology and the scarcity
of corals and crinoids, while there are reasons for believing that
the waters were cool. This facies is typically developed in a
belt crossing the central Sahara, but passes in the northerly
direction into the calcareous bathyal phases of the Atlas region.[2]
The faunas throughout this wide region possess many species
in common, and are strikingly of the facies termed " Austral "
by J. M. Clarke, as discussed below.

During the Middle Devonian the fossa maintained its depth
on the north-west—in West Africa, Amazon Valley, Venezuela
and Bolivia—but shallowed considerably on the south-west and
south.

THE " SAMFRAU " GEOSYNCLINE

Emphasis must be laid at this point on the extraordinarily
similar sequence of events in South America, South Africa and
Australia, stage by stage, from the Silurian up to the early
Cretaceous, as shown in the attached table. Taken in conjunction with a mass of supporting evidence, it favours the
idea of a major geosyncline directly connecting these countries
—traversing Bolivia, north and central Argentina, Cape,
Weddell Sea, passing east of King Edward VII Land and
through Edsel Ford Land, crossing Tasmania and the eastern
part of Australia to New Guinea, its inner margin advancing or
retreating from time to time. On an atlas such a course would
be lengthy and irregular, but under our reconstruction it would
deviate but little from a gentle arc and have a length of only
some 110 degrees in all (Fig. 7).

[1] The hypothetical South Atlantic continent of this time has been called " Flabellites Land " by Schwarz, a name which has been widely repeated, though as in-
appropriate as the expression " Brontosaurus Sea " would be.
[2] Fourmarier (28), 848.

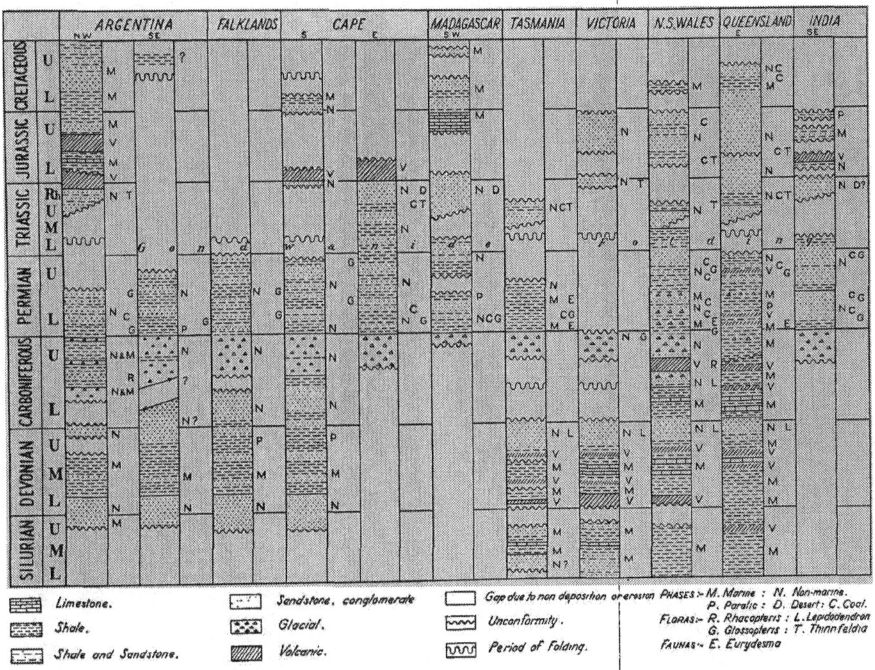
ARGENTINA
NW
SE
FALKLANDS
CAPE
S
E
MADAGASCAR
SW
TASMANIA
VICTORIA
N.S.WALES
QUEENSLAND
E
INDIA
SE
CRETACEOUS
JURASSIC
TRIASSIC
PERMIAN
CARBONIFEROUS
DEVONIAN
SILURIAN
G o n d w a n a l a n d
Limestone.
Shale.
Shale and Sandstone.
Sandstone, conglomerate
Glacial.
Volcanic.
Gap due to non deposition or erosion
Unconformity.
Period of Folding.
PHASES:- M. Marine : N. Non-marine.
P. Paralic : D. Desert : C. Coal.
FLORAS:- R. Rhacopteris : L. Lepidodendron
G. Glossopteris : T. Thinnfeldia
FAUNAS:- E. Eurydesma

This feature, which seems to have played so vital a rôle during the evolution of Gondwana, and which was already in existence in the Ordovician, can conveniently be called the " *Samfrau* " *Geosyncline*—a contraction of the words " South America–South Africa–Australia." Its eastern portion has indeed for long been known as the " Tasman " Geosyncline, to the outer side of which belong New Zealand, New Caledonia and other islands.

Picking up its history at the Lower Devonian we find Tasmania forming land, while in Victoria terrestial volcanoes were discharging the Snowy River porphyries, in places over 700 m. in thickness ; the conditions across Antarctica are unknown, but in the Cape shallowing was certainly in progress. That the sea-way was not entirely blocked is, however, shown by the remarkable development of " Austral " faunal types, significantly in a shallow-water quartzite-mudstone facies, in the Lower Devonian of Reefton, New Zealand (S.I.),[1] Lat. 42° S., Long. 172° E., as will be discussed below. Down-warping continued thereafter in Queensland and New South Wales, which by the Middle Devonian had attained its maximum, reaching northwards almost to the Gulf of Carpentaria and southwards over Victoria and New Zealand and perhaps as far as Tasmania.

The immensely thick strata characterising the deep eastern part of the trough include radiolarian phyllites and red and green jaspers, limestones, tuffs, rhyolites, spilites and basalts with intrusions of keratophyres, etc., igneous activity having been rife in the mid-section. In north Queensland and in Papua massive and largely coralline limestones are dominant.

This Middle Devonian sea was continued along the northern shore of Gondwana through Yunnan, Burma, Himalayas and thence to the Mediterranean region, thus surrounding and isolating Gondwana.

UPPER DEVONIAN REGRESSION

The sea withdrew northwards in the Sahara and probably persisted in shallow form in the Amazon region, but between Argentina and the Cape the Lower Devonian marine succession was followed by fluviatile or lacustrine strata—sandstones, quartzites and shales—with fragmentary remains of plants such as *Haplostigma*, " *Cyclostigma* " and the fucoid-like worm-trail *Taonurus* (*Spirophyton*). In Australia a similar change from marine through estuarine to terrestial facies was general, *Lepidodendron australe* and " *Cyclostigma* " being widespread. Some crustal movement is signalled by the unconformable

[1] Allan (35).

attitude of Upper upon Middle Devonian in Victoria, while thick conglomerates were developed in the Burdekin area of Queensland, but New Zealand, New Caledonia and parts of New Guinea were probably land and presumably formed the outer side of the Samfrau (Tasman) geosyncline.

On the contrary, in Western Australia the Upper Devonian sea invaded the Kimberley district—an overlapping from the general Devonian sea-way with its European fauna that bounded Gondwana on the north.

DEVONIAN FAUNAL PROVINCES

The close of the Silurian brought strong faunal differentiations the world over. Laurasia fell within the *Boreal Province*, which is distinguished by its variety of marine invertebrate life and by the presence of well-defined faunal zones, though, as it happens, there were marked differences also between those of Europe and North America.

The *eastern half of Gondwana* also fell within the *Boreal*, the *western half* within the *Austral Province*. The latter is characterised by uniformity in composition, poverty in gasteropods and corals, presence of pteropods, peculiarities in the trilobites, and above all by the persistence of species through considerable thicknesses of strata with consequent absence of marked palæontological zones. In explanation thereof one can only advance the deduced uniformity of oceanographical conditions throughout a lengthy period, the shallowness of the water, its generally low temperature and just possibly a salinity below normal. Characteristic forms include *Spirifer antarcticus Leptocoelia flabellites*, *Orbiculoidea baini*, *Vitulina pustulosa* and various species of *Conularia, Homalonotus, Phacops, Dalmanites*, etc., as described by Salter, Ulrich, Clarke, Reed, Lake, Kozolowski and Rennie.

While European forms are few, with a general absence of identical species, there are quite a number from the Helderberg and Oriskany of North America, particularly in the section from Bolivia to the Cape, less noticeably in the Maecurú fauna of the Amazon. Even in the Tassili region of the Sahara an austral bias is given, as noted by Haug, by various North American species, though the fauna as a whole is Rhenish, as could be anticipated on geographical grounds. The assemblages in question represent generally the Lower though not the Lowest Devonian.

During the Middle Devonian the affinities with North America become even stronger, when Onondaga and Hamilton species are well represented in the assemblages of Paraguay, Bolivia (Sicasica), Amazon (Ereré), Gold Coast and Sahara

(Ahnet), a relationship that persisted into the Upper Devonian in the Tassili region.

Reverting to the early Devonian, it is striking that the fauna of the Falklands is more closely allied to that of the Cape (with which about one half of the forms are common) than to those of southern Brazil or Bolivia, thereby supporting the view that these islands were then geographically nearer the Cape. The writer has suggested that in view of the shallowness of the sea the Paraná region might have occupied an embayment in which faunal changes could have taken place, an idea advanced independently by Gerth.

Now the Samfrau trough with its austral fauna cannot have come to an abrupt ending off Africa, and the deposits laid down in its presumed easterly extension may perhaps be discovered to the east of the Weddell Sea or the south of Graham Land. It has already been remarked that beyond Antarctica this trough would appear to have been partially obstructed during the Lower Devonian and so cut off from the open geosyncline of Eastern Australia, which harboured a distinct and typically Boreal or European fauna.

Impressively the equivalent neritic fauna of Reefton, New Zealand, includes the characteristic austral form *Leptocoelia flabellites* together with a *Homalonotus* and a *Lingulidiscina* closely allied to, if not identical with, species at the Cape.[1] R. S. Allan[2] rather minimises such austral affinities, possibly because of the great distance separating South Africa from New Zealand and the apparent lack of a suitable connecting sea-way. Under our scheme, however, the latter is provided while the distance is exactly halved—50 degrees of arc as against 100. The palæontological similarities are hence viewed as real and not dominantly due to independent evolution under similar environments, though such was unquestionably an important factor as well. It is worth recording that the discovery of a fauna of this type within Tasmania or New Zealand was visualised by the writer as the logical outcome of the Hypothesis long before Allan's illuminating paper came to his attention.

It is clear that the " African " influence must be fading out at Reefton, for many of the forms obtained there are definitely European, but the lack of close similarities with the corresponding Eastern Australian faunas would suggest an intervening barrier between Australia and New Zealand at that time. Such was, however, over-topped at the close of the Lower Devonian, as shown by Australian corals in the higher strata at Reefton. It is, moreover, noteworthy that precisely during the Middle Devonian some North American forms appear in Eastern Australia and even in Burma, though not in Western

[1] Allan (35), 34·5. [2] Allan (35), 33.

F

Australia or the Himalayas, and in the Upper Devonian in Eastern Australia a fauna allied to the North American Chemung (Dun ; Gürich). This is normally puzzling in that in North America the Chemung fauna did not reach the Pacific coast, but such is explicable under our hypothesis of migration from *eastern* North America along the western and southern shores of Gondwana. The apparent absence of marine Middle and Upper Devonian in the section—central Argentina to Cape—forms no objection, since the equivalent strata within that region are non-marine and must have been deposited along the northern margin of the Samfrau trough with the sea-way situated farther to the south.

DEVONO–CARBONIFEROUS OROGENY

The late Devonian regression heralded an important orogenic period, when the Samfrau geosyncline became intermittently compressed—presumably from the south—with much igneous activity. In a rough way this corresponded with the Acadian Disturbance in Laurasia. On the contrary, the northern margin of Gondwana escaped deformation.

In Peru and Bolivia the Eastern Cordillera was upheaved with much plutonic intrusion — the Alta Planicie escaping — and in Argentina the Pampean ranges were folded along nearly N.–S. axes and injected by granite. Whether the extreme south of Gondwana was similarly affected remains unknown, that sector being today mainly concealed by ocean or Antarctic ice.

In Eastern Australia the *Kanimbla Epoch* of Süssmilch [1] occupies a lengthy period and embraces several tectonic phases, earlier in the south, generally speaking, the first of them late Devonian (Tasmania and Victoria), the second post-Devonian but pre-Kuttung (New South Wales and south-east Queensland), the third post-Lower but pre-Upper Marine (north-east of New South Wales), while still younger episodes may also be represented. These various movements were accompanied by protracted eruptions of lavas and tuffs (dacites, rhyolites, andesites, porphyrites, keratophyres) and by abundant plutonic intrusions (granite, granodiorite, gabbro, alkaline rocks, etc.), the ultrabasics of the Great Serpentine belts of New South Wales (Benson) and Queensland (Bryan) appearing to belong to this epoch also. The later Palæozoic granites of Australia have been termed the " Tin Granites " by Andrews (25). The masses tend to be arranged along or to be elongated in the direction of the foldings, which run N.–S. in the south but veer to N.N.W. in the north-east, a course pursued incidentally by subsequent orogenies. [2]

[1] Süssmilch (14), 80-82.   [2] David (32), 71.

Viewing Gondwana as a whole, igneous activity began about the same time on both western and eastern sides, but was continued longer and was, moreover, more intense on the east. Throughout it was characterised by intense metallisation, which gave rise to immense deposits of gold, tin and tungsten ; indeed, this was the second of the three dominant metallogenic periods in the Southern Hemisphere. Distributed symmetrically on both sides of the continent the ore deposits fade out to the north and probably to the south as well. In south-eastern Peru and in Bolivia gold predominates ; in central Argentina (Pampa ranges) tin, tungsten, lead and some gold and copper ; in Tasmania tin with some tungsten, copper and lead ; in Victoria the gold deposits of Bendigo, Ballarat, etc. ; in New South Wales numerous occurrences of tin ; in Queensland the gold at Gympie. The tin and tungsten lodes of Stewart Island, New Zealand, may perhaps belong to this important epoch.

LOWER CARBONIFEROUS GEOGRAPHICAL CHANGES

Information concerning this period is singularly scanty, since the encircling fossa moved generally outwards, thereby leading to the deposition off the mainland of estuarine or continental deposits or else exposing the sea floor. Within the trough itself deepening went on with the accumulation of thick sediments generally quite conformable with those of the Upper Devonian and accompanied in places by eruptions. In the south, on the contrary, a ridging-up of the centre or else of the *outer side of the Samfrau geosyncline* appears to have been initiated both to the east and west of Antarctica, such developing into the pre-Gondwanide folds that on rising shed detritus northwards, i.e. inwards, into the trough and ultimately upon the mainland itself.

The greater part of South America became land, though not the " Devonian " syncline of the Amazon between " Brasilia " and " Guiana." *Inside* the pre-Gondwanide foldings continental sediments were deposited in the " foredeep," extending from northern Peru through Bolivia (Titicaca), north-western Argentina (La Rioja), west Argentina (San Juan), but *outside* at Paracas on the present coast of Peru. The various floras— in which *Rhacopteris* and *Lepidodendron* are prominent—indicate a Culm or, at latest, Westphalian age. In south-eastern Peru, a little to the north of Lake Titicaca, Douglas[1] discovered what must be a marine equivalent—Upper Avonian, i.e. uppermost Lower Carboniferous—while a similar occurrence is revealed at Leoncito Encima in western Argentina. Near by in San Juan the writer found both these marine and fresh-water

[1] Douglas (20).

phases to be associated with glacials, as will be described later. Recently one of the plants from Paracas, identified by E. W. Berry as *Palmopteris furcata*, but revised by Gothan as *Sphenopteris parasitica*, has been recorded by Oliveira from beds in Piauhý in north-eastern Brazil. In the stretch between central Argentina and Tasmania no marine equivalents are known, only some fresh-water beds with scanty plants having Devonian as well as Lower Carboniferous affinities (Falklands and Cape), and a loose block with fish-scales (Devonian) from near the South Pole.

In Australia the geosyncline moved outwards and marine Tournaisian and Viséan were deposited to the east of a line running from Sydney to Bowen *via* Barraba, Warialda, Texas and Coondarra. In the Queensland section, according to David, this made a long gulf bordered by land on both sides, the apparent narrowness of which is deceptive since allowance has to be made for the considerable subsequent compression. There are traces of another embayment lying to the north-west in the Burdekin Valley (Star Series) and back of Cairns.[1] The Burindi (Viséan) seas, which connected with the Asiatic Tethys, were shallow and warm, as shown by the abundance of oolitic limestones as well as reef-forming corals, but with a scarcity of foraminifera.[2] To the west, i.e. towards the shore, these give place to estuarine and continental strata characterised by *Lepidodendron veltheimianum*, e.g. Drummond Range in Queensland, and to the north also. Massive Viséan limestones mark the continuation of the Upper Devonian seas in the Kimberley district of Western Australia.

During the close of the Lower Carboniferous folding took place locally, while the shore-line of Gondwana migrated outwards generally. In Eastern Australia were formed the Kuttung and equivalent strata with the *Aneimites-Rhacopteris* Culm flora, for example, in New South Wales between Port Stephens and Bingara and in Queensland at Mount Mudge (Drummond Range) and Newelton near Cairns, precisely as in western South America.

The succession in Spiti, India, provides a close parallel with that in New South Wales, the Lower Carboniferous Lipak Series (marine) being followed by the Po' Series with *Rhacopteris* below and *Fenestella* above, succeeded by a conglomerate presumed to be glacial.

In western Sinai strata with *Lepidodendron mosaicum* and *Halonia tortuosa* rest on thin marine beds with *Spirifer mosquensis* and represent, according to Seward,[3] either the top of the Lower or bottom of the Upper Carboniferous. Floras of

[1] See Süssmilch (35), Fig. 1.   [2] Benson (23), 30.
[3] Seward (32), 355.

about the same or slightly later age having European and North American affinities are recorded by Zeiller and Charpentier for southern Oran in Morocco, among which is *Archæosigillaria vanuxemi*, also discovered in the continental strata of 'Uweinat in north-western Sudan,[1] which last are widely distributed and are known to extend westwards into French territory, overlain by the mesozoic " Nubian Sandstone." In the western Sahara the marine Lower Carboniferous is widely developed and, like the Devonian, passes into calcareous phases towards the north, while beyond in eastern Morocco, that is to say, on approaching Laurasia, it changes to a Culm facies and is followed by Upper Carboniferous, which includes coals and carries a typical " Northern " Flora.

UPPER CARBONIFEROUS MOVEMENTS

During this period Gondwana experienced some warping. Marine beds are now known from a boring at Therezina, Piauhý. Deepening of the Amazon syncline took place, where the Carboniferous rests disconformably on Middle and Upper (?) Devonian, and in the pre-Gondwanide " foredeep " from western Venezuela through Colombia, mid-Peru and mid-Bolivia, limestones and shales were deposited characterised in places by *Fusulina*. Bowman states that in northern Peru these limestones decrease in thickness from east to west, while E. W. Berry points out that in Bolivia this calcareous facies changes in the westerly and southerly directions into a littoral or continental one, and he consequently maintains the absence of any westerly connection with the ocean then occupying the Pacific, from which the interior sea must have been parted by the Lower Carboniferous (pre-Gondwanide) fold-ranges.

The strong faunal similarities of this " pre-Andean " foredeep with the Amazon region have led Berry and Gerth to favour a direct connection between them, although a strait continuing northwards to Colombia is not excluded.

During the late Carboniferous and early Permian the Pacific trespassed in north-western Peru upon the western side of the barrier chain in the Amotape Mountains, but failed to cross the latter, as shown by faunal differences with the Amazon Permian. Farther south the Upper Carboniferous fauna of Barreal, western Argentina, and the Lower Permian fauna of central Chile were probably also derived from the Pacific region.

The other South American faunas are of late Carboniferous or Uralian age with *Productus semireticulatus*, *P. cora*, *P. lineatus*, *Spirifer condor*, *Chonetes glaber*, *Seminula*, *Dielasma* and sometimes *Phillipsia*, and have many species in common

[1] Sandford (35), 338.

with those of the Urals and the south-western part of the United States of America.

Of high interest are the occurrences of plant-bearing Upper Carboniferous (*a*) at Gachala in Colombia [1] with *Calamites*, *Cordaites* and *Neuropteris*, and (*b*) in Cochabamba in Bolivia [2] with *Lepidophloios* underlying fossiliferous Uralian—the only localities where such Arcto-carbonic floras are yet known in this continent.

In the eastern and southern parts of South America, Falklands, South Africa, Madagascar and peninsular India the younger Carboniferous is represented by glacials, followed by Permian continental strata with the Glossopteris Flora, and in Eastern and Western Australia and in Tasmania by similar successions with occasional marine intercalations or phases, as set forth in the sequel.

As regards Australia during the Carboniferous and Permian the significance of our palæogeographical reconstruction must be pointed out. Not only has the Samfrau geosyncline a general east-west trend, running therefore roughly parallel to the Hercynian and Altaide geosynclines and foldings (Chapters VIII and XVI), but its history is closely related thereto. The Altaide folding and regression in Central Asia during the mid-Carboniferous was apparently marked by corresponding movements in southern Gondwana, where ridging-up took place, notably in Australia, from Adelaide to Darwin, as David has pointed out, thereby prolonging the Northern Territory outwards —that is to say, eastwards under our scheme—probably for a good distance. During the widespread Uralian transgression the waters of the Himalayan geosyncline spread over Burma, eastern Yunnan, Tonking, East Indies and Western Australia, but were prevented by the aforesaid barrier from reaching Eastern Australia save by a route round it lying well to the east. South Australia remained, as before, attached to Antarctica and hence to Africa.

The Upper Carboniferous and "Permo-Carboniferous" faunas of Eastern Australia and Tasmania, though distinctly of Tethyan facies, accordingly became somewhat specialised, hence the remarkably few species in common with those of Western Australia, as is finely brought out in Benson's review (23) of these regions. It should not be overlooked that migration would nevertheless have been feasible *westwards along the Samfrau trough*, which could readily explain the presence of the New South Wales *Eurydesma* fauna in South-west Africa and of the *Stuchburia* fauna in southern Brazil (Reed)—both puzzles under orthodox ideas.

New Zealand was seemingly land during the early Car-

[1] Gerth (32), 133.   [2] Berry (33).

boniferous, and was possibly situated on the outer side of the Samfrau geosyncline though closer to Australia than at present, but it and New Caledonia were partially submerged in the Permo-Carboniferous with the deposition of sediments carrying faunas of East Australian and Tethyan affinities.[1]

Along the north of Africa there seems to have been only slight uplift during this period.

THE CARBONIFEROUS GLACIATIONS

Early workers such as Koken, with limited data at their disposal, came to the natural conclusion that the Gondwana " System " commenced with a *single stratum of morainal material* of approximately the same age everywhere. Since it underlay strata yielding fossils of either Permo-Carboniferous or Permian affinities, that glaciation was deemed to have been of Lower Permian age.

Each year makes it clearer that this great " Ice-Age " of the Southern Hemisphere was not a single episode, but a complex series of glaciations with milder interludes, embracing in all a *very considerable period of time*. There has already been revealed a whole group of tillites, fluvio-glacials and inter-glacials referable to separate ice centres and overlapping in time as well as space, precisely as during the Pleistocene. Indeed the plurality of tillites is becoming embarrassing, and is rendering precise correlation somewhat difficult. The associated fossils are moreover pushing the beginnings of this multiple refrigeration further and further back, until today the conclusion is unescapable that glaciation began in places so far back at least as the Lower Carboniferous—late Viséan in western Argentina and Namurian in New South Wales. The " post-glacial " strata are generally Lowest Permian, showing that the *main refrigeration*—itself compound—must fall within the Carboniferous. Later recurrences, unquestionably falling within the Lower Permian, are admitted for New South Wales (Lower and Upper Marine Series) and possibly Bolivia. Indeed the time is long overdue for the elimination from literature of the alleged great " Permian Ice Age," which still finds a number of adherents, notable among whom stands Professor C. Schuchert.

So far as our problem is concerned, the absolute time-scale is not a serious matter, since the mutual relationships of the various occurrences are seldom under real dispute, their litho-logical, stratigraphical and palæontological similarities suggesting incidentally the intimate geographical connection of the several areas known to have been ice-capped during the late

[1] Benson (23), 36.

Palæozoic. Our geographical and historical readings will, however, be affected only when closer comparisons have to be made between those dominantly glacial formations and their *non-glacial equivalents* in northern Gondwana or in Laurasia.

The literature is already so vast that only a brief résumé can be submitted ; for details and arguments the reader will have to be referred to the more important contributions by Koken, Holland, David, Süssmilch, Schuchert, Howchin, Coleman, Keidel, Fox, Ward and du Toit. The latest and certainly the best presentation is, however, *Die Permokarbonischen Eiszeiten* by Salomon-Calvi (33), a concise and balanced review in which general conclusions are incidentally reached rather similar to those held by the writer.

In the following pages it is assumed, as argued elsewhere, that save for sporadic ice-action, glaciation had on the whole ceased by the beginning of the Permian, a view which, although strongly opposed by Schuchert, has received wide support from those directly connected with the working-out of the stratigraphy of those deposits.

THE GLACIALS

The following are the principal occurrences known with the direction of ice-movement, as inferred from striated floors or boulder-pavements, given in brackets :

1. *In Pre-Cordillera of Western Argentina*

(*a*) Tillite of Leoncito Encima (N.N.W.) with intercalation yielding *Syringothyris* and *Spirifer*—seemingly Viséan [1]—and presumed general equivalent at Barreal overlain by strata with *Spirifer supramosquensis*—regarded by Reed [2] as probably denoting the lower part of the Upper Carboniferous.

(*b*) Sierra Chica de Zonda in San Juan (N.N.W.) with three tillites having *Rhacopteris* and *Cardiopteris* above the lowest, which is therefore of late " Culm " or Namurian age.

(*c*) Tillite of Jachal (N.W.) and Sierra de Umango—the extension to the north of (*b*).

2. *Sierras of Buenos Aires*

Compound tillite of Rio Sauce Grande near Sierra de la Ventana (N. 65° W.) overlain by the Bonete Series with mollusca and the Glossopteris flora.

3. *Uruguay*

Compound glacials of Fraile Muerto (N. 40° W.) and Gregorio (E.–W.) with unconformity at base.

---

[1] Private communication from Dr H. Harrington.
[2] du Toit (27), 149.

4. *Brazil* [1]

Around and presumably beneath the southern half of the great Paraná Basin between Lat. 21° and 31° S. Generally a single basal tillite resting everywhere unconformably on older rocks, but compound in certain areas, as in the State of São Paulo ; thin or missing to the S.E. in Rio Grande do Sul, which is supposed to have lain near the ice-centre (N.W., as indicated by erratics). Overlain by shales carrying in Paraná *Lingula* and in Santa Caterina *Aviculopecten, Stuchburia* and other forms reminiscent of New South Wales and succeeded by the Bonito " Coal Measures " with a Glossopteris flora. The tillite is not definitely known north of Lat. 21° S.

5. *Falkland Islands*

Thick " Lafonian " tillite (N.–S.) resting disconformably on Devono-Carboniferous and followed by Glossopteris beds.

6. *South Africa*

Vast extent of Dwyka tillite originating from several distinct centres, usually thin and single in the north but thick and compound in the south (S. in west ; S.W. in centre and south-east), followed in South-west Africa by sediments with *Eurydesma, Conularia* and *Myalina,* and elsewhere by lacustrine shales and Glossopteris beds. Not known just north of the Transvaal for certain.

7. *Central Africa*

Tillites of Cassange district of north-central Angola, Lat. 9°-10° S. ; Long. 17°-18° E. (Mouta and O'Donnell [2]) and of eastern Congo in the Lukuga region (N. and N.E.) (Fourmarier), and between Lake Kivu and the Lualaba River (N.W.) (Ball, Shaler, Passau, Horneman and Boutakoff),[3] indicating more than one ice advance and overlain in the Congo by lacustrine beds with Glossopteris. The northerly limit as yet known is about Lat. 1° N.

8. *Madagascar*

Boulder beds of the Sakamena-Sakoa area in the south-west (direction unknown), followed by black shales and coal-bearing strata with the *Glossopteris* Flora and that by an horizon carrying *Productus.*

[1] Summarised by Oppenheim (34), 7-15.
[2] Mouta (33), 49, Pl. VII.
[3] Reviewed by Veatch (35), 118-26 ; 147-57.

9. *India*

(*a*) Talchir tillite occupying wide areas in the Central Provinces, Orissa and Bihar forming the base of the Gondwana System and varying greatly in thickness (N.N.E. to N., though probably variable), overlain by shales and Glossopteris beds.

(*b*) Thin Boulder-bed of the Salt Range followed by strata containing *Eurydesma* and *Conularia* and those in turn by the "Productus Limestone," which includes a layer with the Glossopteris Flora (N.N.W. as suggested by boulders).

(*c*) Boulder-beds of Tanakki and Talhatta in Hazara associated with the peculiar " Agglomeratic Slate " series, and of Blaini at Simla.

10. *Eastern Turkestan*

Reference is here made to the glacials discovered by Norin (30) resting on metamorphosed Lower Carboniferous (?) next the Tarim basin—Lat. 40°-42° N. ; Long. 88°-89° E.—and overlain by marine beds, probably Permian, because, although they are situated far within Asia, they are not very distant from those of Spiti.

11. *Australia*

(*a*) The *five* glacial horizons of New South Wales spreading over a lengthy period : the two earliest in the Kuttung Series (Namurian) (N. 13° W.) associated with *Rhacopteris, Aneimites* and *Lepidodendron*, succeeded by those of the Kamílaroi System, namely, the Lochinvar glacials, the Lower Marine Series, beds with *Gangamopteris*, Allandale glacials with *Eurydesma* (Lower Permian), Greta Coal Measures and the Branxton and Muree glacials in the Upper Marine Series (Middle Permian). Recently the Kuttung glacials, volcanics and flora have been discovered to the west and south-west of Cairns up to Lat. 17° S. in northern Queensland.[1]

(*b*) The glacials of Victoria, South and Central Australia (Finke River) and Tasmania and those of the Irwin, Wooramel, Gascoigne and Lyndon rivers and of the Desert Basin in the Derby district of Western Australia, overlain by marine strata with 'a Tethyan fauna of Upper Carboniferous and Permo-Carboniferous affinities, and probably those near to and to the north-east of Laverton, Western Australia (Ward (25)). (The ice radiated outwards from the Tasmanian region during the late Carboniferous, reaching the sea in Western Australia and northern New South Wales.) The recurrence during the Permian is marked by erratics in the Upper Marines in the

---

[1] Süssmilch (35), 101.

Minilya River area of Western Australia up to Lat. $23\frac{1}{2}°$ S. and in the Bowen River area of Queensland up to Lat. $20°$ S.

ICE RADIATION

The above crystallises our knowledge of the movements of the late Palæozoic Ice, which is a matter for congratulation when it is remembered that the tell-tale floor is usually exposed only within a selvedge, the pavement passing in the one direction beneath the tillite and being destroyed by erosion in the other. Information is nevertheless scanty regarding the northern edge of the regions that were glaciated, and much remains to be learned concerning it.

One of the chief merits of the Displacement Hypothesis lies in its professed ability to explain the present-day distribution of such Palæozoic glacial remnants ; indeed, to many persons the grouping thus afforded constitutes the weightiest argument in support of the theory of Continental Sliding. While such an attitude is a narrow one, there is no doubt that under the new conception the insuperable physical obstacles met with under current views can be enormously reduced, even if all the difficulties cannot effectually be disposed of as yet.

As is well known, ground moraine is commonly scantier in the vicinity of the *glacial centre*, where the rock-floor was more prone to post-glacial erosion, while away from that quarter the glacials usually attain their maxima to fade out beyond into outwash gravels, boulder-shales or " varve-shales." In the heart of territory suspected to have been glaciated the moraine may be missing at the base of the Gondwana succession or only preserved in troughs, as is the case in the Transvaal or Congo. One can well imagine the difficulty of deciphering the glacial history of, say, Canada, eroded for about a million years, stripped of most of its drifts, buried beneath new sediments and once again eroded.

The extreme occurrences in Argentina and New South Wales are today fully 222 degrees apart, those of South and Central Africa 31, those of Tasmania and North Australia 25, and they consequently mark out enormous spaces on our present globe. Under the scheme proposed (Fig. 9), the regions in question become condensed into an irregular oval only 95 degrees across from west to east and 63 from north-north-west to south-south-east, one more comparable with the maximum area known to have been glaciated in the Northern Hemisphere during the Pleistocene, which is set by Antevs [1] at 80 degrees in length by about 55 in width.

[1] Antevs (29), Fig. 1.

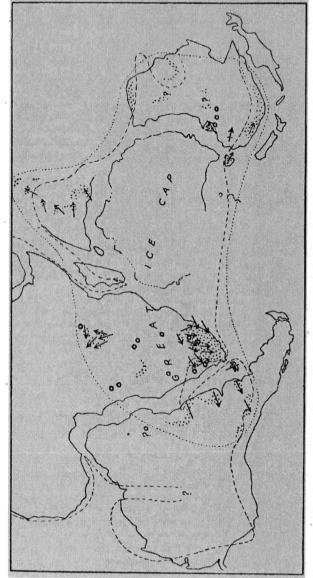

FIG. 9.—Showing by stippling the distribution of the late Carboniferous Glacials in Gondwana. Arrows mark deduced ice-flow; lines indicate that precise direction is uncertain; dotted lines mark limit of ice; broken lines indicate edge of continent; circles denote occurrences of Ancient Tillites.

CIRCUMPOLAR GROUPING

It is worth noting that were one to take the central point from Antevs' map as marking the Pole, such would fall into western Greenland and some 12 degrees away from its true position. The Pleistocene Ice-Cap was therefore *eccentric* to the North Pole, provided that the latter had not shifted. A similar eccentricity could possibly have characterised its Palæozoic representative, which would then affect our geographical reconstruction, perhaps materially. Ignoring, however, such a contingency and placing the Carboniferous South Pole somewhere off the coast of Moçambique, a more or less radial movement therefrom will be disclosed for the several sections of Gondwana (Fig. 9). Coleman's insistence that the South African ice must have formed the southern part of a far greater mass has been brilliantly confirmed by subsequent discoveries in Angola and Congo, though much has still to be learned about those lands. The conditions over Antarctica are as yet unknown, though in our scheme that entire continent is compelled to fall within the sphere of ice-action. Assembled as suggested, the component portions become mutually consistent : separated as they are today they are in certain essentials scarcely intelligible.

So far as can be made out, ice-capping started earliest in Argentina, extending outwards, perhaps in increasing waves, with general easterly migration across the face of the continent, just as in the case of the Pleistocene sheets, the general agreement with the latter being indeed remarkable.[1] Some shifting of the Pole relative to the land would almost certainly have occurred during the lengthy period involved, which would help to explain the large total area glaciated as well as its elongated shape, particularly if one accepts the Tibetan occurrence as part thereof, though such could perhaps have constituted an outlier.

Criticism has been passed that with such a large land-mass insufficient water-surface would have lain near at hand to provide the evaporation needed to feed the enormous ice-fields. On the contrary, over extensive areas the ice-front discharged into the ocean or else the sea lay not far away, as indicated by the marine post-glacials, while at least two gulfs penetrated far into the continent, one on either side of Indo-Madagascar. Indeed, about the middle of the Carboniferous, Gondwana seems to have shrunk to about its *minimum*. Furthermore, as the writer has pointed out, it is unlikely that the entire territory known to have been affected was ever completely covered at

[1] du Toit (21 a).

one and the same time. Again, as a study of Antevs' maps will
bear out, the region in question would not have been less
favourably situated physiographically for ice-capping than that
of the Pleistocene.

POST-GLACIAL BEDS

Proof that the glaciated regions were on the whole rather
low-lying is afforded by the wide distribution of the succeeding
marine, paralic or lacustrine shales, boulder-shales and mudstones with occasional limestones that mark the cessation of
polar conditions at the close of the Carboniferous.   Their
dominantly fine-grained nature is obviously due to the small
relief of the ice-worn continent, the high proportion of bare
rock-surface and the clayey texture of the morainic covering.
Present in western and central Argentina, eastern Congo,
southern Madagascar, Salt Range and Peninsular India
(Talchir), they cover or else underlie huge areas in Uruguay
and Brazil as the carbonaceous pyritic Iratí (Iratÿ) shales
containing the little reptile *Mesosaurus*—with calcareous facies
towards the north—and in South Africa as the similar and
equivalent " White Band " of the Dwyka—also with that fossil
—and have a great development in the Lower Marine Series of
Western and Eastern Australia and Tasmania.

They were deposited in shallow seas, in basins like the Black
Sea or Baltic or in estuaries having a low salinity, where the
bottom muds were in places charged with sulphuretted hydrogen.
While marine fossils have in places been found in the top of
the glacial matter or even in bands carrying erratics, such
remains tend to become more plentiful and wider-spread higher
up, say between 100 and 250 m. above the morainic deposit.
This unique sedimentation was ended by the earlier phase of
the Gondwanide Orogeny.

The presence in particular localities of such forms as
Eurydesma, *Productus*, *Spirifera*, *Protoretepora*, *Parafusulina*,
Paralegoceras or of palæoniscid fishes, dates these " post-
glacials " as more or less contemporaneous and allots them at
latest to the Artinskian, that is to say, to the Lower Permian.

CHAPTER V

Gondwana Systems and their Geographical Setting. Permo-Triassic Sedimentation. Glossopteris Flora. Permo-Triassic Vertebrate Life. Gondwanide Orogeny. Upper Triassic-Rhætic Transgression. Trias-Rhætic Floras. Mesozoic Desiccation. Trias-Liassic Volcanicity. Jurassic. Initiation of " Drifting." Cretaceous.

GONDWANA SYSTEMS AND THEIR GEOGRAPHICAL SETTING

THE concept by Suess of " Gondwanaland " was largely formulated from the widespread distribution in the Southern Hemisphere of those dominantly terrestial stratal assemblages showing the " Gondwana " facies, which have invariably been called " Systems," although frankly admitted not only to embrace strata ranging in age from Carboniferous to Jurassic, but to straddle the Palæozoic and Mesozoic eras. Research has enabled these thick non-marine assemblages to be split up into series and stages, and their inter-correlation to be effected with fair satisfaction considering the immense distances normally involved (see table opposite p. 62). Difficulties admittedly exist, for instance, the precise horizons at which the boundaries corresponding to the European Systems should be drawn, though such are in the way of being settled.

As the writer has elsewhere emphasised, the stratigraphical resemblances between the " Systems " as represented in the several parts of Gondwana are truly remarkable, and constitute a powerful argument for a single and " condensed " unit in the past rather than for several widely separated areas linked together by hypothetical land bridges.

An explanation of these systems can furthermore be found in the intermittent compression and upheaval of the Samfrau Geosyncline that traversed four of the existing components, and a concrete picture thereby got of the physiography of this great southern unit from the Carboniferous to the Jurassic—one vastly different from the unsatisfying impressions otherwise to be gathered from current literature.

The close of the Carboniferous saw renewed movement
within this lengthy trough—which foldings can, following
Keidel, be termed the "*Gondwanides*"—in several pulses,
which attained their maximum during the early Triassic and
which appear to have corresponded approximately with the
Asturian, Saalian and Pfalzian phases in Laurasia. The up-
heaving of such exterior fold-chains, the evidence for which will
be presented later, would have led to the progressive cutting-off
of portions of the encircling ocean, to their conversion into a
chain of great lakes by the damming back of the internal
drainages over the "foredeep," to the silting-up of the latter

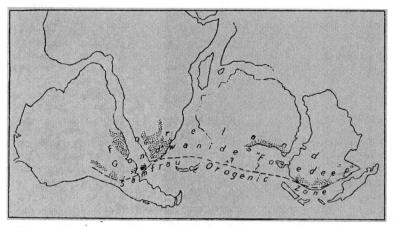

FIG. 10.—Showing general distribution of the Permian Gondwana strata (stippled)
bordering and connected with the Samfrau Orogenic Zone.

from both mainland and rising geanticline and to the subsequent
crumpling of those new sediments by the invading plications.
The general absence after the earliest Permian of marine intercalations from the foredeep is proof of the efficacy of this barrier,
which can well be appreciated on inspection of those folded
chains of hard quartzitic rocks as actually preserved in Argentina,
Falkland Islands and Cape. When the sea reached the interior,
it did so much later, only during the Triassic, and round either
end of the lengthy barrier—in Bolivia and Queensland—or
else from the unprotected north—in Brazil, North Africa and
Arabia. The accumulation of sediments within the foredeep
would have been governed by ever-varying factors such as the
uplifting of the barrier, its down-cutting by the escaping rivers,
the climatic environment, etc. In the interior region secondary
depressions would have developed under tectonic or sedimentary
warping (Fig. 10).

Only under some geographical arrangement, such as here presented, could non-marine sediments have accumulated on the scale and to the thickness found, during which process lakes and swamps played no small part, despite the predominance of " interior basin " conditions.

In each of the components of Gondwana there is a striking absence of marginal Permo-Triassic fold-ranges set at an angle to the Gondwanides, which could have served to complete the sides of the basins in which these fresh-water deposits must have been laid down. On the contrary, over great stretches the little-disturbed Gondwana beds actually reach the present oceans or pass in those directions beneath younger coastal sediments.

Sedimentation attained a maximum and was generally least interrupted in the Gondwanide foredeep that stretched from Bolivia through La Rioja, Sierras of Buenos Aires, Falklands, Southern Cape and the eastern edges of Tasmania, New South Wales and Queensland. The axis of this long depression is accordingly characterised by a conformable or pseudo-conformable succession, while away therefrom the newer groups—markedly the Upper Triassic—tend to transgress across the older ones, so as ultimately to rest upon the pre-Gondwana formations. Over the foreland proper the successions are conspicuously thinner, while overlaps and unconformities are common and sometimes widespread, exceptional being certain troughs in East Africa and Madagascar, which suggestively adjoined the Indo-Madagascar gulf.

Coals are all but absent from the axial portion of the trough probably because of too active a crustal movement, but vast quantities thereof are to be found over the more stable region to the north, although, unlike the Carboniferous fields of the Northern Hemisphere, but few seams occur in superposition. Coal-forming began much later than in Laurasia : first in the Lowest Permian (Bonito) in Brazil, shortly thereafter in the Lower Permian in South Africa (Ecca)—Natal, Transvaal, Rhodesia, Nyasaland, Tanganyika, eastern Congo—southern Madagascar (Sakoa), India (Karharbari), and Australia and Tasmania (Greta). Younger coals were formed in the Upper Permian in Argentina (Paganzo), South Africa (Beaufort), India (Damuda), Australia (Newcastle) and Antarctica. More restricted still were the Upper Triassic and Rhaetic seams, though significantly within the same general areas just specified. As could be anticipated, the floras of the equivalent formations contain stage by stage a high proportion of common or related species belonging to the Glossopteris vegetation.

The influence of the rising Gondwanide chains on the climate of the interior foreland must have been considerable. Even today their much-eroded remnants in the south of the Cape

G

effectively reduce the precipitation over the Karroo despite the warm Agulhas current close by. Thus can be explained the wide development of Permian-Rhætic strata of the " Red Beds " facies, indicative of an aridity that culminated during the closing Triassic and affected no inconsiderable proportion of Gondwana. As will be seen further on, the geographical setting was seemingly the duplicate of that which prevailed in Laurasia during the period—Carboniferous to Triassic—where, however, such desiccation started a bit earlier (p. 167). Noteworthy is the general absence of salt and gypsum-bearing strata, the exception being those in North Africa, Eritrea, Abyssinia and Somaliland belonging to the Triassic, which represent a littoral or lagunar facies that preceded the Liassic submergence.

PERMO-TRIASSIC SEDIMENTATION

The non-marine strata bridging the Palæozoic and Mesozoic belong to one or other of two fairly distinct facies : (a) deltaic, lacustrine or swamp, and (b) basin, playa or eolian, deposits.

The former embraces conglomerates with worn pebbles, current-bedded or laminated grits, whitish sandstones, grey or black laminated shales, coals, fireclays, pyritic and ferruginous concretions and plants, but a scarcity of vertebrate remains. The climate at the time was normally humid and either warm or cold.

The latter is characterised by breccias, unsorted conglomerates with angular material, yellow or red felspathic sandstones—sometimes with strong cross-bedding—more or less unstratified arenaceous mudstones and mudstones of blue, green, red, purple or variegated tints, calcareous or dolomitic concretions, occasional vertebrate remains, but scarcity of plants, save silicified wood, and a general absence of coals. The climate was sub-normal, semi-arid or arid, giving rise to intermittent rivers and temporary lakes, while wind-action usually played an important part. Because of its dominant coloration such is generally known as the " Red Beds " facies, though it must be recognised that red sediments can equally well be produced under a humid tropical environment.

The particular facies represented varied naturally both in space and time, but in a rough way there was a progressive increase in such basin types in passing up through the Permian into the Triassic ; further reference will be made to this in Chapter XIII. A most illuminating review of the conditions under which these various formations were accumulated in Gondwana will be found in Case's monograph " Environment of Tetrapod Life in the late Palæozoic of Regions other than North America " (26). As he has well pointed out, such change in the direction of aridity influenced profoundly the evolution of all biological elements (p. 204).

THE GLOSSOPTERIS FLORA

Up to the close of the Culm—Lower Carboniferous—period the vegetation was more or less uniform throughout the world ; thereafter, during perhaps the Westphalian, typically European assemblages were represented in the northern part of Gondwana in western Sinai, Oran, Sahara, Bolivia and, just possibly, North-western Australia,.for which a wider distribution will doubtless be discovered in the future. Even during the succeeding Stephanian the *Arcto-carbonic* or *Northern Flora* existed in Colombia, Bolivia, perhaps Piauhý, Malay States and Sumatra, though the last two areas may perhaps have formed marginal portions of Laurasia. The northern tree-fern *Psaronius* is not uncommon in Brazil.

To the south of a line, which for the present can be drawn from northern Argentina through São Paulo, north-eastern Congo, Uganda, northern Tanganyika and Kashmir, the *Antarcto-carbonic* or *Southern Flora* is alone known to be represented. Some intermingling took place in the beginning in Argentina, southern Brazil and South Africa in the shape of *Sigillaria, Lepidodendron, Cordaites* and *Sphenopteris*, but not in India, Australia or Antarctica, though at a slightly later date a high proportion of Eurasiatic elements such as *Pecopteris* and *Sphenophyllum* is found in Southern Rhodesia and some in northern Queensland.

All this confirms the idea that the Glossopteris Flora originated during the Carboniferous far to the south, perhaps, as Seward has suggested, in Antarctica, possibly, as the author fancies, in Argentina. It should be noted too that this vegetation, so far as our limited information goes, occupied at least four-fifths of the Gondwana of that time, whence it spread during the mid-Permian into Russia, presumably *via* Persia, while certain elements reached Central and Eastern Asia. The uniformity of the vegetation generally, from Argentina across to Australia, from the Carboniferous to the Rhætic, stage by stage, is proof first of the remarkable evenness of climatic and other conditions, and secondly of the absence of geographical hindrances to spreading, such as mountain ranges or seas. The Glossopteris Flora is nevertheless characterised by a poverty in genera and species in strong contrast to the contemporary vegetation of Laurasia.

Although it was closely associated with glacials in the basal part of the Gondwana System, there are no adequate grounds for supposing, as has so often been done, that such was essentially a very cold-climate flora. The genus *Gangamopteris* is more restricted both in space and time than *Glossopteris*, disappearing after the Lower Permian, whereas the latter genus survived in a few places into the Triassic and mingled with the Thinnfeldia Flora. For details the illuminating writings of Seward (24 ; 31) should be particularly studied.

PERMO-TRIASSIC VERTEBRATE LIFE

The Karroo is as yet the one great region in the world wherein is preserved a practically unbroken chain of vertebrate life from the middle of the Permian to the close of the Rhætic as revealed in a host of genera and species. Away from this type region similar forms are, however, being obtained in ever-increasing numbers, and it is striking that under our reconstruction that expanding life province forms a fairly well defined belt trending east-north-east from the Karroo through Nyasaland, Tanganyika, Madagascar, India into Indo-China and Mongolia, which, moreover, coincides pretty well with the area showing a "*red beds*" *facies* of sediments and incidentally with the desert region of the late Triassic. Case, von Huene, Haughton and Nopsca have all commented upon the profound influence that such a dry environment would appear to have had upon the course of evolution. Outside the tracts in question no typical Permian forms have yet been discovered.

By the close of the Lower Permian there were already established in the Cape two dominant groups, the *Pareiasauria* and *Therapsida*, confined to the south-western corner of the Karroo basin, from which they spread with strong evolution. The L. Beaufort (lowest Upper Permian) reptiles are closely related to those of the presumably equivalent " Copper Sandstone " of the Kazan Stage of the Urals—with a Northern Flora, be it noted—but, as all palæontologists agree, are distinct from those of the North American and European Permian with the exclusion of the Russian, though the two faunas must undoubtedly have sprung from a common stock.

It will in the sequel be shown that during the Carboniferous North America was united to Western Europe, and with much probability to north-west Africa, so that migration could have been possible from this North Atlantic land-mass southwards into Gondwana just as was the case with the Northern Flora. During the late Carboniferous the great southern ice-cap would have barred the way, and so migration would have been possible only down the western side of South America until Gondwana had been freed from ice. Haughton (30) has given reasons for supposing that the ancestors of the Karroo animals inhabited an area to the south or south-west of the Karroo, which in accordance with our view would correspond to the early Gond-wanide up-warpings between Argentina and the Cape. If so, similar or earlier forms may ultimately be discovered in the strata of the Sierra de Pillahuincó to the south of Buenos Aires. It is true that no Permo-Carboniferous amphibia or reptilia have yet been found in the pre-Cordillera or in north-eastern Brazil, but in the limited state of our knowledge such negative evidence is decidedly unsafe.

The Therapsida seemingly originated in Gondwana, being unknown in North America, though found in Russia and sparsely in Scotland. The L. Beaufort fauna is rich in Anomo-dontia, Therocephalia, Dinocephalia and Gorgonopsia, certain forms of which migrated northwards *via* Tanganyika into the Ural region, presumably round the eastern end of the Zechstein (Upper Permian) sea by way of Persia, a route that seems to have been revived during the later Triassic. The uppermost L. Beaufort fauna (uppermost Permian) is as widely spread, being represented in Nyasaland and Tanganyika and in the Dwina area of northern Russia, the " Scutosaurus Beds" of which contain Pareiasauria, Dicynodontia and Therocephalia with strong Karroo alliances, though forms having North American affinities are also present and, strikingly too, repre-sentatives of the Glossopteris Flora. The " Gordonia " fauna of Elgin, Scotland, is furthermore allied to that of the Karroo.

The Lower Triassic shows the genus *Lystrosaurus* set forth by a wealth of species in South Africa, in India (in the Panchet, associated with Dicynodon), at Laos in Indo-China and in Tibet (Tsingling area)—together with that strange Karroo reptile *Chasmatosaurus* ; also the aquatic *Hovasaurus* of Mada-gascar allied to *Tangasaurus* of East Africa and *Youngina* of the Cape.

Typical Triassic European amphibia, such as *Trematosaurus*, *Capitosaurus, Cyclotosaurus* are found in the Karroo, *Mastodon-saurus* in India and *Capitosaurus* and *Cyclotosaurus* in Australia, while the South African *Batrocosuchus* belongs to a family only known from India and Australia.

During the Upper Triassic the community of forms is even more remarkable. The Rio do Rasto " red beds " of Brazil contain Rhyncosauria and Parasuchia, known from Europe, India and East Africa, Cynodontia and Dicynodontia—includ-ing the large *Stahlekeria* (von Huene) closely allied to the South African *Kannemeyeria*, and *Scaphonyx* closely related to the Karroo *Erythrosuchus*—while India has yielded *Hypero-dapedon* and *Gomphodontosaurus*, and South Africa numerous Cynodontia to mention only a few examples. Finally the Rhætic of the uppermost Karroo has furnished numerous Saurischia (dinosaurs) and Pseudosuchia, among which are various well-known German genera such as *Thecodontosaurus*, *Plateosaurus*, *Gresslyosaurus* as well as allied forms, the first-named in Australia also.

These lists could be considerably amplified, but should suffice when taken in conjunction with the associated floras and the fresh-water fishes and mollusca, to prove the continuity of the various parts of Gondwana from the Permian to the Rhætic and the consanguinity of their terrestial life, as well as the close connection of that mass on more than one occasion with the lands of Laurasia.

THE GONDWANIDE OROGENY

(1) The agreement between the southern Cape fold-ranges and those of the Sierras of Buenos Aires, stratigraphical as well as tectonic, was brilliantly established by Keidel (16), while by his discovery of the Glossopteris vegetation in the Sierra de Pillahuincó Harrington (34) has upheld the writer's contention that those strata correspond to the Ecca Beds of the Cape. Several eminent geologists have nevertheless denied any close geographical connection between these two tectonic belts on the grounds that in the south-western corner of the Cape the nearly north-south Cedarberg flexures actually mark the western end of the east-west Zwartberg foldings and that the latter are not prolonged into the Atlantic—a weighty argument, if correct.

To this it can be said that :

(*a*) The Indian Ocean does not mark the southern limit of the Zwartberg foldings, which limit must be situated much further to the south—beyond the edge of the Agulhas Bank maybe.

(*b*) North-east of a line drawn diagonally from Mossel Bay to Karroo Poort the strata are intensely compressed and often overturned to the north, due in part to renewal of pressure during the Cretaceous : in the northerly direction these folds ripple out.

(*c*) South-west of that line these post-Permian compressions trespassed upon the field of the pre-Devonian folding with south-south-easterly to southerly trend—from Van Rhyns Dorp to Hermanus. The stiffened basement was less amenable to renewed flexing across than along the older grain, whence the development of a weaker set of " posthumous "—Cedarberg—foldings in the latter direction and the production of a generally open cross-folding giving basins, diagonal and pitching folds, etc. It is a belt of such a nature fully 150 km. wide that passes west-south-westwards into the ocean between Cape Town and Cape Agulhas with unknown southern extension.

(*d*) In extraordinary fashion an identical tectonic arrangement is found in the Falklands involving similar strata over a zone 200 km. across, openly folded in two directions set at right angles to one another.

(e) In eastern Argentina the structures resemble those to the east of Mossel Bay, isoclinal folding and overfolding and overthrusting to the N.N.E. being pronounced with some signs of cross-folding and of renewed movement during the Cretaceous. The various exposures forming inliers in the Tertiary indicate a belt at least 80 km. wide, while, if the Sierra de Tandil be included, the width of the fold-system becomes 230 km.

Keidel's views are therefore confirmed and also the author's synthesis in which the Falklands are included. Towards the north-west this orogeny is traceable into the pre-Cordillera of San Juan and La Rioja, where it has become involved in younger disturbances, though it seems to have affected Bolivia and possibly even Peru.

A Lower Triassic age is given in Argentina by the strong plication of Lower and in places of Upper Permian and the unconformable superposition of the Upper Triassic in Mendoza and elsewhere : in the Cape by the fact that in the Zuurberg the Rhætic (Stormberg) rests discordantly on the tilted and much eroded Lower Permian, with which Upper Permian is involved farther to the north.

(2) Furthermore, the Cape Foldings could scarcely have died out immediately to the east in the Indian Ocean. Unfortunately their hypothetical continuation through Antarctica is concealed by the Weddell Sea and by ice, but may possibly be represented among the mountains recently observed by Lincoln Ellsworth [1] —such as the Eternity Range or Sentinel Peak—and is hence presumed to pass to the east of the South Pole, Queen Maud Ranges and King Edward VII Land, Tasmania and Victoria and then to traverse the eastern part of Australia from Grafton to Cape Yorke—in a generally north-westerly direction—with possible prolongation into New Guinea. The maximum width exposed is fully 150 km., but the outer limit is concealed by the Tasman Sea. Those fold-chains may indeed have constituted part of the hypothetical land of " Tasmantis," which is supposed to have connected Australia intermittently with New Zealand between the Permian and Jurassic, but which sank during the Cretaceous.

Since the importance of this orogeny does not seem to have been stressed by Australian geologists, the following details are submitted. From the New South Wales boundary to beyond Cooktown the pre-Silurian basement schists have been brought to the surface and upon their eroded edges the Triassics have been deposited at various points. The Mesozoic makes several

[1] Ellsworth (36), 7-9.

belts—partially detached from one another by later warpings—
e.g. one cropping out along the edge of the Great Artesian Basin
from Dubbo along the main divide to Taroom and north-
westwards, and the narrow strip beginning on the coast at
Grafton, proceeding inland a little west of north past Brisbane
with forking towards Gladstone and Bundaberg. These show
the base of the Triassic repeatedly transgressing across formations ranging in age from pre-Silurian to Lower Marine, while
the higher divisions tend to overstep, a phenomenon noticeable
also around the main Triassic basin of New South Wales,
evinced in the wedging out of the Narrabeen shales. Within
the Bowen River syncline the Triassic rests with apparent conformity upon Upper Coal Measures, though only because that
trough persisted as such during the younger movement. David[1]
states that the strong plication of the Upper Bowen Coal
Measures must have been completed before the Upper Triassic.

The pressures along this Australian arc of over 20 degrees
in length were directed inland, as indicated by the asymmetrical
nature of certain of the folds and particularly by the steep and
occasionally inverted north-eastern edge of the Great Bowen
syncline. Folding and upheaval therein began so far back as
the Middle Permian, while the principal phase took place
probably about the Middle Triassic, but was continued in
certain areas into the early Jurassic. It is noteworthy that
this shove seems to have been propagated right through the
mass of the continent to Western Australia, where the Permian
Irwin River Series was folded in pre-Jurassic times with a
similar north-north-westerly strike and the Kamilaroi strata of
Mount Wynne and the Wooramel River up-domed. Such inter-
Triassic movement was represented in New Caledonia also, and
perhaps in New Zealand.

The relationship of the Australian Triassic orogeny to that
of Indo-China is dealt with in Chapter XVI.

(3) The marvellous parallelism shown by the stratigraphies
and tectonics of South America, South Africa and Australia
during the early Mesozoic, following upon those previously
recorded for the Palæozoic, must indicate, when expressed
mathematically, an enormously high probability that the
*phenomena in question were due to the same controlling orogenic
forces operating throughout that lengthy period.* That, step by
step and in full detail, such could have happened with those
continents stationed at their present distances apart and with the
tectonic trends of South Africa and Australia set at a right
angle, is asking too much of human credence, and the author
finds himself compelled by sheer weight of evidence to accept
some form of continental drifting as explanation therefor.

[1] David (32), 73.

With the continents properly reassembled, the Gondwanides in their course from Bolivia to the East Indies are found to constitute an arc of some 110 degrees in length following a small circle fairly closely, Figs. 10 and 11. On its outer side lay the relatively narrow segment made by Chile, Patagonia, Graham Land, New Zealand, New Caledonia and probably New Guinea—all " border lands " in the sense of Schuchert.

Remarkable, too, is the way in which in the South Atlantic region the Gondwanides cross at right angles the " grain " of Gondwana, which runs nearly parallel to, and has seemingly determined, the opposed fractured margins of Southern Africa and the mid-section of South America, as more fully set out in Chapter VI. Down the eastern side of Brazil run the foldings, probably of early Cambrian age, called the " Brasilides " by Keidel (16) with general south-south-westerly trend traceable into Uruguay and detectable beyond in the Sierra de Tandil. The youngest structures in South Africa are the post-Nama plications, seemingly of early Cambrian age also, called the " Afrizides " by Krenkel, running from Lüderitz with south-south-easterly course. Both terrains are moreover cut by bodies of pre-Devonian granite related to such foldings.

If South America be brought into its proper position next South Africa, these two ancient fold-systems are discovered to form a *single wide orogenic belt* with nearly meridional trend embracing both shores of the Atlantic. Thereupon, as Holmes (29) has aptly remarked, the *crossing of the Brasilides by the Gondwanides, begun in Argentina, is completed in the Cape*, a relation expounded by Maack (34) as well. It will later be shown that in the North Atlantic a similar crossing of older tectonic belts by the Hercynian orogeny can be made out that discloses a second and northern bond between the New and the Old World during the past.

Noteworthy is it to find that during the late Permian or early Triassic, while compression was taking place in the south of Gondwana, a state of tension had actually developed in the north, one doubtless connected with the deepening of the Tethys, that led to faulting, under which blocks of Gondwana beds were lowered into the basement and so preserved, and in some instances became covered unconformably by late Triassic strata, as for instance in the eastern Congo, peninsular India and Western Australia.

THE UPPER TRIASSIC-RHÆTIC TRANSGRESSION

The erosive period of the earlier Triassic, following the Gondwanide folding, left an expanded and apparently low-standing continent composed of the following major elements :

(1) The broad peneplained region in the N.E., N. and N.W., dipping in those directions beneath the Tethyan ocean, with a gulf on either side of India just as during previous epochs, the western arm thereof reaching to Madagascar and probably just trespassing upon East Africa ;

(2) The great central and moderately diversified interior possessing over its greater part a dry climate ;

(3) The southern Gondwanide " cordillera " hundreds of kilometres wide and still experiencing compression, trending roughly E.–W. with ends deviating northwards in Bolivia and Queensland—a great barrier through which certain drainages passed southwards to the Triassic ocean from

(4) The gentle foredeep with its extensive lakes and wide " pans " in process of being silted up ; and

(5) The external portion of the continent situated to the south of (3), already much eroded, represented by the Pacific margin of South America, Patagonia, Graham Land and parts of New Zealand, New Caledonia and New Guinea, which, during the succeeding Jurassic, proceeded to develop into an exterior geosyncline.

In the north overspilling took place on either side of the Tethyan geosyncline that ran almost straight from the Mediterranean to Tonking. Upper Triassic and Rhætic marine or paralic strata were deposited upon the margins of both Gondwana and Laurasia, unconformably upon older formations for the most part, as in Malay States, Tonking, Burma, Assam, India, Afghanistan, Baluchistan, eastern Persia, Oman, Syria, Abyssinia (Adigrat), Somaliland (Lugh) and northern Madagascar. In North Africa the germanic " Keuper facies " was developed over the subsequently folded region of Tunis-Morocco (Solignac).

Along the western side of South America marine Upper Triassic is known at various points from northern Peru to southern Neuquen, commonly within a few degrees of the present Pacific and normally *outside* the bordering Gondwanide folds. The sea, nevertheless, overtopped that barrier in northwestern Argentina (Famatina) and extended far to the east as shown by the faunas with *Pachycardia* and *Myophoria* described by Reed and Cox from various localities within the Paraná basin, while a recent discovery has shown its former presence in Sergipe as well. This brief, though widespread, incursion, during which much of central South America may have been temporarily cut off from Africa, was followed by continental conditions with pronounced " red beds," the southern limit thereof being Uruguay. Outside of these occurrences marine conditions are unknown in eastern Patagonia, Falklands, South Africa, southern Madagascar, Antarctica and Australia.

During the earlier Triassic, New Zealand and New Caledonia were possibly united in a land-mass, but thereafter, along with Ceram, Rotti, Timor, etc., became submerged to reappear in the Rhætic. Connection between New Zealand and Chile—*via* Graham Land—is suggested by their respective Thinnfeldia Floras, which include the uncommon genus *Linguifolium*. It was along this southern land that the geosyncline developed that came to link those countries faunally during the Jurassic.

The foredeep, situated just inside the Gondwanide chains, was conspicuously marked out by more or less unbroken sedimentation from the Permian onwards, as in southern Bolivia and north-western Argentina (Bermejo) and in parts of La Rioja, in Cape–Natal (Beaufort), Tanganyika (Tangà), southern and western Madagascar (Sakamena), New South Wales (Narrabeen-Hawkesbury) and Queensland in parts of the Bowen syncline (Carborough). Away from the axis of this lengthy trough overlapping with unconformity is normally found, hence outside this restricted zone a *break occurs within the continental Triassic, and the middle division at least is usually missing.* Dixey (35) has given an instructive and detailed review of the stratigraphical, lithological and palæontological development of the Upper Karroo and its equivalents throughout Gondwana, and of its dominantly transgressive relations, to which the reader must be referred.

In both the Northern and Southern Hemispheres the later Triassic marks out an almost universal negative movement, which usually continued until well into the Jurassic.

TRIAS-RHÆTIC FLORAS

The early Triassic saw a great impoverishment in the Glossopteris vegetation, followed by the establishment of the " Thinnfeldia Flora," consisting of some Permian relics, locally evolved types and immigrant Eurasiatic forms as represented by various species of *Thinnfeldia (Dicröidium), Callipteridium, Pterophyllum*, etc.

Significantly it was best developed in the south and east of Gondwana—including Antarctica—which must largely have been due to the higher rainfall in those sectors—and even penetrated into the Malay States and Tonking. It was, furthermore, remarkably uniform in its composition ; localities so far apart as Chile, India, Tasmania and New Zealand are actually represented by certain identical species. In addition, allowing for the immigrant forms, this flora displays important differences from the contemporary vegetation of Laurasia. Interchange became freer, however, during the Jurassic, when northern species invaded East Africa, India, Australia and New Zealand,

and made their way, presumably *via* Antarctica, into Graham Land and Patagonia.

MESOZOIC DESICCATION

During the Triassic the warm and semi-arid conditions of the late Permian, leading to "red beds" phases, became extended, to attain their maximum during the Rhætic. From the wide distribution of this kind of deposit and its tendency to overlap upon the crystalline basement one can deduce a progressive levelling-up of the landscape, in part through wind action. Significant is the fact that the late Triassic torrential and deltaic phases are usually marginal to the Gondwanide ranges, from which their material was derived, and, furthermore, tend to carry coal-seams, e.g. the so-called "Rhætic" of Argentina, Molteno Group of Cape–Natal, Knocklofty of Tasmania, Hawkesbury of New South Wales, and Ipswich of Queensland. Conspicuous further away from such elevated sources occur the varicoloured deposits of the "red beds" type, that is to say, "claystones" with occasional calcareous concretions and associated yellow or red sandstones—generally of Keuper age— as represented in the Mearim of north-eastern Brazil, Rio do Rasto of the Paraná basin, Stage V of western Argentina, Red Beds and Bushveld "marls" of South Africa, Manda of Tanganyika, part of the Adigrat of Abyssinia and the Maleri and Parsora of India, in some of which vertebrate remains are not uncommon.

More conspicuous are the succeeding, rather massive, yellow or reddish sandstone formations, in places hundreds of metres thick, e.g. the Botucatú and Areiado of Brazil, Cave, Bushveld and Forest Sandstone of Southern Africa, upper stage of the Lunda Series of Angola, Lubilash (Lubilache) of the immense Congo basin, Adigrat of Abyssinia, Isalo of Madagascar, Pachmarhi of India and Bundanba of Queensland—all of them for the most part of Rhætic age.

They are fine to medium in texture with the grains often finely rounded, coarse only sporadically, without good stratification, but with frequent strong false-bedding. Over large areas æolian conditions must have prevailed, though much of the constituent sands may have been water-laid, and a semi-arid or arid climate is strongly suggested, namely, over the central part of the Paraná basin, Transvaal, Southern and Northern Rhodesia, Congo, central Moçambique, Tanganyika, Kenya, Madagascar and central India. Unsorted stream-gravels typical of a dry climate occur locally at the base of the Congo Lubilash,[1] while wind-faceted pebbles occur in the Areiado Sandstone of Minas

[1] Veatch (35), 48-61 ; 159.

Geraes, Brazil,[1] in the Isalo sandstone of Madagascar (Wade) and in the Upper Duruma or the Shimba Grits of East Africa (Dixey).

Floral life was of the scantiest, mainly trees of coniferous type, and rare dinosaurian remains. Stockley[2] has called attention to the abundance of silicified wood throughout the Permo-Triassic deposits of Central and Southern Africa, and has suggested that such silicification of woody tissue may have been effected by the alkaline carbonates contained in saline waters—as could be anticipated under an arid environment.

Under our reconstruction this drier region forms a belt trending E.N.E. (Fig. 11). To the south thereof the loess-like

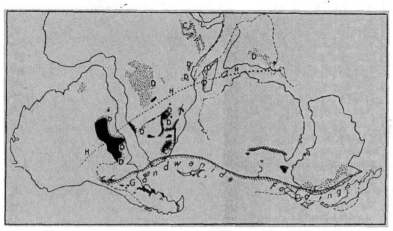

FIG. 11.—Showing the distribution of the late Triassic and Rhaetic Gondwana Formations. Stippling indicates continental sediments; black, volcanics; cross-hatching, areas of intrusive dolerite; D, semi-arid to desert phases; C, coals; HH, axis of southern high-pressure belt; broken lines, continental margin (where determinable).

Cave Sandstone of the Union of South Africa extended and to the north the Lubilash of Angola and Congo which were partly water-laid. The widely spread Bundanba of Queensland is considered by Bryan to be lacustrine. It is highly suggestive that this more arid territory should have occupied the central-west and centre of Gondwana precisely as is the case in each fragment thereof today. It is important to observe that *this desiccation began later and also ended slightly later than that in Laurasia.*

The lagunar facies of this period with salt and gypsum in North Africa is referred to in Chapter XIII.

[1] von Freyberg (32), 196-200.   [2] Stockley (36), 27.

TRIAS-LIASSIC VOLCANICITY

The bonds holding Gondwana together were first loosened at the close of the Triassic with the pouring out of magma, or its consolidation underground in tension-fractures, over a gigantic area, mainly inside the Gondwanide ranges, but outside them in Patagonia (Fig. 11).

The *Basalts* of the Paraná Basin (Serra Geral) occupy an oval area 1900 km. long ; of Basutoland (Stormberg) 500 km. ; of the Kalahari perhaps 1000 km. ; of the Lebombo a belt 1000 km. long ; together with minor areas in north-eastern Brazil and western Argentina, Zuurberg (Cape), the Kaokoveld (South-west Africa), Southern Rhodesia, upper Zambezi valley, Angola-Rhodesia border, Tete area (Portuguese East Africa), Nyasaland and possibly north-west Tanganyika. Actually these are merely remnants of much vaster lava fields, the total extent of which could well have exceeded 4,000,000 sq. km. Thicknesses of over 1000 m. have been measured in certain places and, since the underlying strata have over wide areas been riddled with the doleritic intrusive facies, the total volume of magma erupted must have been colossal. The Basutoland region is characterised by numerous volcanic pipes, but the bulk of the lavas there, and all those elsewhere, have manifestly been poured out from fissures, which are represented by the abundant dykes that cut the effusions as well as their basement.

Forming sills and dykes these Dolerites significantly avoid the Gondwanide fold-belt, though occupying a wide region parallel thereto in Uruguay, Brazil, Southern Africa, Antarctica and Tasmania and just touching New South Wales. To the north of that territory they become scarce, i.e. Angola, Congo and western Tanganyika, while eastern Tanganyika, Madagascar, India and Australia escaped. Their co-genetic nature is borne out by their wonderfully uniform petrological composition and character. Crustal tension at the period of their intrusion is indicated by the dyke-swarms of South Africa, which through their orientations—nearly N.–S. on west, E.–W. on south and N.N.E.–S.S.W. on east with a complementary set trending W.N.W. to N.W. over the Basutoland region—show that the shape of Africa was being roughed-out even at this early date.

Remarkable is the eruptive belt of the Lebombo trending north-south, which is only a section of the monoclinal structure, dipping eastwards, traceable from the mouth of the Bashee River in the Cape to the Sabie River on the Rhodesian border, a distance of over 1250 km. developed *during the eruptive phase.* It is marked by vast outpourings not only of limburgites and

basalts, but of *rhyolites* (quartz-porphyries), intruded by gabbro,
granite and granophyre in elongated masses stretching N.–S.
and by a dolerite dyke-swarm with similar trend. A maximum
thickness of some 9000 m. of volcanics has been measured. The
writer[1] has suggested that this belt of flexure marks the position
of a deep-seated magmatic wedge in which strong differentiation took place. In the Rhodesian section granites cutting these
basalts occupy a wide area. From farther north in the Lower
Zambezi-Shiré area Dixey (30) has described an identical basalt-
rhyolite-basalt succession with associated intrusions. It is
striking that in the coastal region of South-west Africa there
should be similar large intrusions of granite and granophyre
into sediments and lavas regarded as of late Karroo age.

It is hence not surprising to find in the district of Mendoza,
Argentina, not only Triassic porphyrites, but intrusive into them
granite and quartz-monzonite-porphyry of pre-Liassic age. In
Neuquen and Patagonia,[2] resting on the ancient platform
though widely covered by younger formations and certainly
occupying an enormous stretch between the Atlantic and the
Cordillera southwards to Santa Cruz at least, extend little-
disturbed quartz-porphyries, keratophyres, tuffs and occasional
sediments. In places they overlie the earlier-erupted porphyrites.
As shown by the presence of the fossil crustacean *Estheria
draperi*—a Stormberg species—and some plants, effusion began
in the later Triassic, but continued in Patagonia into the Jurassic,
merging further to the west with the great Andean Jurassic
volcanicity that affected South America down to Cape Horn
with the prolific emission of porphyrites. The significance of
the acid volcanic phases in Patagonia and the Lebombo-Shiré
regions will be pointed out later.

Right at the opposite end of the Samfrau geosyncline
igneous activity occurred during the Upper Triassic and
Liassic in New South Wales (Gunnedah) and Queensland
(Brisbane and Esk Series), the Bundanba Series was invaded by
large masses of grano-diorite and syenite, while rhyolites and
tuffs were poured out in New Zealand in the Coromandel and
Canterbury Plains areas.

The Liassic Rajmahal basalts of peninsular India can
further be mentioned, and also those curious occurrences of the
Tethyan geosyncline, largely of submarine origin, known as the
" Panjal Traps " in Kashmir and Afghanistan, some of which
are so late as Triassic though others are distinctly older.[3]

Later it will be shown that the closing Triassic was an
important eruptive period for Laurasia also in the Alpine,
Acadian and Cordilleran regions.

[1] du Toit (29 a), 212.   [2] Windhausen (31), 199, 218, 273.
[3] Wadia (34), 158-05.

JURASSIC

Marine deposits of this epoch are conspicuously absent from both sides of the South Atlantic, for that ocean only proceeded

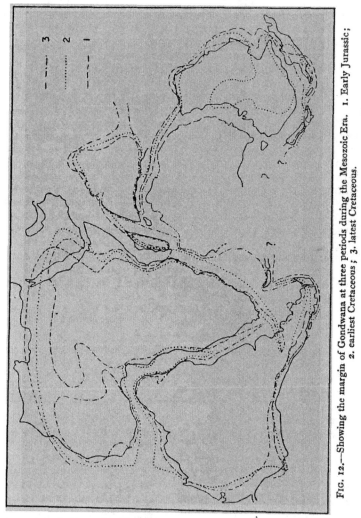

FIG. 12.—Showing the margin of Gondwana at three periods during the Mesozoic Era. 1. Early Jurassic; 2. earliest Cretaceous; 3. latest Cretaceous.

to open during the Cretaceous—Fig. 12. The Jurassic sea, however, covered the north-north-western margin of Africa. It

furthermore invaded the western side of South America from
northern Peru downwards, the controlling syncline running not
exactly parallel to the present coast nor to the Tertiary Cordillera
between northern Chile and Patagonia, but obliquely thereto—
a little east of south—and so cutting across the Gondwanides,
its *outer side* corresponding with the coastal Cordillera (Windhausen). The rest of the continent was land.[1]

Throughout this lengthy arc sedimentation (including littoral
and deep-water phases) was much interfered with by eruptions
(mainly porphyrites and tuffs)—really a continuation of the
Triassic volcanicity—coupled with the injection of masses
chiefly of grano-diorite, and hence typically " Andean." A
minor movement occurred during the mid-Jurassic on the
western side of the trough, which cut the latter off from the
deeper sea and led to extensive gypsiferous deposits.

Excluding local species, which are numerous in the shallow-
water phases, the faunas show strong affinities to those of
the Mediterranean with a proportion of Indian, East Indian
and Madagascan types, especially during the later stages. Of
unusual importance are beds in South Georgia with probably
Jurassic radiolaria and the mid-Jurassic plant-bearing strata of
Hope Bay, Graham Land (Lat. $63\frac{1}{2}°$ S.) with the flora described
by Halle, associated with Andean rocks and folded prior to the
Cretaceous.

A syncline must therefore have connected Patagonia with
western Madagascar, almost touching the Cape–Moçambique
coast, to trespass on the latter during the early Cretaceous.
Another one, or else the open sea, passed on the Pacific side of
the South Pole to New Zealand—where the Jurassic fauna is
typically Andean (Gregory)—and thence to New Guinea. Land
was made by New Caledonia (for some time), Tasmania and all
Australia, save where the late lower Jurassic sea had drowned
the fresh-water Jurassics that had been deposited over two
lengthy sections of the western and north-western coasts of
Australia, and brought in European and Himalayan elements.

This gulf between Western Australia, Antarctica and India
seemingly did not penetrate farther than Ceylon even during
the later Jurassic, when the sea lapped the eastern edge of India.
The Indian peninsula terminated north-eastwards as heretofore
at the long-established Himalayan geosyncline, in which were
being deposited the deep-water Upper Jurassic Spiti Shales and
younger sediments with a fauna of Mediterranean affinities but
with conspicuous south Andean elements as well.[2] On the
opposite side of India the second and longer synclinal gulf
turned off south-south-westwards, overlapping on its eastern side
Afghanistan, Baluchistan, Sind, Cutch and the western part of

[1] See Gerth (35), Pl. 27.   Reed (21), 277.

H

Madagascar—with extension to Patagonia—and on its western the margin of Tanganyika, Kenya, much of Somaliland and Abyssinia, great portions of Arabia and Persia and· part of Syria, whence its waters extended westwards to the Atlas region and south-westwards to the Cape Verde Islands. In this quarter of Africa the marine facies is replaced in the southerly direction by a sandy littoral phase.

In connection with the Madagascar Gulf the progressive overlapping by younger beds in a southerly direction along the east side of Africa, beginning with the Triassic in Abyssinia and ending with the Danian near Beira, would indicate the steady propagation of this furrow south-westwards between Africa and Indo-Madagascar, a view confirmed by the successive changes of facies shown by the strata in that island. The faunas of this furrow are throughout Tethyan in composition.

Arresting are the equivalent fresh-water deposits that must have been laid down in enormous lakes and which incidentally denote a radical change of climate from that marking out the Triassic. The Lubilash of the Congo—which seemingly ranges up into the Lower Jurassic—with its soft red sandstones (and a shaly division) and occasional ostracod remains, occupies a continuous area of more than 1800 km. by 1200 km. and has a thickness of over 450 m. It was " formed by stream and wind action, for the most part, on a semi-arid land surface and, to a lesser degree, in more or less temporary dry climate lakes." [1]

" Lake Walloon," [2] some 1000 km. from north to south and 900 km. across, was developed behind the Gondwanide barrier of Eastern Australia—thereby determining the Great Artesian Basin—its south-western rim characterised by numerous rocky islands. From the channel leading seawards through the D'Aguilar Range near Brisbane two arms, occupying synclines, led off respectively northwards toward Rockhampton and south-wards toward Grafton, while minor basins were formed in the remainder of Australia, o far to the north even as Cape Yorke. In them were deposited the Walloon and Clarence River Series with coals, more especially in central and eastern Queensland. They strongly resemble the corresponding formations of Indo-China.

It is curious that characteristic Jurassic floras have so far been yielded only by the great area situated to the south-east of the Afghanistan–Madagascar–Patagonia trough, these show-ing a marked uniformity in composition.

Noteworthy, in contrast to Laurasia, is the absence from Gondwana during this epoch of orogenic movements save within the Chilean trough.

[1] Veatch (35), 163.   [2] David (32), 80.

INITIATION OF " DRIFTING "

Although the complete break-up of Gondwana seems not to
have been effected until the Cretaceo-Tertiary, one can deduce
from the distribution of the marine mesozoics that the bonds
were already relaxing in the Jurassic and that appreciable
drifting was in progress during the Cretaceous, a state of affairs
that incidentally was being duplicated in Laurasia.

The precise mechanism is not readily visualised, even the
protagonists of the Hypothesis having generally evaded the
question, which must nevertheless be properly answered.   The
author's conception is briefly as follows :

" The several gulfs leading inland from the encircling ocean
became extended toward the heart of the continent with the
production of wide rift valleys, and sliding began.   The rate
of movement being necessarily extremely slow at all times,
though inferentially rhythmic and sometimes spasmodic, *such
rifting need never at any stage have produced abyssal seas*, because of the concurrent ascent of magma and its differentiated
products from below and the continuous infilling sedimentation
from either shore.   The sea floor would furthermore always
have had a mantle of unconsolidated bathyal material that could
have become stretched without experiencing true fracture or
that could have slid sideways, as in the case of the Atlantic
earthquake of 1929 (p. 225).   Deposition *pari passu* with displacement would appear to be a *sine qua non*.

" Indeed, during the earlier stages at least, biological isolation
of the detached mass is not necessarily implied, since sedimentation within the tension-rift might conceivably have been competent, especially between crises, to overtake sinking and so
provide temporary land-connections, as, for example, during
the early Cretaceous (Wealden).   Furthermore, while centrifugal movement was active, bulging-up of the sea floor in
advance of two or more diverging masses could well have
linked them together for a time, an aspect that will be developed when dealing with the Tertiary Mountain System of
the Earth.

" Considering the stupendous masses involved, the varying
degrees of tension in the several sectors, the differences in the
thermal condition of the supporting sima, the fact that erosion
and deposition were always active, the ever-increasing resistance
met with in the piled-up frontal foldings, etc., the dispersion of
the continental fragments must have been a highly complicated
business and not one lending itself to full mathematical analysis.
If the thrusting of one block upon another be found by structural
geologists to prove immensely complicated, why should the

phenomena of rifting in the concealed rear of the mass have to be conceived as simple."

The problem is raised at this stage because from the later Mesozoic onwards the several components of Gondwana must be viewed as no longer maintaining their relative geographical fixity, but as experiencing progressive and radical changes in their spacing, orientation and shape.

CRETACEOUS

The first half of this epoch was marked out by general transgression and with extension of the gulfs that bounded India (Fig. 12) : the second by tectonic movements of far-reaching character.

The African section experienced sweeping advances by the Mediterranean sea over Baluchistan, Arabia, Syria and North and West Africa, which submerged most of the peneplained territory, residuals such as the Central Sahara and Tibesti escaping. Not only did younger stages overstep in the southerly direction, but there was developed at the base a peculiar clastic facies, which, just as in the case of the Siluro-Devonian of the same region, was not of precisely the same age everywhere, passing up with apparent conformity into the succeeding marine marls and limestones, whatever their age might be. In the east such constitutes the well-known *Nubian Sandstone* of brown, red or variegated colour with saliferous and gypsiferous layers and containing silicified wood but rarely marine fossils —a formation usually considered to be wave-worked terrestrial detritus. In the Sudan it has become continental. It is of Lower Cretaceous age in Spiti (Giumal), Hazara, Somaliland, Eritrea, Syria and Atlas region, but of Upper Cretaceous over most of Egypt and in Nigeria, where it passes into marine phases and is overlain by fresh-water beds with coals.

By the beginning of the Eocene most of Arabia and North and West Africa had been submerged, and the Nummulitic Limestone was being deposited conformably or disconformably upon the Cretaceous with northward retreat of the shore-line thereafter. Basic lavas were erupted from widely scattered centres.

It is surprising to find in north-eastern Brazil a close parallel to the North African terrains in the thick red and white continental facies, shown by fish remains to be of Cenomanian age along the boundary of Ceará and Piauhy and in Maranhão with wide extension into the interior.

The *opening of the South Atlantic* was effected during this epoch by means of rifting and stretching, which began in the north during the Neocomian, was propagated southwards and was completed by the Campanian (Upper Cretaceous) at least, perhaps a trifle earlier (Haughton) ; Windhausen would set it at about the Senonian or Danian.[1]   This is shown by the Cretaceous littoral deposits ranging in age from the Albian upwards between Rio Grande do Norte and Bahia in Brazil on the one side and between Cameroon and Mossamedes (with a tiny patch south of Lüderitz) on the other.   Around the promontory of Africa the older Cretaceous faunas of the west coast are characterised by Mediterranean elements ; those of the south and east coasts by Indian elements, but by the Senonian intermingling had occurred.   Turning to north-eastern Brazil one finds faunal elements from the Mediterranean, Cameroon and Angola regions as well as from the Pacific.   On the contrary, Atlantic affinities are not found in Patagonia until the Danian (latest Cretaceous).

The southern Cape furnishes valuable evidence regarding the manner of opening of the South Atlantic.   According to our view the Gondwanide fold-ranges were still linking Argentina and the Cape at the beginning of the Cretaceous and thereby shutting off the southern seas from the enlarging Atlantic basin. In strike valleys in and on the southern side of the Cape chains were laid down terrestrial and estuarine Wealden beds (Enon) followed by Neocomian (Uitenhage) strata with a fauna of Indian aspect characterised by *Trigonias*, one that reappears just outside the Gondwanides in north-western Neuquen in Argentina.   The mid-Cretaceous saw renewed crumpling of the Cape Foldings along east-west lines—and hence in places slightly oblique to the older structures,—probably with some thrusting towards the north as well, which may have been repeated in the Argentine section (though such is not yet capable of proof because of the greater degree of erosion) for near by in Neuquen and farther south were produced during the Cenomanian the *Patagonide Foldings,* with occasional overthrusting towards the north-east, that are suggestive of a slight twist of Patagonia with reference to the Brazilian shield.

During the later Cretaceous the bond formed by the Gondwanides was snapped, South America proceeded to drift westwards, taking the Falklands with it, thereby giving rise to the First Andean Orogeny, while the Atlantic " Rift " was continued southward to meet at right angles the important trough that had been extending itself from India past Madagascar and the Cape to Patagonia (Fig. 12).

Crustal tension at this time is revealed by the many strike and diagonal faults in the southern Cape with downthrow nearly always towards the southern ocean, one of them having a drop of over 4 km.   The rough outlines of the African pro-

[1] Windhausen (31), 540.

montory and the eastern side of South America were clearly
determined by the ancient grain and the mesozoic foldings.
Interesting is the fact that these three, sub-parallel, fractured
continental margins mark zones characterised by suites of
igneous rocks of late Cretaceous or early Tertiary age in which
alkaline types play a part.

Turning now to the northern and western sides of South
America, the Cretaceous makes an almost continuous curved
belt from Trinidad to Terra del Fuego—much compressed by
younger movements of course—showing normally the phases
indicative of transgression by the Caribbean Sea and Pacific
Ocean, with terrestrial facies on the inner side. Volcanicity was
rife on the outer side, save in the central section, and was
presumably connected with the squeezing-up of the ocean floor
by the westerly drifting block. Marine Neocomian, with much
porphyrite, follows the Jurassic conformably in the established
syncline that extended from northern Chile into Patagonia.[1]
Into the fauna thereof *austral* elements entered and spread
northwards, the Uitenhage assemblage with its Trigonias having
already been referred to.

The inter - Cretaceous Patagonide movements drove the
Neocomian seas of the west off the foreland and led to the
deposition thereon of the famous terrestrial " Dinosaur Beds "
with various estuarine and marine interludes and unconformities.
This was connected with the great marine late Cretaceous
transgression that affected the territory northwards from Graham
Land and brought in an Indo-Pacific fauna with such forms as
Lahillia and *Kossmaticeras*, the large number thereof having
indeed induced Wilckens[2] to postulate a continuous shore-line
from Patagonia to New Zealand. The pre-Upper Cretaceous
plications of that island, which pass south-eastwards out to sea
in the Otago district, may possibly be the direct continuation
of the Patagonides.

The increased drift of South America with the piling-up of
material in front overloaded the block and so enabled the
waters of the South Atlantic to invade its rear at the close of
the Cretaceous with the consequent deposition of mixed brackish
water and marine phases that pass in a north-westerly direction
—i.e. towards the pre-Cordillera—into fresh-water deposits
that occupy the " foredeep " through an immense distance.

A digression is needed to discuss the Cretaceous tectonics of
the Persian Gulf region, where the southern section of the highly
folded Zagros system has been discovered to strike south-
eastwards through southern Persia and then southwards through
Oman into the Indian Ocean with palæozoic and mesozoic
sediments intensely crumpled and thrust south-westwards and

[1] Gerth (35), 300, 387.   [2] Benson (23), 52.

westwards over the Arabian foreland in pre-Maestrichtian and probably pre-Cenomanian times. Lees[1] has pointed out the difficulties in accepting the views of either Suess, Argand, Wegener or Kober ; Krenkel[2] also refuses to admit the scheme of the last named. Under our hypothesis, however, strong disturbances could be anticipated in such very position, seeing that :

(*a*) The region formed the north-western side of the syncline extending from Afghanistan to Moçambique, initiated back in the Permian at least and lengthened and widened thereafter ; and

(*b*) India is presumed to have begun its north-eastward drift of perhaps 1500 km. during the early Cretaceous.

Squeezing could therefore have taken place between the Arabian and Iranian blocks and the obliquely moving mass of India. If so, then the Oman structures should after proceeding southwards beneath the Indian Ocean double back sharply towards Cutch, where indeed Lees asserts that "Jurassic and Lower Cretaceous are considerably folded and are unconformably overlaid by Upper Cretaceous." Possibly, to follow a suggestion from Holmes, their continuation may lie buried beneath the Alpine foldings that subsequently overwhelmed Sind.

It is noteworthy that in Madagascar the Jurassic and Cretaceous possess low dips save in the extreme north of the island, where they have been considerably disturbed with the intrusion of syenite and other alkaline rocks and effusion of basalts, etc., similar igneous types occurring opposite in East Africa. It is highly suggestive that the north-north-easterly prolongation of this zone of alkaline rocks would pass by Socotra into the Oman region with its curious " Semail " igneous assemblage of keratophyres and associated types, while beyond that, in the Kathiawar Peninsula of India, will be found rocks intrusive into the Deccan lavas including alkaline kinds.[3] On *a priori* grounds strong igneous activity could be expected in proximity to a gigantic zone of crustal shear such as continental sliding would demand here.

Presumably connected with such crustal rupture is the enormous spread of late Cretaceous or early Eocene Deccan " traps " of western India that occupy over 500,000 sq. km.— mainly basalts erupted from fissures now indicated by numerous dykes ranging from basic to acid that cut these lavas or their basement. These tend to follow older lines of fracture that trend between E. and E.N.E. mostly.[4] With such volcanicity can also be linked the corresponding as well as the younger lavas

[1] Lees (28), 629-33, Figs. 6, 7.   [2] Krenkel (25), 34-6, Fig. 5.
[3] West (34).   [4] Holland (15), 357.

of Abyssinia and of the Red Sea, Aden and East African rifts —all manifestations of considerable internal tension.

The particular distribution of land here favoured receives strong support from the discovery in north-western Madagascar of the sauropod genera *Titanosaurus* and *Laplatasaurus* known from Argentina and India, though not yet from Africa, and of *Antarctosaurus* represented in Argentina. Their presence finds readier explanation thereby than that advocated by von Huene (29) under which the more roundabout connection is made *via* Australia. Under his view Antarctica occupied its present position and must have been cold, whereas under the author's that region lay well to the north of the pole during the late Cretaceous.

With the Indo-Madagascan block drawing away from Africa while Asia was approaching from the north, the parting Tethyan geosyncline that stretched westwards through northern Persia must have become narrowed. Such agrees with the change in sedimentary deposition and in the following of the neritic " Hippurite Limestone " by shallow-water " Flysch " deposits as, for example, in Baluchistan. Volcanicity was well marked, the material being largely of andesitic character. As shown on Fox's maps [1] and argued by Grabau,[2] the triangular mass of India was for a time connected to China by its eastern corner, a relationship disclosed by the faunal differences of its eastern and western littoral deposits (Fig. 12). The invading late Cretaceous seas, on the contrary, severed India from the remainder of Gondwana in a submergence that progressed well into the Tertiary.

During the early Cretaceous Australia was still linked to the other masses, though not to Africa. Pushing between it and New Zealand the Aptian sea trespassed over the Queensland coast and poured into " Lake Walloon," presumably from the north rather than the south-west, thus bisecting Australia. The coldness of the waters of this basin is indicated by the absence of reef-forming corals or rudistids and by the ice-borne erratics contained in its sediments in its southern section.[3] Some peculiar mollusca, like *Maccoyella*, are restricted to this epi-continental sea. After the Albian the sea retreated, the interior lake was restored and coals proceeded to form there and also in the coastal Maryborough area. Throughout the Cretaceous, New Guinea and New Zealand formed the margin of Australia and consequently shared in its vicissitudes with occasional though partial submergence. Only during the Tertiary did this belt of " border lands " proceed to drift off from the mainland towards the Pacific.

The breaking away of Australia from Antarctica and its

[1] Fox (31), Pl. 10.   [2] Grabau (28), 530, 619.   [3] David (32), 84.

easterly drift with slight anti-clockwise rotation led to the production of foldings along its eastern margin.   Bryan's analysis [1]
discloses four parallel equidistant anticlines trending N. 30° W.,
each with short steep western. limb, with tendency towards
over-turning, the most disturbed one on the east, developed by
pressures directed from E.N.E.   In south-eastern Queensland
that compression was followed by strike faulting.   This, the
latest important folding in Australia, was post-Cenomanian and
probably late Cretaceous, but may have been continued into
the early Tertiary.

[1] Bryan (25), 37, Figs. 2 and 3.

CHAPTER VI

COMPARISONS BETWEEN THE FRAGMENTS OF GONDWANA

Criteria. South America and Africa. Patagonia. Falkland Islands. Madagascar. India and Australia. Antarctica. Boundary relations between Gondwana and Laurasia.

CRITERIA

In such inquiry we must take into account collectively and weigh up all the physiographical, stratigraphical, phasal, climatic, tectonic, palæontological and volcanic relationships. Negative evidence must not be relied upon to any extent because of the imperfections of our knowledge or of the geological record, but may become weighty in cases where support can be got from other directions. For instance, the conclusion that there was no close connection between North and South America during either the Palæozoic or Mesozoic is pointed to by the absence from among the various fossils already known of any closely related terrestrial forms not only of vertebrate but of plant life. Such biological alliances as have been observed are quite explicable by indirect linking *via* Europe and North Africa. It is vital to note that not only does the recorded evidence favour in unmistakable fashion our special reconstructions, but no important relations have thereby become disclosed that would appear adverse to such an interpretation. Indeed each new discovery merely serves to reveal another bond between the now-scattered fragments of this great " Southern Continent ".

SOUTH AMERICA AND AFRICA

The concordance between the opposed shores, incidentally pointed out and discussed by others long before Wegener, has consistently been extended by each fresh geological observation until at present the amount of agreement is nothing short of marvellous. When it is recognised that such linkages cross from coast to coast, not only directly but diagonally as well, and are furthermore of widely different ages, it will have to be conceded that such probability of closer continental union must, when expressed mathematically, be extremely high.

106

The fact that in the Southern Hemisphere these two masses provide the most convincing test for such long-distance correlation is the excuse for the somewhat detailed nature of the following comparisons. The general agreement was set out by the author in 1927 (Fig. 13), which proved surprising even to Wegener himself. These views are summarised again with additions based upon more recent work by Windhausen, Walther, Harrington, Falconer, Oppenheim, von Huene, Maack, von Freyberg, Washburne, Beetz, Mouta and others.

Of the major controls the most obvious is the rounded corner of north-eastern Brazil with its counterpart in the Gulf of Guinea, due allowance being made for the projecting youthful delta of the Niger. The second is that long ago pointed out by Keidel (16) between the Sierras of Buenos Aires and the Cape Ranges. The third is that between Santa Caterina (Catherina) and the Kaokoveld, for, as Maack [1] has recently stressed, the Gondwana rocks of that part of Africa would seem to have constituted the actual eastern rim of the great Paraná Basin.

The distances between Bahia Blanca and Cape São Roque and between Cape Town and the Niger mouth are almost identical, while the intervening coast-lines show an agreement that extends even to their various inflections.

Wegener erred slightly in his reconstruction in making no allowance for the continental shelves, as he subsequently admitted.[2] The writer, on the contrary, used a distance between the respective coasts of some 400-500 km. in 1927 and one of 500-600 km. in 1928 largely because of the observed phasal variations within the sedimentary groups. It has since been recognised that such variation takes place not always perpendicularly but sometimes obliquely to those shores (Fig. 15), while the mid-Atlantic Rise is now viewed as more probably a recent structure and not a relic of fractured Gondwana. This has served to bring down the distance to some 300-400 km. usually, which not only gives a better fit, but becomes identical with the scheme of Maack (Fig. 14)[3] and almost the same as that of Baker.[4] The rotation of South America relative to Africa is about 50 degrees. The hypothetical strip between the two present shores would of course have been occupied essentially *by land*, down to the Jurassic at least, after which the ocean proceeded to enter from either end as the rift of the South Atlantic widened.

In our comparison the section south of Bahia Blanca will be considered separately below.

(1) The Gondwanide Belt north of Bahia Blanca corresponds with that of the southern Cape, showing in each case

[1] Maack (34), 210.   [2] Wegener (29), 72.
[3] Maack (34), Maps 1 and 2.   [4] Baker (32), Fig. 20

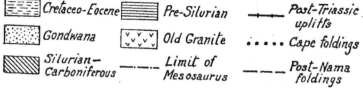

FIG. 13.—Suggested Continental Restoration for the South Atlantic Region under the Displacement Hypothesis : A, Agua Suja ; B, Burnier ; Bv, Boa Vista ; Ki, Kasai ; Ko, Kaokoveld ; Ks, Klein See ; L, Lüderitz ; N, Neuquen ; P, Postmasburg ; S, Salobro ; Sc, Santa Caterina ; Sv, Sierra de la Ventana ; U, Uitenhage.

(*a*) intense folding with overturning—and in the case of Argentina of overthrusting—towards the foreland, strata up to the Permian being involved ; (*b*) the quartzites of the Sierra de la Ventana correspond lithologically with the Table Mountain sandstone of the southernmost Cape (of Mossel Bay) ; (*c*) the fossiliferous

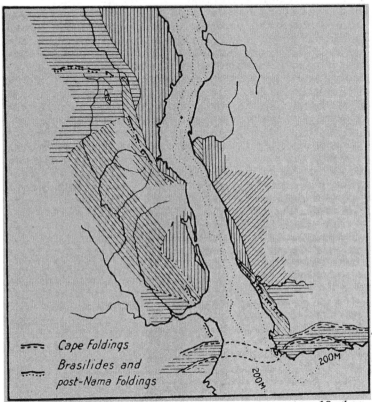

Cape Foldings

Brasilides and
post-Nama Foldings

200M

200M

200M

FIG. 14.—Showing Tectonic Correspondences between South America and Southern Africa—after Maack, with simplification. The course of the ruling indicates the dominant trend of the particular folding. Scale 1 : 30 million.

Devonian beds are identical ; (*d*) the Witteberg quartzites of the Cape are, however, either represented by greywackes or cut out by thrusting in Argentina ; (*e*) the thick composite and cleaved glacials in Argentina duplicate the Dwyka tillite, and may be conformable with the beds below just as in the Cape south of Lat. 33° S. ; (*f*) dark shales follow, overlain by identical strata—Bonete and Ecca—with a Glossopteris flora ;[1] (*g*)

[1] Harrington (34).

dolerite intrusions are practically absent in each case ; (*h*) consolidated ferruginous " gravels " rest upon bevels cut across the folded beds on the north-eastern flank of the Argentine ranges and duplicate in extraordinary fashion the early Tertiary " high level gravels " and silcretes of the south of the Cape, having similarly been raised and dissected (though long after the supposed separation of the continents) ; (*i*) traced north-westwards into San Juan and Jachal the Siluro-Devonian quartzites become darker, the fossiliferous slates greener, the presumed equivalent of the Witteberg represented by grits and greywackes, while the glacials become discordant ; (*j*) traced eastwards from the Cape into Zululand or northwards into Van Rhyns Dorp the Table Mountain sandstone becomes thinner, develops a reddish and in places conglomeratic basal facies, the marine strata are missing, while the glacials transgress with unconformity.

(2) The pre-Silurian stratigraphies and tectonics north of (1) are equally striking, for both coastal regions show, resting upon the Archæan, widespread developments of unfossiliferous systems of the general type—quartzite, slate, limestone (sometimes with volcanics). In particular regions two such systems are present, one seemingly of middle the other of late pre-Cambrian age, each folded within certain zones and intruded by younger granites of more than one age (see Fig. 7) : (*a*) in South Africa the edge of the post-Nama folding running N. 35° W. from near Worcester past Van Rhyns Dorp and Lüderitz to the shore in Lat. 25° S., crossing the Archæan grain, which is variable but ranges generally between W.–E. and W.N.W.–E.S.E. ; (*b*) in the Sierra de Tandil the corresponding strata are little disturbed and then mainly by the Gondwanide movements, but the older transverse disturbances can still be detected. In Uruguay, however, these strata—the Minas Series—are folded and run in a curve from Maldonado through the departments of Minas, Treinta y Tres and Cerro Largo to the Brazilian border with strike varying from nearly N.–S. to N.N.E. and even N.E. with local deviations, due mainly to intrusive granites. This ragged belt constitutes the southern end of the " Brasilides." On the contrary, the Archæan grain is commonly W.–E. and N.W.–S.E. The parallelism with South Africa is almost perfect.

(*c*) The Windhoek region reveals a powerful older structure at least 400 km. wide trending N.E.–S.W. along which compression was renewed in post-Nama times, one that formed mountain chains during the Carboniferous and was seemingly up-arched during the early Triassic. Its prolongation into Brazil coincides with a great embayment in the Paraná Basin, wherein the granite and gneiss show varying foliations (at Caçapava some crystalline limestones trend about W.–E.), while close by in Uruguay the crystalline floor has been domed-up in the Riviera inlier. This is the only part of the basin wherein such a reversal of dip takes place. Significantly, too, the basal Gondwana is poorly developed or missing over this particular region just as in central South-west Africa, while the Upper Gondwana overlaps on to the older rocks in both countries ; (d) in the Kaokoveld and the coastal belt of most of Angola the Archæan strike is nearly N.–S., which agrees with that in Brazil opposite, where the direction according to Maack is normally N. 50°–60° E.—that is to say, both coasts follow the foliation closely ; (e) on the other hand, the post-Nama plication in the former region trends about N. 30° W., while that affecting the various belts of the Assunguy Series in Paraná and São Paulo is generally N. 20° E. The latter beds, much intruded by granite and quartz-porphyry, are currently regarded as Ordovician, but the evidence therefor is not convincing, and they are more probably late pre-Cambrian ; (f) the important Lunda axis (Beetz) that runs south-westwards from Katanga to Mossamedes will, when prolonged, coincide with the up-domed part of the basin and its embayment in Paraná ; (g) the inflection of the coast north of Rio is to be correlated with the marked change in strike of the Minas Series from N. 60° E. in the south to N. 30° E. along the prominent quartzite " back-bone " of the Espinaço Range past Grão Mogul into Bahia. Further inland and within the curve made by this ancient formation the presumed lower palæozoic Bambuhy Series has been thrown into folds along axes directed N.W. and N.N.E. respectively. On the Angola side the quartzitic Oendolongo Series is represented by a chain of outliers curving roughly with the change in coast line ; (h) in the regions further north the older systems usually lie well inland though parallel to the coasts, and comparisons become less reliable. North of the Cuanza in Angola the Bembe System strikes N.N.W. to N.W. far across the Lower Congo River with cross-folding and is hence in general accord with the corresponding formations in Bahia (Tombador-Caboclos-Paraguassú Series) with strong development of limestones and dolomites in both cases.

(3) Dealing with the post-Silurian strata within the regions stretching from Uruguay to Minas Geraes and from Clanwilliam to the Cunene River respectively we find : (a) the almost horizontal Carmen Sandstone of Uruguay and Furnas Sandstone of Paraná are similar to the equivalent Table Mountain Sandstone of Van Rhyns Dorp ; (b) each is thinner and softer than in the south and carries small white quartz pebbles ; (c) each is succeeded by marine Lower Devonian shales with calcareous nodules—Rincon de Alonso, Ponta Grossa and Bokkeveld shales—with closely allied " austral " faunas ; (d) the soft

Tibagí sandstone with *Spirifers*, recently found by Oppenheim to form a local phase within the Ponta Grossa shales, duplicates the "Fossiliferous Sandstone" of Ceres and Clanwilliam; (*e*) the higher Barreiro sandstone is correlated by Maack with the Witteberg quartzites, but is stated by Oppenheim to represent the Itararé; (*f*) the base of the Itararé glacials is unconformable and transgressive, exactly as is the case of the Dwyka tillite from Clanwilliam northwards; (*g*) the glacials are scanty or absent in Rio Grande do Sul just as next the Windhoek axis; (*h*) an easterly or south-easterly source is presumed for the glacials in Uruguay and Brazil, but, on the other hand, northerly and north-easterly sources for those in South-west Africa; (*i*) marine interludes are indicated by associated bands with *Stuchburia, Aviculopecten, Chonetes*, etc., in Brazil and with *Eurydesma, Conularia, Myalina*, etc., in South-west Africa; (*j*) the Iratí shales are identically lithologically and palæontologically with the carbonaceous "White Band" of the Dwyka, each containing the reptile *Mesosaurus*—though the species are said to be different—not known in other parts of the world. The first formation extends from Uruguay to São Paulo and westwards into Paraguay, but is unknown in Argentina west of the Paraná River or in Bolivia: the second from Robertson across to Grahamstown and northwards to Mariental with an outlier in the southern end of the Kaokoveld, but is not represented in Natal, Transvaal or about Palapye. Thus the basin within which this uncommon type of organic sulphuretted mud was deposited is fairly well defined and must have formed an estuary opening presumably to the north, where the facies becomes calcareous.

(*k*) The deltaic phase of the Brazilian Bonito Series or "Coal Measures" underlies this horizon and reaches Africa only in the Kaokoveld and just touches the western Karroo. The centre of coal-formation is Santa Caterina, whence there is deterioration towards São Paulo and Uruguay respectively; (*l*) the succeeding Lower Permian strata show several contrasted phases (Fig. 15), namely, (i) a green-blue lacustrine facies in the Argentina–Cape section, (ii) a red and variegated facies to the north-west thereof in Central Argentina (Paganzo), Uruguay (Melo), Brazil (L. Estrada Nova and Corumbatahy), South-west Africa and western Kalahari (Ecca), and (iii) a deltaic "coal measure" facies beyond that in western Argentina, eastern Kalahari, Natal, Transvaal and Rhodesia; (*m*) the Upper Permian is absent in Brazil, the Kaokoveld and Hereroland; (*n*) over the greater part of these regions a disconformity is widespread beneath the Triassic, which shows transgression and occasionally an angular discordance; (*o*) the brilliant red Rio do Rasto and Piramboia groups parallel the Stormberg Red

Beds and Bushveld " Marls ", well developed in Hereroland though only feebly in the Kaokoveld ; (*p*) the Botucatú Sandstone agrees with the Cave-Bushveld-Forest Sandstone, the " Plateau Sandstone " of Hereroland and the " Hauptsandstein " so widespread in the Kaokoveld, all of which have been

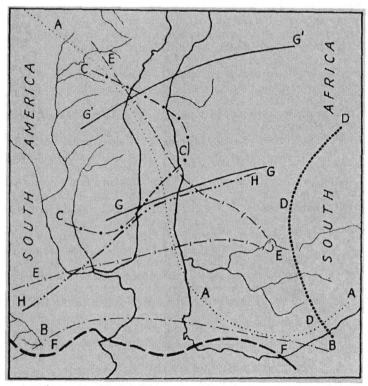

FIG. 15.—Tectonic and Phasal Comparisons between South America and South Africa. Siluro-Devonian sandstones to W. and S.W. of AA ; Glacials unconformable to N. of BB ; Brazilian " Coal Measures " (Bonito) within CC ; South African " Coal Measures " (Ecca) to E. of DD ; Ecca (blue-green) between DD and EE ; Ecca (red) to N.W. of EE ; Gondwanide Foldings FF ; Gondwanide Upwarpings GG, G'G' ; Triassic unconformable to N. of HH.

largely formed through wind action under a dry climate, secondary silicification being not uncommon ; (*q*) the succeeding volcanics build the actual coast only near Torres and in the Kaokoveld ; (*r*) the lavas are interbedded in their basal portions with thin sandstone bands or tufaceous sandstones, and are basalts with little olivine or else basaltic andesites with intersertal and porphyritic textures, while tuffs are rare ; (*s*) these

I

effusions terminate the Gondwana successions in both countries ; (*t*) widespread sills and dykes of mesozoic dolerite connected with (*r*) penetrate the flat-lying strata of the Paraná and Karroo basins, but not the Gondwanide belt ; (*u*) Maack has noted the north-west course of the lengthy dykes in Brazil, which accords with their east-west strike along the lower Cunene River.

(*v*) Marine Jurassic is absent from both coasts ; (*w*) various divisions of the Cretaceous, commencing with fresh-water Wealden and characterised by Mediterranean–North African faunas, occur at wide intervals from Ceará to just south of Bahia and from Senegal to Porto Alexandre, with a patch of late Cretaceous just south of Lüderitz ; (*x*) kimberlite and melilite-basalt of presumed late Cretaceous age form pipes in Namaqualand and South-west Africa and in Brazil, for example, Serra das Lages, 100 km. from Rio, and at and to the east of Patos in the western half of Minas piercing Triassic sandstones ; (*y*) as Brouwer has pointed out, there are centres of alkaline rocks both plutonic and effusive on both sides of the Atlantic, for instance, in South-west Africa. the foyaites of the Granit Berg and the phonolites of the Klinghardt Mountains near Lüderitz, the various necks near Windhoek and the occurrences in Angola near Tiger Bay and east and north-east of Mossamedes,[1] and in Brazil the Poços de Caldas north of São Paulo, the occurrences in the State of Rio de Janeiro such as the Serra de Gericino, Serra de Tingua, Cabo Frio and the laccolithic body penetrating Gondwana beds near Lagēs in Santa Caterina ; (*z*) the diamondiferous deposits of the coastal region between Rio and Bahia recall those of the Atlantic shore between Van Rhyns Dorp and Angola.

(4) More to the north there runs from Bahia to the Rio São Francisco the trough of the Estancia Beds, the upper part seemingly of Permian age with plications generally directed a little east of north, while a tiny patch of marine Triassic has recently been discovered 70 km. up from the mouth of that river. Similar strata may just possibly be present in Gabon, a region which is but little known. The littoral in each case is much covered by Cretaceous and Tertiary strata, the former somewhat disturbed in places. Reference might here be made to the well-known volcanic line of the Cameroon that extends south-westwards into the ocean through San Thomé, etc., with alkaline lavas, in view of the presence off the corner of Brazil of the analogous phonolite island of Fernando Noronha. Well inland in both continents Permo-Triassic Gondwana deposits have a wide development—in the Congo Basin and in Maranhão and Piauhý—but too far apart for effective comparisons.

(5) Between the north-eastern side of Brazil and the Guinea

[1] Beetz (34), sect. viii.

coast closer agreement can once again be established due, first, to the reappearance on this northern side of Gondwana of palæozoic strata and, secondly, to the important fact that *the ocean there cuts acutely across the structures of both lands* (Figs. 7 and 8).

(*a*) The roughly N.–S. strike of the older systems of French Congo and Cameroons is repeated in northern and southern Nigeria and southern Dahomey, changing to S.S.W. and even to S.W. in Gold Coast, as finely displayed by the Birrimian and Tarkwaian Systems, to S.S.W. and S. in Sierra Leone and to N.W. in French Guinea. In Brazil the strike is S.S.W. in Sergipe, changing according to Waring to E.–W. with local deviations in Parahyba and southern Ceará, to N.E. in northern Ceará, to N.N.W. on the River Gurupy, N.–S. near Para and E.–W. in Dutch Guiana. Despite the exceptions that have been taken to Wegener's comparison of these basement rocks, the accord between Brazil and West Africa is distinctly good ; (*b*) erosion of the very flat arch of Siluro-Devonian sandstones of West Africa has removed them from the structural axis, which runs W.S.W. from about Tassili through Sierra Leone. Northwest of that line they form a continuous sheet in Niger, Senegal, French Guinea and north-western Sierra Leone dipping faintly northwards. If prolonged, this axis would pass into the Guianas, where the similar horizontal Roraima Sandstone caps the crystallines and forms lofty territory running north-westwards into Venezuela. South-east of this axis, in Togoland and Gold Coast, lie the presumably equivalent Voltaian Beds, dipping faintly southwards and reaching almost to the Atlantic, and the downfaulted strips trending south-westwards of marine Devonian and probably early Carboniferous between Accra and Sekondi. Directly opposite extends the wide shallow Amazon syncline with its marine Silurian, Devonian and Carboniferous strata resting on crystallines and trending a little south of west. Possibly the continental facies of some of these beds may be represented in the similar gently inclined Serra Grande Series of Piauhý trending N. by E. and reaching almost to the sea. Marine Carboniferous has lately been found in a boring inland at Therezina, Piauhý. A reason is now disclosed for the marine palæozoics of the Gold Coast at a point where their presence would never have been anticipated under current theories. As the northern margin of Brazil becomes better known, further similarities and congruences might well reveal themselves. It is, furthermore, significant that the several diamondiferous areas in British Guiana, Sierra Leone and Gold Coast fall on a nearly straight line.

Our reconstruction shows a large triangular gap between the Guianas and Senegal, which, as Gagel has suggested, might have originally been filled up by a block that drifted northwards to form the rise on which stand the Cape Verde Islands, just as in the south the Falkland Islands block is presumed to have escaped to the south-west. The latter, being part of a mountain structure, could be expected to have maintained itself above sea-level ; the former, being part of the shield, could, on the contrary, have become submerged. Suggestive thereof is its strong volcanicity.

(6) Finally come the Alpine foldings of the Atlas Ranges, trending W.S.W. till near the Atlantic and then S.W., continued beneath the Canary Islands, and corresponding precisely with those of Trinidad and Venezuela, which have an E.–W. strike, each involving not only Cretaceous and Tertiary but older groups of very similar facies respectively, all overturned towards their continental masses, as discussed in Chapter IX (a). Gentil suggests that the base of Madeira is the extension of the Moroccan Meseta, that of the Azores of the Spanish Meseta ; Gregory's [1] opinions on this region are worthy of study.

(7) When allowance is made for the missing " inter-con-tinental " material, lost through erosion or sunk beneath the ocean within a zone at least several hundred kilometres wide, it will have to be admitted that the resemblances between the coastal regions of South America and Africa are both numerous and surprisingly close. The correspondences are far more than could be ascribed to mere " coincidence," being, indeed, precisely such as could be predicted under the Displacement Hypothesis.

PATAGONIA

Using the Gondwanide fold-belt as our base-line and restor-ing the lands as in Fig 7, Cape Horn is now brought to within 1000 km. from Port Elizabeth, and a test of the validity of our reconstruction is thus afforded.

(a) As set forth by Windhausen, the Patagonian platform consists of granite, gneiss, schists, phyllites, greywackes and occasionally crystalline limestones of unknown age, though early Palæozoic or older ; an E.–W. strike has been noted therein. These are comparable with the pre-Devonian basement—Malmesbury and Cango Series—of the southern Cape, which forms the shore at many places between Cape Town and Port Elizabeth. Involved in the Cape foldings, its strike runs more or less parallel therewith, i.e. about E.–W. with deviations to the north or south, but at its eastern end it passes out to sea with nearly south-east course ; (b) pre-Triassic erosion reduced Patagonia to a peneplain while the southern Cape ranges experienced considerable denudation during the early Triassic ;

[1] Gregory (29), ci.

(c) the Patagonian platform over a distance of at least 15 degrees of latitude received a covering of Upper Triassic and Lower Jurassic volcanics—mainly quartz-porphyries (rhyolites) and keratophyres, tuffs and conglomerates, with subordinate sediments largely of continental or estuarine character, the oldest of which contain abundantly the crustacean *Estheria* (*Cyzicus*) *draperi*, a Stormberg species. The Triassic sea lay to the west, in Neuquen. Comparison can be made with (1) the Rhætic-Liassic basalt-rhyolite-basalt outpourings of the Lebombo belt of South Africa and (2) the equivalent Sandstone-basalt remnants next the Zuurberg just north of Port Elizabeth resting unconformably upon tilted palæozoics, just as in Mendoza, Argentina, the plant-bearing Rhætic followed by basalts rests with discordance on Permian and older strata; (d) it is suggestive that the important meridional Jurassic monocline of the Lebombo, traceable for 1400 km. and running out to sea south of Port St. Johns, would, if extended, coincide with the ridge crossing Patagonia south of the Gulf of San Jorge that separates the two main areas of late Cretaceous deposition.

(e) At the close of the Jurassic the south-westerly prolongation of the Moçambique trough to Mossel Bay (Cape) and thence westwards *outside* the Gondwanides enabled the Lower Cretaceous Uitenhage fauna—itself derived from India *via* Madagascar —to reach Neuquen (western Argentina); (f) inter-Cretaceous movements involving overthrusting—the Patagonides of Keidel and Windhausen—curving around the Patagonian nucleus, striking E.N.E. at the Atlantic coast and affecting strata up to Lower Cretaceous, can be compared with the similar folding and faulting of the Uitenhage beds in the southern Cape, where the downthrows are usually on the seaward side with displacements that attain a magnitude of from 3 to 4 km. with general E.–W. direction; (g) the widespread Upper Cretaceous transgression brought the shallow sea upon the coasts of Natal and Pondoland and a vast proportion of Patagonia southwards to Cape Horn and over parts of Graham Land; (h) continuity of these southern lands is indicated by the Indian and Australian elements in the Liassic flora of Chubut, the Jurassic flora north-west of Cape Horn and that of North Graham Land, and the Wealden assemblages of San Martin in Patagonia and the Uitenhage Series of the Cape. As Seward has remarked, it is likely that certain " Northern " Wealden plants originated in this southern continent. As Halle has pointed out, the climate of North Graham Land could not have been frigid during the late Jurassic, though today that island is ice-bound; (i) the Patagonian sauropod genera *Titanosaurus* and *Laplatasaurus* occur in Madagascar and India and *Antarctosaurus* in India; [1]

[1] Von Huene (29), 107, 193; Nopsca (34), 89.

(*j*) the presence of strong Indian Jurassic and Cretaceous elements in Argentina argues for a continuous continental shore or shelf between those lands by way of the Cape (see Fig. 12).

(*k*) The close of the Cretaceous saw the marginal fractures around Africa widen, the southern ocean with its Indian, Australian and New Zealand faunal elements spread northwards into the South Atlantic and trespass over northern and southern, though not over central Patagonia, with repetition during the Oligocene ; (*l*) Argentina, Antarctica and Australia nevertheless remained connected down to the Oligocene at least, as indicated by the diprotodont marsupials and giant horned tortoises of Patagonia and Australia that are not represented in Africa ; (*m*) marked resemblances are shown by the neritic Oligocene faunas of those two regions, as first emphasised by Ortmann and furthered by Hedley and von Ihering ; (*n*) Versluys (34) has commented on the resemblances between the living corals, primnoids, etc., on the continental slopes of Australia, South America and some intermediate islands, but differences from those of South Africa ; (*o*) E. C. Andrews[1] has stressed the floral resemblances between Australia and New Zealand and South America, but has not favoured spreading *via* the Antarctic on climatic grounds. One could, however, urge that even by the later Tertiary Antarctica might not yet have drifted to near its present position.

The reconstruction in Fig. 7 admittedly leaves a triangular space off South Africa. When, however, one makes allowance for the continental shelf of Patagonia—300 km. broad in places—the Agulhas Bank, South Georgia, North Graham Land, the pedestals of Bouvet, Prince Edward and the Croiset Islands, the submarine ridge discovered by the *Meteor* at 560 m to the south-west of the Cape, and that off Natal, such hiatus becomes practically filled and the balance could readily be accounted for by slight distortion of Patagonia in its westerly drift. The fit thus obtained between South America, Africa and Antarctica is hence good and far better, indeed, than might have been expected.

THE FALKLAND ISLANDS

This remarkable group constitutes a most striking link between South America and South Africa despite the fact that it is situated only a few degrees from Patagonia and well down below the fiftieth parallel. Our knowledge of it has been enormously extended through the work of Andersson, Halle and H. A. Baker and the descriptions of the fossils by J. M. Clarke, Halle, Seward and Walton.

[1] Andrews (16), 219.

(*a*) The dominant Devono-Triassic succession has been affected by two main tectonic structures, the more important one in the north trending W. to N.W., the other crossing it nearly at right angles. These can be compared with the Gondwanides of Argentina and the Cape as discussed on p. 87; (*b*) the thick Siluro-Devonian sandstones correspond with the Table Mountain Series of the Cape and are succeeded by shaly marine beds with a fauna having some twenty species in common with that of the Bokkeveld and some twenty-five with the equivalent Lower Devonian of South America, followed by quartzitic strata just like the Witteberg Series. Baker has stressed the surprising similarities with the Cape System not only lithologically but in regard to thickness, which reaches a total of some 3000 m.; (*c*) the equally thick Gondwana beds, known as the Lafonian System, have at their base thick glacial boulder-beds, usually folded and cleaved, identical in their characters with those of the Dwyka tillite, succeeded by a great mass of mudstones, shales and sandstones with a Glossopteris flora, the upper part of which may perhaps range into the Triassic.

The position to be allotted to the Falklands in our restoration (Fig. 7) can be approximately determined by a series of rather pretty comparisons : ·(1) the folding is along two axes but of fairly open character as a rule and thus comparable with that of the Ceres-Clanwilliam region and neither with that of the Caledon-Riversdale section nor with Argentina, where the first-mentioned set of foldings is dominant and intense ; (2) the Lower Devonian sandstones are gritty and pebbly, as in the Cape-Clanwilliam region or in Uruguay, and not quartzitic as in Mossel Bay or the Sierra de la Ventana, moreover, the glacial horizon of the western Cape may, according to Baker, just possibly be represented in West Island ; (3) the Devonian fauna of the Falklands is very closely related to that of the Bokkeveld, less so to that of Paraná ; (4) the Devono-Carboniferous quartzites of the Falklands are identical with the Witteberg Series of the Cape and not closely comparable with strata in Argentina ; (5) the strongly cleaved tillite of the islands is slightly disconformable to these quartzites, as in northern Ceres ; (6) the tillite is of great thickness with some intercalated shales and hence like that of the southern Karroo and the Rio Sauce Grande ; (7) the Lafonian sandstone just above the tillite has no lithological representative in the Cape, but finds its equivalent in the Itararé series of Uruguay and Brazil ; (8) the absence of the " Mesosaurus zone " links the islands with Argentina ; (9) the great development of lower Lafonian sediments devoid of coals recalls the facies of the Sierra de Pillahuincó and of the southern Karroo ; (10) the upper Lafonian beds resemble the

Upper Karroo and not the Triassic of South America; (11) the few dolerite intrusions suggest again the Cape-Clanwilliam region.

" If we admit the displacement hypothesis, these affinities would collectively assign the Falklands to a position along the northern edge of the broad belt of foldings that linked the Cape with Argentina, but would place them somewhat nearer to Africa. Despite the fact that these islands rise from the Patagonian coastal shelf, and that the late Mesozoic foldings (Patagonides) strike south-eastward near the Gulf of San Jorge, and therefore toward them, as remarked by Windhausen, the Falklands, if we ignore the mere tectonic parallelism, display not the slightest affinities with Patagonia, where the late Triassic volcanics and sediments rest upon the crystallines, with no representatives of the Palæozoics. It must accordingly be concluded that this outpost of Gondwanaland belongs to the north-east rather than to the west and actually affords very striking support to the displacement hypothesis." [1]

Their distance is about 5500 km. from the Cape, but only 1500 km. from their supposed Argentine continuation.

MADAGASCAR

Past and present opinion has been unanimous that Madagascar formerly constituted part of Africa and connected furthermore with India. Indeed such union formed the keynote of P. L. Sclater's " Lemuria " and W. Blanford's " Indo-African continent " of the Tertiary era and Suess' " Gondwanaland " of still earlier date.

There have nevertheless been considerable differences of opinion as to just where those several masses were joined and how and when they became parted. The valuable work done by H. Besairie (30) and other members of the Geological Survey during recent years is, however, contributing towards this problem as well as emphasising the geological resemblances between Madagascar and its neighbours, particularly East Africa, an aspect which has been enlarged on by J. W. Gregory [2] and F. Dixey (35).

Influenced by the parallelism of the opposed shores of the Moçambique Channel, the writer had for long imagined—like Wegener, and indeed most other persons—that Madagascar had formerly lain next to the Moçambique-Inhambane coast and that it had moved off for some degrees in the north-easterly direction, following in the wake of India. Further consideration in the light of new data has, however, shown up the improbability of such a viewpoint, and has, on the contrary, suggested an original position opposite Kenya and Tanganyika

[1] du Toit (27), 102.   [2] Gregory (21), chap. xxvii, 367-73.

with subsequent displacement a little to the east of south for about 1000 km. accompanied by slight rotation.   On the north-east lay originally the Seychelles-Mauritius block with India and Ceylon immediately beyond ; not so far away to the south-east and east stretched Antarctica.

The criteria for such a replacement of Madagascar are as follows :

(*a*) Allowing for fringes of younger Tertiary beds the coasts of mainland and island show a clear parallelism throughout the distance of 1600 km. ; (*b*) the southern end of the island then forms the extension of the north-eastward trending coast line between Quelimane and Moçambique, which has been regarded by most geographers as determined by an immense fault.   Not only does the southern shore of Madagascar transect all ancient structures there, but no marine Mesozoic or early Tertiary occur along it facing southwards.

(*c*) The strike in the pre-Cambrian basement of the main-land between Kenya and Moçambique and along the full length of Madagascar forms regular curves, concentric about a point situated within Tanganyika (Fig. 7).   In the south the direction is N.E. to N.N.E., changing in the centre to N.–S. and then N.W., and in the north to N.–S. with deviations to N.N.W., but run parallel to the coast in the eastern side of the island.   More-over, in Ceylon and southern India (Mysore and Vizagapatam) the strike is from N. to N.N.W. with gentle folding in Ceylon, and is in accord with the replacement-scheme postulated.   Such a persistence of grain over so enormous a region cannot be dis-regarded.   Furthermore, the petrological resemblances of these three main areas are extraordinary, for, in addition to wide-spread gneisses, mica-schists, quartz-schists and quartzites, are found granulites, pyroxenites (para-pyroxenites), crystalline limestones and cipolins (marmorites) extensively developed—representing highly altered sediments—together with an abun-dance of types exceptionally rich in graphite, spinel, garnet, corundum, sillimanite, kyanite, cordierite, sapphirine, etc., which are scarcely known from other parts of the world, but have significantly been signalled from Enderby Land, Antarctica. The workable graphite and monazite deposits and the rarer thorianites of Ceylon parallel those of Madagascar, the first two minerals being also present in the Uluguru Mountains of Tanganyika (accompanied by uraninite), while in the Usambara Mountains behind Tanga are rocks of the charnockite suite reminiscent of southern India and Ceylon.   A peculiar new type from Enderby Land called " enderbite " has now been identified from Ceylon and Madras.[1]   In Madagascar, too, are younger alkaline-granites and -syenites recalling those near Madras.

[1] Tilley (36).

(*d*) There is an absence of lower palæozoics of the Waterberg or Table Mountain sandstone type, such as could be expected had the southern end of the island lain due east of, and close to, Zululand ; (*e*) the development of the Madagascan Karroo is more or less intermediate between those of the Zambezi-Shiré area, eastern Tanganyika, Tanga and peninsular India, each sub-division showing a facies closely resembling that belonging to one or other of those four regions. The latter seem therefore to have formed continuous land—" Gondwana "—before the Upper Carboniferous ; only during the Permian did the ocean invade the region and enter a trough that proceeded to develop between the Afro-Arabian and the Indo-Madagascan masses— the forerunner of the Moçambique Channel ; (*f*) the basal glacials of the Onilahy Valley of south-western Madagascar with their overlying dark shales have no known representatives in eastern Africa, but agree with the Dwyka tillite of Zululand and the Talchir of India. The direction of ice movement in the island is regrettably unknown ; (*g*) the succeeding coal-bearing Sakoa Series with a Glossopteris flora accords equally well with the Ecca of the Transvaal, Rhodesia, Nyasaland and Tanganyika and with the Karharbari of India. This division is, moreover, only present in the south, thus paralleling Tanganyika, where coals are not yet known north of Morogoro ; (*h*) the seemingly Lower-Permian marine incursion of south-western Madagascar, with its scanty fauna just above the Coal Measures, has no equivalent on the mainland. It could, however, correspond to the deep-water Upper Ecca shales of South Africa and Nyasaland and the Ironstone Shales of the Raniganj of India.

(*i*) The break at the base of the trangressive Upper Permian Sakamena Series does not fit with that in South Africa or Nyasaland, where the gap occurs on a higher level, but accords more with that in the Tanga area and with that in the Gondwana System of many parts of southern India ; (*j*) the mollusca of the Kidodi area near the Uluguru Mountains—the only place where marine late Permian has been found in southern Africa— include two species characterising the Upper Permian of the Sakamena of southern Madagascar, which in the north is represented by a distinct fauna with *Xenaspis*, *Cyclolobus*, etc., closely related to those of the Lower and Middle Productus Limestone of north-west India and the Urals ; (*k*) the higher Sakamena beds have yielded remains of the reptile *Tangasaurus* and the amphibian *Rhinesuchus*, the former known from the Tanga area, the latter from South Africa, also typical Karroo fish remains and fresh-water mollusca. The floras of both Madagascar and East Africa include the European genera *Ullmannia* and *Voltzia*, unknown elsewhere in Africa. In the north the upper part of the Sakamena has developed a marine—Lower Triassic—facies.

(*l*) The thick Isalo Sandstone, which traverses the island from south to north, is Upper Triassic to Middle Liassic in age and is transgressive over the older divisions, which in the north includes marine Lower Triassic not represented on the mainland. The Isalo can be paralleled precisely with the Escarpment Grits and succeeding red " desert " sandstones of the Zambezi-Shiré area (Dixey), in part with the Manda beds of the Ruhuhu area (Stockley) and more fully with the thick assemblage of the Mariakani Sandstone, Shimba Grits and Mazeras Sandstone of the Tanga area and southern Kenya. Silicified wood is abundant in each of these sandstone groups. Comparison can also be made with the equivalent strata in India—Pachmari, Parsora and other Upper Gondwana beds. Wade [1] has recorded wind-faceted pebbles either in the top part of the Isalo Sandstone or its upward continuation, while similar pebbles have been reported from the Shimba Grits; [2] (*m*) it is suggestive that these equivalent strata reach closest to the present coasts a little to the east of Cape St. André and in the Shimba Hills south-west of Mombasa respectively, in each case through up-doming during early Jurassic times along axes directed generally north-westwards, as mentioned by Wade and by Parsons respectively.

(*n*) Estuarine and marine Jurassic follow either conformably or with slight break in both Madagascar and the Tanganyika-Kenya territory ; (*o*) there is accordingly an absence of the Mesozoic volcanicity so marked in the region from the Cape to Nyasaland. Such is held to be a highly important criterion. Eruptive rocks are admittedly abundant in Madagascar, but are of post-Liassic and mostly of post-Jurassic age, though some dolerite dykes, like certain in Tanganyika, may perhaps be of Stormberg age.

(*p*) Parsons [3] has described a considerable amount of faulting and low-angled thrusting in the region just west of Mombasa, in which the strata have been pressed partly from the north, but mainly from the east in pre- and post-Callovian (mid-Jurassic) times ; this compression he regards as an eastern manifestation of the Great Rift System. Wade [4] has pointed out similar structures in the region of Cape St. André in Madagascar along two axes, the main one directed almost N.–S., the other N.W.–S.E., the former giving rise to anticlines, to overthrusts from the west or to normal faults, the latter to up-doming of the Triassic and Jurassic. These structures are of importance in the search for petroleum. All these disturbances

[1] Wade (14), 13. [2] Gregory (21), 48.
[3] Parsons (29), 70-82. [4] Wade (14), 25.

strongly suggest the result of a slight anti-clockwise twist of Madagascar whilst still attached to Africa, due, it can be hinted, to an eastward drag on its southern end by the mass of Antarctica. At this very time the meridional monocline of the Lebombo was being produced, while warping was taking place in Ceylon.

(q) Marine Jurassics border the opposed coasts of Madagascar and East Africa, dipping seawards at low angles in each case, but, while they run the full length of the island, they have not yet been found on the mainland south of Lindi. Both sequences contain Indian forms like those found in Madras, Cutch and Spiti ; (r) on the other hand, while the Cretaceous succeeds the Jurassic right through the whole length of Madagascar, it is missing in Kenya but extends along the coastal region of Tanganyika and occurs at various spots in Moçambique, always with faunas having Indian affinities ; (s) regressions occurred during the Jurassic and Cretaceous on both mainland and island, to which is due the preservation of vertebrate remains, finely exemplified in the famous " Dinosaur Beds " of Tendaguru, Tanganyika, that contain, among other forms, the colossal *Brachiosaurus*, known from North America though not from Europe, and *Tornieria*, related in some way to *Titanosaurus*. Remarkable, therefore, is the discovery in Upper Cretaceous lagunar deposits in north-western Madagascar of the two genera *Titanosaurus* and *Laplatasaurus* known from Argentina and India, neither of which is as yet known from East Africa, while the Argentina sauropod *Antarctosaurus* is represented in India. To explain the presence of these three genera Nopsca[1] follows Gregory in postulating a migration right across the Pacific. Von Huene,[2] on the contrary, makes the round-about connection *via* Australia and Antarctica, which latter must under his views have been a cold region, since it must throughout have remained fixed in its present position. These difficulties are removed by our hypothesis, which assumes India as connected to Patagonia *via* Madagascar, a route that would at no point have passed within the polar circle of the time ; (t) India was still in attachment to Madagascar during the mid-Cretaceous, but in the latter part of that epoch not only were marine beds being deposited on the eastern as well as the western side of the island, but European faunal elements now made their appearance therein and on the eastern side of the Indian peninsula also, showing that water-ways had been established between the various masses (Fig. 12).

(u) The Eocene sea in turn transgressed over the edges of East Africa and western Madagascar with minor submergences

---

[1] Nopsca (34), 90.
[2] von Huene (29), 179 ; see, however, Schuchert (30).

during the Miocene. The break represented by the Oligocene suggests that Madagascar became reunited to Africa temporarily. Only thus can be explained the well-established faunal relationships that are exemplified by the lemurs, monkeys, tortoises, hippopotami, etc., as cited by Gregory; (*v*) the Moçambique Channel was developed by the movement of Madagascar to the east of south, while the crystallines of the Seychelles—perhaps the nucleus of Mauritius also—are visualised as fragments left behind in the rear of India. Such would accord well with Bailey Willis' [1] clear picture of the displacements that mark the coastal belt of East Africa, namely, the sinuous meridional Lindi–Mafia–Zanzibar fault converging upon the Ruvu-Mombasa fracture which defines the north-eastward trending continental shelf. Within the included angle is set the horst of Pemba Island, repeating on a tiny scale what is presumed to have been the arrangement with Madagascar during the immediate past; (*w*) prominent along the eastern side of the latter is a swarm of basic dykes indicative of tension; (*x*) the profuse volcanicity of the Great Rift System—with Moçambique as well—is repeated in Madagascar—mainly in the west and centre—and the islands of Great Comoro, Mauritius, Réunion and Rodriguez, initiated in the Jurassic and in places continuing today. An immense variety of types ranging from acid to basic—many of them rare—is represented, belonging largely to the alkaline class. The north-western region of Madagascar, and with it the volcano of Great Comoro, belongs to the petrographical province of East Africa. Taken along with the respective alignments of the various faults and volcanoes, this provides weighty testimony for an intense fracturing under which the basin of the Indian Ocean has originated; (*y*) lastly, the physiographical evolution over East Africa, Madagascar, Ceylon, India and Western Australia during and since their presumed parting has been wonderfully similar, and has resulted in the development of elevated laterite-capped peneplains now in course of dissection. Whether such profiles have been preserved anywhere in Antarctica as well is a problem for future research.

INDIA AND AUSTRALIA

It has already been urged that the Indian block must have been much larger than the existing surface of India, since its leading edge had deeply under-run Tibet and had also been sliced-off in the overthrust mass of the Himalayas. Overturned coal-bearing Gondwanas are, for example, found wedged in beneath this range within the great bend of the Brahmaputra River. Furthermore, on the east the right-angle bend made

[1] Willis (36), 99-110, Pl. VI.

by Burma would need to be straightened out, thus giving a total former extent of at least 35 degrees in length with a frontage of some 30 degrees, as shown by the dotted lines in our reconstruction. In addition, the evidence suggests that a wedge-shaped fragment that fitted between India and Western Australia has become involved in the Tertiary crumplings of Burma, Siam and Malay States with its counterpart in the resistant massif of Annam (Cathaysia), though the details have admittedly still to be worked out. The suggestion made by Wing Easton [1] of the derivation from Antarctica west of Wilkes Land of the basement of the islands composing the Sunda arc in the East Indies should be recollected.

Even with such extensions this palæozoic India only just reaches opposite to Western Australia, and for that reason any close resemblances in the structures of these two lands would scarcely be anticipated. Nevertheless there are several significant points of agreement :

(a) The Archaean rocks of India and Western Australia are not only similar lithologically, but show the same general strike —between E.N.E. and E.S.E. in the Central Provinces and Orissa and E.S.E. in the opposed corner of Australia, changing to S.S.E. in the interior and north of that territory (Fig. 7) ; (b) these formations are covered unconformably by at least two thick pre-Cambrian systems having enormous distribution, which lie flat and are only locally disturbed, of the general type " quartzite-shale-limestone-volcanics (acid and basic)." Such are the Cuddapah of Madras, Vindhyan of Central India and equivalent strata in Assam and Burma that have been caught up in the later plication, and the Nullagine and Elvire systems that occupy the northern half of Western, Northern and Central Australia.

(c) The Cambro-Ordovician geosyncline of the Himalayas, Burma, Yunnan and Tonking finds its continuation southeastwards through the Kimberley district to South Australia, Tasmania and the Ross Sea. Striking is the Lower Cambrian *Redlichia* fauna throughout this lengthy trough, with persistence of the North American facies until the early Ordovician, when European forms appear ; the north-western end of the Australian " channel " was then upheaved and the seas transferred to the east. Peninsular India and the south-western part of Australia were of course united to East Antarctica ; only first in the later palæozoic did an arm from the encircling ocean interpose itself temporarily between India and Australia ; (d) during the later Devonian the sea overflowed upon the north-west of Australia as well as over Burma and Yunnan, but its readvance during the Lower Carboniferous failed to reach that part of Australia

[1] Wing Easton (21), 507.

although covering lower Siam, Malay States and Sumatra with its shore-line situated somewhat to the south-west or south.

(*e*) The late Carboniferous and early Permian seas not only inundated the previously mentioned areas, but pushed southwards along the western side of Australia to Perth at least, the shore-line running meridionally in that land and hence parallel to the Samfrau trough on the opposite side of the continent. The faunas disclose Salt Range and Himalayan species, but show marked differences to those of Eastern Australia; (*f*) associated therewith are the glacials of Western Australia, the product of ice-sheets originating seemingly to the south-east followed by the Irwin River and Collie Coal Measures and overlying marine beds with which can be compared the Talchir glacials derived from a centre probably to the south or south-west succeeded by the Karharbari Series in Bihar, Orissa and Madras, also with a Glossopteris flora; proximity to the sea is nevertheless indicated by the marine intercalation at Umaria in the Rewah coalfield. Again in Tenasserim, Lower Burma, Gondwana coal-bearing strata overlie the Permo-Carboniferous Moulmein limestone.[1] In contrast are beds of Stephanian age with *Pecopteris*, etc., in Malay States and Sumatra that reveal the nearness of the Northern or Asiatic continent.

(*g*) The late Permian and early Triassic regression converted into land a wide tract between India, Australia and China, thereby extending Gondwana to the north-east; much of Burma, Yunnan, Siam, Tonking, Annam, Malay States, Sumatra and the margin of Western Australia were raised save where certain troughs persisted or else proceeded to deepen during the Middle and Upper Triassic. This permitted the interchange of terrestrial life between Gondwana and Asia (Cathaysia), such accounting for the Karroo reptiles *Dicynodon* and *Lystrosaurus* in India and Tibet (Sinkiang) and the second-named in Indo-China (Laos), and for members of the Glossopteris flora in the Triassic coal-bearing formation of central Yunnan and eastern Tonking. Tension is evinced by block-faulting in India and Western Australia with the lowering into the basement of areas of coal-bearing Lower Gondwanas; (*h*) the wide Upper Triassic transgression with its *Myophoria* fauna flooded much of Indo-China and also Assam, and probably converted Cathaysia into an island; (*i*) the close of the Triassic was marked by strong compression in Indo-China and the Malay States [2]—the only quarter of the world where powerful movements of this date are known—the overfolding being directed toward the north-west and hence, after allowing for some Tertiary distortion, trending approximately parallel to the slightly earlier compression-zone in north-eastern Queensland.

[1] Fox (34), 42.　　　　　　　　　　[2] Fromaget (29).

The massif of Annam seems to have played an important part in this squeeze between the opposed masses of Gondwana and Asia, which became once again united.

(*j*) Unconformable upon these folded beds lie Liassic and Jurassic deposits, generally of a non-marine facies, developed over Indo-China, in Assam, at intervals along the eastern coast of India and more continuously along the north-western and western sides of Australia, the floras being closely comparable. In the succeeding Jurassic submergence the Indo-Australian gulf must have attained a considerable size, for that sea reached so far south at least as Ceylon and Perth respectively (Fig. 12) ; (*k*) during the early Cretaceous Australia was apparently connected to China *via* Madagascar and India, but this long chain was broken by the later Cretaceous, when the ocean entered between those parting blocks. The date of separation of Australia was seemingly effected thereafter, probably during the Tertiary

ANTARCTICA

The rôle of the Antarctic is a vital one. As will be observed from Fig. 7, the shield of East Antarctica constitutes the " keypiece "—shaped surprisingly like Australia, only larger—around which, with wonderful correspondences in outline, the remaining " puzzle-pieces " of Gondwana can with remarkable precision be fitted. The angle of Crown Princess Martha Land conforms to the indentation at Beira : Princess Ragnhild Land lies next the south-eastern corner of Madagascar, Enderby Land opposite Ceylon and the eastern side of India : the great bulge of Wilkes Land fits into the Australian Bight, while Tasmania and the submarine bank to the south of it discovered by the *Aurora* project into the Ross Sea. New Zealand, restored to its former place next Australia, is brought into line with King Edward VII Land, while the Antarctic Archipelago falls off the Cape and in addition makes connection through a much-shortened South Antillean arc (p. 195) with Patagonia.

The geological agreements thereupon become not less remarkable :

(*a*) The length of coast line between Coats Land and South Victoria Land is some 8000 km., of which the greater part has merely been sighted at intervals but never trodden, while large stretches are ice-covered. At the relatively few places where the rocks have been examined, they have proved to consist of granite, gneiss, amphibolite and occasionally phyllite, schist and limestone, that include, more particularly in the Enderby quadrant, gneisses in which garnet, pyroxene, cordierite and sillimanite figure, together with varieties of norite and charnockite. These in striking fashion recall the Archæan complexes of the shields situated to the north—of Moçambique, Madagascar, Ceylon and India—but unfortunately it has rarely been possible to obtain detailed information concerning the dominant strike of these rocks, for on that vital question Antarctic literature is generally silent. The trend is, however, N.–S. in the eastern part of Wilkes Land and perhaps in the western as well, thus agreeing with the normal direction in South Australia and Victoria, with which the geology can be compared, save that the Tertiary of the Eucla Basin is only doubtfully reproduced in Antarctica.

(b) The Lower Cambrian *Archæocyathus* Limestone discovered as blocks on the Beardmore Glacier, possibly too the limestone *in situ* at Mount Nansen nearer the Pole, has its nearest representatives in that of South Australia, where similar beds occur in the long Cambrian syncline that crosses Australia from north to south and is therefore directed towards Antarctica. A similar lump was dredged on the opposite side of the continent, in the Weddell Sea. Furthermore an Archæocyathid has been recorded from the basal Nama of South-west Africa ; (c) the Upper Devonian fish-scales obtained from erratics near Mount Suess suggest the equivalent formation in Victoria, the only spot in our reconstruction where comparable strata are developed.

(d) From the rocky coast of Adelie Land for a distance southwards of about 2000 km.—to Long. 145° E. and to within 300 km. of the Pole—resting generally horizontally on the shield, lies the Gondwana Beacon Sandstone, exceeding 1000 m. in thickness in places. The transported fragments of sandstone obtained off Enderby and Coats Lands suggest that the ice-sheet conceals an enormous cap of such beds. The formation consists of pale, massive, cross-bedded sandstones with felspathic grits and conglomerates, yellow or grey, thin-bedded sandstones and in its upper parts of dark shaly beds with carbonaceous shales and coals. The existence at any point of a basal tillite is not yet assured ; (e) these lithological characters and the presence of *Glossopteris indica* suggest the Upper Permian Newcastle Series of Tasmania or the Raniganj of India, but the discovery of the fossil stem *Rhexoxylon* in a loose block indicates the existence of still higher horizons, since that unique genus characterises the Middle and Upper Triassic of Southern Africa ; (f) the constantly associated dolerite sills, from Adelie Land to the Queen Maud Ranges, agree precisely with the mesozoic intrusions of Southern Africa and Tasmania that fade out in the northerly direction.

(g) Marine mesozoics are absent from the curving margin of East Antarctica, and are scarcely developed along the shore-lines opposite ; (h) as set forth later, the Antarctandes con-

K

stitute the prolongation of the Andean system ; (*i*) the fracture-pattern of South Victoria Land and of the Ross Sea (Fig. 35) was doubtless produced during the later Tertiary, judging from the associated volcanism, and is comparable with the graben-structure of South Australia and the faulting that has determined Tasmania ; (*j*) in view of such faulting the rapid deepening of the surrounding ocean to 4500 or 5000 m., together with the regular semicircular sweep of both coast and submarine contours, strongly suggests that East Antarctica is bounded, and has been determined, by tension-fractures. The abnormal depth of the continental shelf—from 250 to 650 m. generally—is probably due to the weighting by the ice-cap.

(*k*) The Ross Sea volcanoes and that of the Gauss Berg (Long. 89° E.) have erupted lavas of alkaline type (kenyite, etc.) closely related to those of East Africa. The Sjövold cones of Long. $72\frac{1}{2}$° E. have only been sighted so far ; (*l*) as a link with Africa must be reckoned the Kerguelen Archipelago with its striking fiord-system, a group largely composed of probably Miocene basalts resting upon Oligocene lignitic beds ; phonolite, syenite and diorite also occur.[1] These Tertiary rocks have yielded carbonised trunks of *Dadoxylon* and *Cupressinoxylon* ; (*m*) of high significance is the presence there of the living *Phreodrilus*, intermediate between the terrestrial and aquatic worms, found in pools on mountain peaks in New Zealand, Tasmania, Australia, South Africa, Falklands and Patagonia, just as the isopod *Phreatoicus* occurs under a similar environment in these first four countries. Neither organism could have crossed the ocean.

(*n*) The part played by Antarctica in marine faunal migration is referred to more than once, and it would appear that such a rôle must have been of extreme importance. While a route always lay open along its western or Pacific shore from the Palæozoic onwards, an alternative one came into being during the late Cretaceous along its eastern or Indian Ocean side, thereby providing a more direct passage between the Malay region and Patagonia. The precise date of the initiation of the latter is a problem for research. These two routes naturally traversed regions possessing rather different oceanographical conditions ; (*o*) biological evidence indicates a connection between Australia and South America down to a late date in the Tertiary (Chapter XIV). Presumably the link was broken first at the Australian end. The enormous distance between these two continents, well brought out in our reconstruction, suggests that the biological affinities to be anticipated between them before their separation from Antarctica should be generic rather than specific in character ; (*p*) of significance is the hypothesis

[1] Mawson (34).

favoured by Gaudry, Moreno and Windhausen [1] of an *Antarctic "Biological Asylum"* from which the neighbouring attached lands derived new forms of life.

THE BOUNDARY RELATIONS BETWEEN GONDWANA AND LAURASIA

The mutual relations of these bodies must obviously have varied considerably from time to time. For relatively lengthy periods they were separated by wide ocean deeps—comprehensively referred to as the "Tethys"—while on rarer and briefer occasions they became united along one or more stretches of margin to form a unit. Although such palæogeographical studies are most fascinating, it must suffice merely to refer to those few though important junctures, when connection was thus established and interchange of terrestrial life became feasible. Incidentally the remarks concerning biological migration (p. 294) should be studied.

North and South America were on the whole parted by ocean till the Tertiary, but it is thought that the much-folded regions of Appalachia, Iberia, Venezuela and North-west Africa must in their *uncompressed state* have extended far out into the present Atlantic region towards one another, and could well under our hypothesis have been thus united during certain times within the Palæozoic, as supposed by Wegener, Baker and others, the disjunctive Mediterranean basin having not yet opened.

The post-Dinantian (Suedetic) movement thereupon enabled the "northern" vegetation, accompanied by reptiles and amphibians, to enter northern Africa, whence it spread to the north and west of South America until opposed by the "southern" flora, with which it mixed slightly in Brazil and South Africa. It also reached Sinai and Persia, but failed to enter India or Australia, possibly because of the gulf that led from Afghanistan towards Madagascar, which may have introduced certain climatic difficulties. Nevertheless the European tetrapod *Actinodon* has been found in Kashmir associated with the southern plant *Gangamopteris*.

The westerly spreading of the Permian Tethys from the Mediterranean region across to Central America cut off North Africa not only from North America but Western Europe, though in Asia Minor and Persia connections seem to have remained or else to have been initiated with Russia, a link that broke down before the Triassic. Further east the "Productus Ocean" delayed exchange of life between Indo-Australia and China until the beginning of the Triassic. During the Upper

[1] Windhausen (31), 344, 548.

Triassic a temporary union was re-established in the Persia-Pamir sector and during the Liassic in the Indo-Chinese region. In the Lower Jurassic Central Europe may briefly have been joined to Armenia, while in the Upper Jurassic the extreme west of Europe became linked to Algeria and Western Asia to Afghanistan and India. In the Lower Cretaceous Western Europe was still joined to Algeria, while the easterly connection was now made *via* Austria, Armenia and Arabia. In the Upper Cretaceous the westerly link was broken and one established between India and China, while the early Tertiary saw the westerly connection restored.

The rhythmic nature of the connections between Laurasia and Gondwana is well brought out by this epitome, which could go to support Staub's idea of the alternating poleward " drift " and poleward " flight " of those masses. This regular " make and break," alternately in east and west, and that over so long a period, strongly favours, on the other hand, the author's view of the repeated impact of the one continent on the other with some slight rotation of the masses each time, as is pointed to by the progressively changing orientation of the belts of compression developed through such conflicts, an aspect that will be elaborated in Chapter XVI.

CHAPTER VII

LAURASIA—SECTION I

Northern Continent. Reassembly. Comparisons between Sides of North Atlantic. Late pre-Cambrian. Cambro-Ordovician. Faunal Provinces. Taconian Orogeny. Silurian. Caledonian Orogeny. Devonian. Acadian Orogeny. Carboniferous Sedimentation and Orogeny.

THE NORTHERN CONTINENT

WHILE the lands of the Northern Hemisphere could be treated in the same comprehensive fashion as those of the Southern, they happen to have been less widely dispersed. Our essential object being to establish a case for closer geographical union between North America and Eurasia, it merely suffices to concentrate upon the North Atlantic and Arctic regions for which the evidence is quite strong, though it would clarify matters to push our inquiries so far back as the Cambrian. The remoter parts of North America and Asia will only be touched on lightly, chiefly to show that their histories are in harmony with our general ideas of earth evolution. Here again the *study of the polar region provides the key to the elucidation of Laurasia.*

While Gondwana was essentially a stable shield which suffered little save marginal compression, Laurasia was, on the contrary, a composite mass built up of several minor shields parted on occasions by sea-ways, which during various stages in its evolution became squeezed and welded together or else dissolved into similar or into other components. It consequently experienced an inordinate amount of bodily distortion, crumpling following upon crumpling, sometimes along the same lines, but more frequently at angles to the earlier trends, and often in complex fashion. The restoration of its framework, attempted in Fig. 16, is therefore only a first approximation. It is curious to observe that in almost every respect, but markedly through the presence of a persistent marginal fossa, a lesser counterpart to Gondwana was formed by that section known as " Laurentia," which indeed for lengthy periods constituted a sub-continent.

Notable, however, is the general symmetry of the continent, as brought out by the three orogenic belts of the far east, far west and centre (Urals), and by that along the southern margin, as will be amplified in Chapter XVI. It must be pointed out

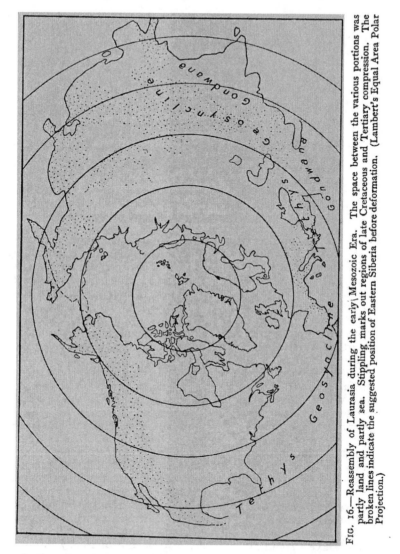

FIG. 16.—Reassembly of Laurasia during the early Mesozoic Era. The space between the various portions was partly land and partly sea. Stippling marks out regions of late Cretaceous and Tertiary compression. The broken lines indicate the suggested position of Eastern Siberia before deformation. (Lambert's Equal Area Polar Projection.)

further that during much of geological time Laurasia must have lain farther to the south with its southern margin extending across the Equator, as will be argued in Chapter XIII.

REASSEMBLY OF THE PARTS

The accurate " fitting " of North America to Eurasia presents much difficulty because of the shortness of available undistorted " base line." A key is provided nevertheless by the steep edges of the Greenland and Scandinavian masses, at between the − 1000 and − 2000 m. contours, and by the wonderfully straight physiographical feature, taken to be a stupendous crustal break, that runs from Tromsö across the Arctic to the McKenzie River, as discussed in Chapter X. The Greenland block can then be apposed remarkably well against that of Norway both as regards distances and inflections, the north-eastern corner of the former falling opposite the southern end of Spitzbergen, which must be displaced somewhat to the east, the angle of Cape Brewster at Scoresby Sound entering the indentation off Christiansund, and the slightly concave coast-line down to Cape Farewell paralleling the drop of the sea floor that runs evenly south-westwards from Norway outside the Hebrides and Ireland. The distance apart at which the present coast-lines would be set can be taken at between 350 and 500 km.

The partial closing-up of Baffin Bay and Davis Strait and of the several waterways between Baffin Land and Ellesmere Land, as indicated by Taylor,[1] would bring Newfoundland nearly opposite Ireland, though at some considerable distance therefrom, a point which is discussed below. Some of that space could be accounted for by the rather wide extensions of the continental shelves off Newfoundland and the British Isles (the latter including Rockwall), by other sunk lands or perhaps by Iceland, though the last can rather be viewed as a mass built up by volcanic action during the process of displacement.

Under such a scheme, which is admittedly approximate, the amount of sliding relative to Europe of the northern end of Greenland would have been about 800 km. ; that of its southern end and of the United States much more owing to the deduced rotation of the continent about some centre in Alaska as well as its north-westerly displacement. During the opening of the Arctic Basin, Spitzbergen and Franz Josef Land may have suffered some slight north-westerly movement also. Allowance would have to be made for the anti-clockwise twist during the Tertiary of the Iberian peninsula, which would bring Galicia a bit nearer to Brittany and Portugal opposite the eastern corner of Newfoundland. Not improbably the British Isles, because of the strength of the Alpine push in Central Europe, experienced a similar though weaker rotation, and Newfoundland an appreciable anti-clockwise twist, but such likelihood is here ignored.

[1] Taylor (28a), Fig. 1 ; Bucher (33), Fig. 99.

Van der Gracht has commented on the probability of crustal distortion affecting not only the continental masses, but the adjacent shelf margins.[1]

In his replacement scheme Baker brings Newfoundland right against the south-western corner of Portugal with Cape St. Vincent entering into White Bay and thereby obtains an astonishingly close fit between the two coasts, one which he claims is supported by the stratigraphies and tectonics of the respective countries. The argument is, however, ineffective, since the formations and structural lines of the two lands are widely different. Actually on the one side Archæan and Palæozoics are concerned involved in pre-Carboniferous movements, on the other Mesozoics and Tertiaries (some extremely young) affected by the Alpine disturbances. Again, where the Palæozoics are exposed in the Algarve area of Iberia, their dominant strike is W.N.W. to N.W., or at right angles to that of the equivalent systems in Newfoundland, while their respective facies are, so far as can be made out, rather different.

In our reconstruction, as in that proposed by Wegener, Newfoundland is set to the west of Ireland and the north-west of Iberia, though at a greater distance apart, while the folded Palæozoics of southern Portugal are correlated with those of Nova Scotia, and those of northern Portugal—bordering the Meseta—with probably those of eastern Newfoundland. Fairly wide spaces make themselves evident between the projecting masses of America and Europe. The importance of such spaces in connection with the crossing of the trans-Atlantic orogenic belts has been lucidly expounded by van der Gracht. Lake [2] has pointed out that Murray and Hiort's bathymetric map of the North Atlantic of 1911 shows in mid-ocean about Lat. 50°–52° N. parallel ridges and troughs running a little south of west with the difference between ridge and trough sometimes exceeding 5000 feet, and that these lie almost on a great circle passing through Halifax and Cork. One is tempted to regard them as dismembered portions of the palæozoic fold-bundles of Laurasia. As forming the counterpart of the Iberian block van der Gracht [3] pictures not Newfoundland or Nova Scotia but the sunken Bahamas block, but the latter with greater probability belongs to the submerged mass of central or southern Appalachia.

Only brief allusion need be made to the opposed masses of the North American Archipelago and Siberia. The first-named has been broken into a mosaic by regional tension, while the second has suffered wholesale deformation, only limited parts having escaped, such as the Angara platform. Striking is the

[1] van der Gracht (33), 927.   [2] Lake (33), Fig.
[3] van der Gracht (33), 928.

vast U-shape of the foldings around that nucleus and the wider spread arcs outside the latter, when compared with the opposed D-shape of the Arctic basin.

COMPARISONS BETWEEN THE SIDES· OF THE NORTH ATLANTIC

From our reconstruction a number of outstanding correlations emerge, most of which will receive discussion below.   Incidentally it might be stressed that the North Atlantic region presents a magnificent field for the systematic assembling of data on such relevant problems as the constitution and grain of the pre-Cambrian basement and its intrusions, the thickness, source, mineralogical composition, facies and faunas of the palæozoic and mesozoic strata, the separation of the various orogenic pulses, the location of their particular fronts and advance-folds, the relative movements of the masses concerned, the determination· of the directions of prevailing tension as evidenced by dyke-swarms, etc., all researches which should be carried out with the co-operation of specialists.   The views submitted regarding this region are only put forward with considerable diffidence through the writer's lack of personal acquaintance with any save a rather limited area.

Some of the principal relationships disclosed are the following:

(1) The Archæan complexes of Labrador, Newfoundland, southern Greenland, north-western Scotland and southern Norway are brought within a much condensed region.

(2) The extensive belt of intensely granite-injected and granitised lower palæozoics of eastern Greenland becomes directly opposed to those of Scandinavia and Spitzbergen.

(3) The pre-Cambrian sandstone formations of the same areas as in (1) are collected within a smaller compass.

(4) The American faunas of the Cambro-Ordovician of north-western Newfoundland, north-western Scotland and eastern Greenland now occupy a gentle curved belt which fails to touch Ireland.

(5) The Taconian thrusting—which was towards the Laurentian foreland—follows the very same belt.

(6) The alkaline syenites, porphyries and other allied intrusions fall on a line extending from Ontario through eastern Maine, southern Greenland, north-western Scotland, southern Norway and Kola Peninsula.

(7) The facies of the Ordovician of north central Newfoundland accords with that of the British Isles (excluding north-western Scotland).

(8) The Caledonian structures of central Newfoundland fit with those of southern Ireland and south-western England.

(9) The Old Red Sandstone facies of the Devonian is developed universally over this region.

(10) The Carboniferous has largely been stripped from Newfoundland but still covers most of Ireland, while each is relatively little disturbed in the northern and central parts of those countries.

(11) The Appalachian-Hercynian fold-bundles run evenly through New England, Maine, Nova Scotia, southern Ireland, southern England with fairly well marked front, with the area to the south showing the folds arranged in a wide S-shaped structure that .passes through France, Spain and Portugal with overturning in the main towards Laurentia.

(12) The several phases as worked out by van der Gracht synchronise almost exactly in America and Europe.

(13) The Triassic of eastern North America and western Europe agree closely.

(14) The older strata of the British Isles have been derived from a continent situated to the north-west, which Gilligan (31) suggests might have been Labrador, and also, it can be added, southern Greenland.

(15) The areas of Tertiary basalts and rhyolites with their intrusive granites, gabbros, etc., fall on a belt trending roughly N.-S. from the Irish Sea past the Faroes to eastern Greenland, one that is set obliquely to the present ocean borders.

(16) Geodetic measurements hint that the westerly drift of Greenland is being continued today.

Other points of congruence, as well as of difference, will be mentioned shortly, but sufficient has been set down to indicate the general correctness of Taylor's and Wegener's ideas. Schuchert [1] has tabulated the geological relations between Newfoundland and Ireland, but has unduly magnified the apparent discordances through not recognising that the Irish lower palæozoics belong essentially to the South-eastern or Acadian Geosyncline, and should therefore correspond, as indeed they do, with those of central Newfoundland. Hence, too, the absence from Ireland of typical American Cambro-Ordovician limestones and also Archæan gneisses, anorthosites, etc. He has furthermore failed to notice that these two lands must have been situated a considerable distance apart originally—by at least twice the width of Ireland itself (600 km.)—which would alone account for appreciable facial and faunal differences over such intervening space.

Allowing for such factors and discriminating more carefully between the several palæozoic orogenies, a far greater amount of geological agreement than Schuchert has so far conceded is to be found between these outposts of America and Eurasia.

[1] Schuchert (28), 131-32.

Indeed, as van der Gracht [1] has aptly remarked, there is nothing in the structures to oppose the conviction that those crustal blocks are portions of one original land-mass, and a whole series of facts and deductions to make the strongest plea therefor.

LATE PRE-CAMBRIAN

Favouring our reconstruction is the distribution of those little-disturbed clastic formations known as the Jotnian of Scandinavia (including part of the so-called " Sparagmite "), the Torridonian of north-western Scotland, Igalko of southern Greenland, Avalon (Signal Hill Series) of Newfoundland and those of Labrador, which pointedly recall the subsequent Old Red Sandstone deposits of those very regions.

Noteworthy, too, are the Eo-Cambrian tillites of the Varanger Fiord in Finmark (Reusch ; Holtedahl), Kola Peninsula (Wegmann), Spitzbergen (Kulling), eastern Greenland (Poulsen ; Kulling) and the probably Cambrian one of the Yennesei (Nicolaev), which denote an important glacial tract situated well within Laurasia.

CAMBRO-ORDOVICIAN

Entering between the more stable elements—Cascadia, Laurentia, Appalachia, Caledonia, Baltica, etc.—the epeiric seas generally widened and deepened up to about the close of this period, our concern being specially with the great compound geosyncline that stretched north-eastwards along a great circle from Mexico into Siberia, a structure that persisted till the later Palæozoic (Fig. 17). Practically all the *Archæocyathus* limestones of the Northern Hemisphere, from California by way of Labrador, Spain and Scotland to New Siberia, were formed within this lengthy trough. From the deepening Appalachian geosyncline the sea spread north-westwards by way of the Cordilleran geosyncline, inundated the northern edge of Canada, Ellesmere Land, northern and eastern parts of Greenland, Spitzbergen, Novaya Zemlya (with extension into New Siberia), Scandinavia and Baltic provinces, much of the British Isles, France and Spain, and so back to Newfoundland and Acadia. The stratal development in eastern Greenland is of enormous thickness—over 9000 m., according to Wegmann.

The large oval sub-continent thereby created, embracing Labrador, Greenland and the western margins of Spitzbergen, Scandinavia, Scotland and Ireland, was probably flattish, but despite big diastrophic movements along its borders, persisted throughout the Palæozoic. It can conveniently be referred

[1] van der Gracht (33), 929.

to as " Laurentia " or " Eria " (Suess). Close to it lay " Appalachia," which with " Acadia " must have constituted an
important land-mass extending far to the south-east, in which
direction it may even have been co-terminous with North-west
Africa as supposed by Wegener and Baker. The great furrow
parting Laurentia and Appalachia was in western Vermont

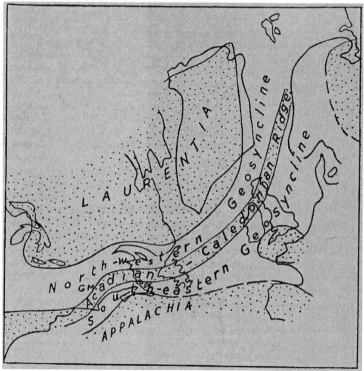

FIG. 17.—Showing the Geosynclines at the close of the Cambrian Epoch in the
North Atlantic Region. (Some allowance has been made for the subsequent
Palæozoic compressions.) GM marks the position of the Green Mountain
" Wall."

markedly constricted by the Green Mountain " wall " or axis,
which right up to the Devonian seems to have played a vital
part in determining, by its upheaval or depression, the interchange of marine life between the seas situated to the north-east
and north-west respectively. It is here that the " American "
and " European " faunas either contrasted or intermingled.

Furthermore, whereas from this point onwards in the south-
westerly direction the Appalachian geosyncline was single and

purely " American " in its life, the trough stretching north-
eastwards toward Scandinavia seems to have been a double one,
divided by a fairly persistent medial land barrier into a North-
western and a South-eastern portion with contrasted facies and
faunas.   Broadly speaking, these are (1) the Quebec–western
Newfoundland – north - western Scotland – eastern Greenland –
Spitzbergen–Novaya Zemlya–Verkhoiansk, and (2) the southern
New England–Nova Scotia–eastern Newfoundland–Ireland–
southern Scotland–Scandinavia troughs.   It must be admitted
that certain geologists have pictured a single furrow of such
depth that neritic forms of life were able to migrate only along
its shores but not to cross it, thereby accounting for the faunal
differences noted in the fossils from its north-western and south-
eastern margins.

The medial ridge is, however, traceable in America through
central New England and thence through the west-central part
of Newfoundland as the New Brunswick geanticline, parting
the " St Lawrence " from the " Acadian " geosyncline.[1]   In
northern Ireland and central Scotland it is marked out by the
broad massif of Dalradian rocks flanked by palæozoic clastics,
which can indeed be specifically referred to as " Caledonia,"
while it must have extended to at least a point off the coast of
Scandinavia, as shown in Jamtland by the thicker and more
arenaceous phase of the palæozoics when traced north-westwards towards the Atlantic.   This " Acadian-Caledonian "
ridge (Fig. 17) not only formed a more or less effective barrier
between the two faunal provinces—at least during the earlier
stages—but determined the positions of the subsequent Taconian
and Caledonian orogenic fronts.

The North-western trough must have been narrow, even when
compressional shortening is allowed for, though a considerable
depth is indicated by the prevailing calcareous facies during
greatest submergence, as well as the great thickness of those
sediments, well manifested in the Greenland sector.   The
South-eastern, on the contrary, was wide and certainly extended
far to the south in Western Europe, where it included a number
of large islands.   Its farther shore is definable in Europe and
reached to Morocco, but is only conjectural off America.   This
trough was marked out by shallower water sediments generally,
which included radiolarian cherts, graptolitic shales, greywackes and in places limestones, lavas and tuffs.   Striking is
the development of such cherts in a zone extending from the
Hudson Valley through Newfoundland to the British Isles.

[1] See Schuchert and Dunbar (34), Fig. 2 ; Heyl (36), 13.

FAUNAL PROVINCES

" American " faunas characterise the North-western trough, and in fact the entire circum-Laurentian fossa, as pointed out by Holtedahl (20) ; " European " ones, the South-eastern one, with further sub-division into a *Northern* facies—Spain to Scandinavia—and a *Southern* one—Sardinia to Bohemia. All this is best brought out by the Trilobite genera.

During the Lower Cambrian the American genus *Olenellus* (*sensu stricto*) flourished in the Appalachian geosyncline west of the Green Mountain wall, in north-western and northern Canada and in the North-western trough. The corresponding *Callavia-Holmia* fauna largely, though not entirely, characterised the South-eastern trough right to New York and on to Mexico. In another geosyncline extending through Meso-potamia and South-eastern Asia to Australia the *Redlichia* fauna existed. During the mid-Cambrian, with the same general distribution of land and sea, *Olenoides* and *Paradoxides* respectively replaced *Olenellus* and *Callavia-Holmia*, the first-named occurring in the eastern prolongation of the North-western trough through north and north-eastern Siberia into northern China, which was not in free communication with the " Red-lichia Sea." During the Upper Cambrian they were respectively replaced by *Dicellocephalus* (scarce in Wales) and *Olenus*, while the American fauna spread through China and so far as Australia and Tasmania, where *Olenus* is absent and *Dicello-cephalus* common. The Chinese form *Huenella* is present in Novaya Zemlya.

The Ozarkian (Tremadoc) *Dictyonema* and *Euloma-Niobe* faunas characterise the South-eastern trough, i.e. eastern Spain, southern France, England and Scandinavia. Of interest is the development on the southern side of that trough of the clastic " Grès Armoricain," extending from Germany into Brittany and occupying much of Spain, with the " Calciferous Sand-stone " as its probable American equivalent. Stromatolite and Cryptozoan dolomites mark out, as described by Holtedahl, the circum-Laurentian sea, i.e. eastern North America, Ellesmere Land, north-eastern Greenland, Spitzbergen, Bear Island, Finmark and Kanin Peninsula, associated with an immense thickness of limestones, slates and quartzitic slates.

While the seas were generally deepening during the early Ordovician, thick terrigenous deposits were being laid down from the erosion of Appalachia, thinning out to the north-west and being replaced in that direction by black graptolitic shales and limestones, which during the Middle Ordovician spread widely over the Laurentian platform. The Canadian (Beekmantown (fauna) is recorded in north-western Newfoundland, north-western Scotland (Durness), Isle of Smolen off Norway, eastern Greenland (Poulsen) and in northern China (Grabau), which last is in contrast to the European fauna of Szechuan and Tonking.

In the South-eastern trough of New England and central Newfoundland the strata—which include numerous graptolite horizons—carry a Western European fauna ; indeed, as Heyl has remarked, the succession in the Bay of Exploits "more closely resembles that of the British Isles than that of the rest of North America."[1] In the mid-Ordovician, however, the sea overtopped the Green Mountain wall and the "Atlantic" waters joined those of the "Trenton Sea." In Ireland over the broad region south of a line drawn from County Mayo to County Down the graptolitic strata, as long recorded by Murchison, contain trilobites abundant in America, but seldom found in Wales or England, e.g. *Remopleurides, Bronteus, Bathyurus,* etc., while American elements together with Scandinavian ones occur even in the Girvan district, especially in the Stinchar limestone. In central Newfoundland and in the British Isles from County Mayo across, *via* County Down, to Aberdeen, the clastic facies deposited during the Tremadoc-Llandeilo off the margin of Caledonia included cherty rocks, shales and "pillow lavas," which passed south-eastwards into deeper-water grapto-litic muds and limestones. Volcanicity was rife, particularly during the mid-Ordovician (Llandeilian) over the lengthy stretch right from Gaspé through central Newfoundland across northern and southern Ireland, northern Wales, Westmorland and southern Scotland to the western side of Norway between Stavanger and northern Trondhjem, and also gave rise along that zone to a wealth of intrusions of various kinds such as diorite, diabase, gabbro and serpentine—a clear warning of impending crustal instability. In Spain and Brittany the Ordovician faunas are, however, of the Northern European type, and contrast with those of Sardinia and Bohemia. It is highly significant that the American elements appearing in the South-eastern trough are confined to those deposits bordering "Caledonia."

This study of the Cambro-Ordovician faunal provinces shows that the distribution of the strata concerned is precisely *such as could be expected were the great space of the present Atlantic to have been non-existent with America situated much closer up against Europe*, as shown in Fig. 17.

Whereas in the South-eastern trough sedimentation went on continuously into the Silurian, i.e. Ireland, much of England, Spain, Scandinavia, Baltic, etc., with some shallowing, in the

[1] Heyl (36), 13.

North-western, on the contrary, the Taconian Orogeny developed within that geosyncline before the close of the Ordovician, vastly modified that trough and in most places destroyed it, which disturbance is recorded in the temporary withdrawal of the sea from the main Appalachian geosyncline.

TACONIAN OROGENY [1]

Many persons have unfortunately failed to discriminate between the two widely parted periods of deformation during the mid-Palæozoic—the Taconian and the Caledonian—the former of late Ordovician, the latter of late Silurian age. Their trend-lines are indeed sub-parallel, but in Europe their over-thrustings are in opposite directions, although paradoxically the second movement would appear to have largely been due to repetition of the same crustal pressures.

A gap above the Beekmantown stage is not uncommon, as in south-eastern Pennsylvania, where the " Mine Ridge " anti-cline shows Upper transgressing across up-arched and eroded Lower Ordovician, while a break occurs in a slightly inferior position (Llandeilian) in southern Scotland and in Spain.

The main north-western Taconian front runs from above New York at first nearly northwards a little to the east of the Hudson and Lake Champlain, thence north-eastwards along the south of the St. Lawrence, east through Gaspé and to the south of Anticosti Island and then within the north-western margin of Newfoundland with north-easterly trend into the Atlantic. Its southerly extension is less easily traced, since it becomes involved in, and is deflected south-westwards by the subsequent Appalachian foldings. The Silurian overlaps the Ordovician in south-eastern Pennsylvania and in the New Jersey highlands, so that the broad orogenic zone is certainly continued south-south-westwards into central Maryland, but the western parts of that state, Pennsylvania and New York together with eastern Canada, formed the Taconian *foreland* (Fig. 18).

Its front is throughout marked by intense overthrusting—great slices of pre-Cambrian and later rocks having been pushed north-westwards over almost undisturbed Cambro-Ordovician in horizontal displacements of considerable magnitude, shifts of perhaps as much as 60 km. having been recorded. Bailey (29) has pointed out that the locus of this long curving front was determined by the obstruction made by the " Logan line " immediately to the north-west that marked the shallowing of the Laurentian sea. Behind this curve strata ranging from the Upper Ordovician downwards were folded along axes

[1] This term was suggested by Schuchert to avoid confusion with " Taconic " in the stratigraphical sense (32).

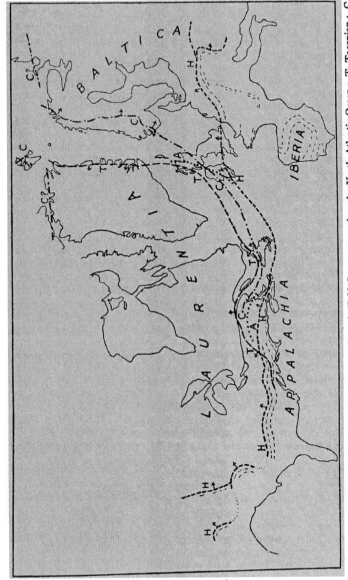

Fig. 18.—Showing the Tectonic Fronts of the Palæozoic Fold Systems crossing the North Atlantic Ocean: T, Taconian; C, Caledonian; A, Acadian; H, Appalachian-Hercynian (with associated foldings mainly in the rear indicated by thin broken lines). Arrows mark the direction of pressure.

roughly concentric thereto in eastern New England and New Brunswick with strong overthrusting directed to the north-west in central and west-north-west in south-eastern Newfoundland (Heyl (31)), while the backland must have lain farther to the south-east.

Proceeding to north-western Scotland, an identical orogeny marks out the belt stretching north-north-eastwards from south-eastern Skye to the northern coast of Sutherlandshire, wherein the little-disturbed palæozoics (of American facies be it noted) have been thrust from the south-east along relatively flat planes, bringing forward slices of Archæan, Moine, Torridonian and Cambro-Ordovician with much imbrication, horizontal displacements of up to over 15 km. having been discovered. Its probable continuation with more northerly trend and with higher-angled thrusting is reported by Read in the Shetlands. Perhaps contemporaneous therewith is the belt of strong crushing along the entire length of the inner side of the Outer Hebrides (Jehu and Craig), the thrusts in the Fort William area and some of the big fault-lines in the central Highlands, e.g. Loch Tay and Tyndrum, wherein the movements were to the north-west (Elles). This orogeny has almost universally been mistermed "Caledonian" although no strata younger than Ordovician are definitely involved therein. The inner margin of the disturbed belt must lie far to the south and south-east, for the Silurian rests unconformably on Ordovician or older rocks in Galway, Lake District, western England and southern and south-eastern Wales.

Passing now to eastern Greenland, enormously thick Cambro-Ordovician of similar facies to that of northern Scotland has suffered in exactly the same way over a distance of at least several degrees trending N.–S. and over a breadth of up to 35 km.,[1] with overthrusting directed towards the west and hence towards the source of those sediments. Throughout this lengthy arc—of at least 45 degrees as measured upon a globe—from Virginia to eastern Greenland, strata belonging to the interior of the North-western trough have everywhere been strongly thrust upon the Laurentian foreland and must in large degree have replaced the late Ordovician furrow by a piled-up mass of material, thereby welding Caledonia to Laurentia, which weld acted as a buttress during the subsequent Caledonian compression and profoundly influenced that movement. Deposition nevertheless went on uninterruptedly in the South-eastern trough—Nova Scotia–Scandinavia—and southwards over France and Spain.

[1] Wegmann (35).

SILURIAN

Erosion of the Taconian chains was followed by the entry of the eastern sea by way of New England into the Appalachian geosyncline, whence it spread not only westwards but northwards through Canada, over northern Greenland and thence into northern Russia and southwards into the Irkutsk region as well. Appalachia continued to shed detritus north-westwards into the main geosyncline. The Silurian seas transgressed over tilted Ordovician in New Brunswick and occupied central and south-eastern Newfoundland, but failed to reach the north-western part of that island, where the older North-western trough had presumably been obliterated. Although the Silurian faunas became almost cosmopolitan, that of Nova Scotia is more European than American.[1]

In the British Isles, save as noted above, sedimentation was continuous from the Ordovician, neritic sediments bordering Caledonia from Galway across to County Down, passing south-eastwards into the graptolitic facies, while the same was the case to the east, but with calcareous phases in Scandinavia, Baltic, Bear Island, Spitzbergen, Urals, Lena River area, New Siberia, etc. Renewed eruption took place along the older line in Scotland, but there was a striking shift in volcanicity to a new belt that ran oblique to the tectonic trend, through Maine, Kerry, Bristol, Brabant and Bohemia, as pointed out by J. S. Turner

The initiation of the Caledonian Orogeny saw extensive displacement of the oceans and the enlargement of Laurentia and more particularly of Asia.

CALEDONIAN OROGENY

Called so by Suess in 1887, this was in places a double one— pre- and post-Downtonian—of Middle to Upper Silurian and Upper Silurian to Lower Devonian age respectively—foldings which are sometimes spoken of as the " *Caledonides.*"

In North America it does not seem to have been differentiated from the subsequent Acadian, though it could well have been of importance in the northern and central Appalachians. In the Matapedia valley,[2] south of Gaspé, the basal part of the Devonian is missing, in New Brunswick and Nova Scotia the Lower Old Red Sandstone (Knoydart) rests discordantly on Upper Silurian,[3] while in Newfoundland a great change in sedimentation took place between the highest known Silurian and the lowest known Devonian (Heyl).

In the British Isles strong pre-Old Red Sandstone folds,

[1] Reed (21), 219.   [2] Crickmay (32).   [3] Reed (21), 219, 222.

along axes directed S.W.–N.E., affected a wide area bounded
on the north-west by a line through northern Ireland and the
central Highlands and on the south-east by one from southern
Wales and Northumberland (at least) and thence north-eastwards through Scandinavia (Fig. 18). In County Kerry the
strata involved included not only the Salopian (Wenlock), but
the supposed Downtonian " Dingle Beds."

Further south stretched the important and roughly parallel
belt of the Ardennes, represented in Portugal and central Spain
with a W.–E. course (Carbonell), in Brittany with N.E. trend
(Kerferne ; Bigot) passing E.N.E. through Belgium and E.
through Thuringia (where the Devonian rests unconformably
on Silurian, whereas to the south thereof these systems are
conformable).

In Europe the pressures were normally directed from the
south-east and were moderate in the west though increasing in
intensity north-eastwards, as shown by the close isoclinal folding in the southern uplands of Scotland, where a great " Fan
structure " is discernible, e.g. Wigtownshire, and the still more
intense deformation in Scandinavia. In central Scotland the
position was much complicated by the sinking of the " Lowlands " region or trough, as the Highlands beyond were folded
and upheaved. Along the border zone from Arran to Stone-
haven a pre-Downtonian movement crumpled up both the
Dalradian and the Margie (Lower Ordovician) series, which
was followed by intense erosion, the deposition of Downtonian
and Lower Old Red Sandstone (each including materials derived
from the igneous suite characterising the Caledonian disturbances (Campbell) in the trough to the south-east, with renewal
of movement during the mid-Devonian (" Acadian "), so that
the " border-rocks " became in places overturned to the south-
east over what can now be viewed as the " backdeep."

In Scandinavia the pressures were—following conventional
interpretation—from the north-west, that is to say, directed
towards Baltica, save in the Hardanger-Jotun region, where
displacements to the north-west are found, which suggests that
" Caledonia " formed a pivot around which Laurasia became
twisted during this period of intense crustal compression
(Chapter XVI).

In Scandinavia the zone of disturbance, first elucidated by
Törnebohm in 1888, divides that country lengthwise into two
nearly equal portions.[1] The north-western margin lies hidden
by the ocean ; the south-eastern runs nearly N.N.E. with some
curving and re-entrants from the Hardanger Fiord, west of Lake
Mjösen, through Jamtland to the Tana Fiord in Finmark, a
distance of 1600 km. with a maximum width in south Jamtland

[1] See Holtedahl (21).

of 200 km., tapering northwards. The overthrusting has been from north-west to south-east along flat planes slicing through the Cambro-Silurian—in places the Jotnian as well—bringing forward vast wedges and nappes of highly metamorphosed rocks—the " Seve Group " of Törnebohm. In the coastal zone between Bergen and Trondhjem pre-Cambrian crystallines are also involved, and in Finmark rocks of the Scottish Moine type, but over the rest of the region the overthrust material consists of what is generally taken to be altered Cambrian to Silurian, highly injected and metamorphosed by deformed eruptives, pushed forward upon the horizontal or slightly disturbed palæozoic foreland of Baltica. The front is bordered by outliers (klippes), while large eroded " windows " (fensters) disclose the floor in the heart of the thrust-zone. Movements of over 30 km. are assured, while some of over 100 km. have been presumed. As Holtedahl (20) has pointed out, the undisturbed Scandinavian " shield " must extend far to the north-west beneath its crumpled and intensely altered covering. Presumably, as Holmquist has suggested, this is more properly an instance of " under-thrusting."

Stages up to the mid-Silurian have been affected normally, but in the central area presumed Downtonian has also been involved (Bergen district), while in the marginal area, where the Downtonian lies conformably and undisturbed upon the Salopian (Wenlock) (Ringerike district), it suggests by its clastic facies tectonic movements during the latest Silurian.

In the extreme north, according to Holtedahl,[1] the little dis-turbed palæozoics sweep eastwards to the Varanger Fiord with some crumpling against the Russian shield, that appears to be continued south-eastwards off the Murman coast (Polkanov) towards the Kanin Peninsula and Timan, while the tectonic front passes out to sea with north-easterly trend at the Tana Fiord. Its manifest continuation is to be found in western Spitz-bergen, where the compressed and metamorphosed Cambro-Silurian Hecla Hook Series, highly injected with granitic matter, strike north-north-west strongly overfolded and over-thrust to the east. The belt of deformation, its western side covered by sea, occupies a width of nearly 300 km.—from Prince Charles Foreland in the west to the centre of North-east Land, where Sandford[2] has found the plications with north-north-easterly course rippling out into the undisturbed eastern fore-land. To the latter, too, can probably be ascribed the nearly horizontal block of Bear Island to the south with its similar strata.

If this interpretation be correct, then Spitzbergen and Bear Island have experienced considerable westerly displacement

[1] Holtedahl (20), 19; (21), Fig. 2.   [2] Sandford (26), 633.

together with some anti-clockwise rotation, and the Barents Sea, despite its shallowness, would be a " stretch-basin," a view to which the shapes of Greenland and Scandinavia lend strong support. The age of this diastrophism, which may possibly have extended to eastern Greenland and affected the Taconian belt there as well, is set in Spitzbergen through the Downtonian Red Bay Series resting on the eroded edges of the folded and metamorphosed strata, followed in places with a lesser break by the Wood Bay Series (Lower Devonian).[1]

In northern Greenland and in Grinnell Land marine lowest Devonian was affected by foldings directed E.N.E.–W.S.W. upon which Lower Devonian follows unconformably. These very early Devonian movements are viewed by Holtedahl as a " delayed Caledonian Orogeny " (see Fig. 18).

The extension of the Scandinavian diastrophism is perhaps represented in the V-shaped belt of folding, with overturning towards its interior, around the archæan and palæozoic platform of Angara, termed by Suess the " vertex of Asia."

Arresting right along the lengthy Caledonian belt is the abundance of plutonic intrusions ranging from isolated batholiths at either end to the intense and wholesale granitisation of the mid-section of Scandinavia and Spitzbergen (and probably eastern Greenland within N.–S. zones, described by Wegmann and Koch,[2] as is suggested by the composition of certain of the granites).

The plan here formulated differs from those usually advanced largely through reference of the diastrophism in north-western Scotland to the Taconian and its extension to Greenland, where it was seemingly repeated by the Caledonian disturbance. While it still leaves various points unsettled, it will meet certain well-merited objections raised by Schuchert and dispose of many of the difficulties experienced by Frödin in his comparison of Scotland with Scandinavia. Holtedahl's [3] illuminating contributions on the evolution of the Laurentian mass should be consulted with the above in mind.

DEVONIAN

Following the Caledonian Orogeny temporary communication was established in North America between the Acadian and Appalachian geosynclines, while later the ocean spread westwards from the latter to Cascadia and connected with the long-established waterway between Ellesmere Land, northern Greenland and Eurasia, i.e. Timan Ridge, Novaya Zemlya, Vaigatch Island, the Urals and Central Asia, but does not seem

---

[1] Holtedahl, as given in Frebold (35), 22, Figs. 11 and 65.
[2] Wegmann (35); Koch (35), 137, Fig. 10.   [3] Holtedahl (20; 25).

to have reached the Pacific generally. Northern Eurasia was broken into horsts and the deposition thereafter of marine strata was confined to the gräben between them. The faunas of this Arcto-Asiatic ocean are closely related to those of North America and Europe. On the south side of Laurentia the shore stretched from New Brunswick to Devon and Brabant (just missing Ireland), made an embayment over north Germany and proceeded along the Harz and Carpathians and eastwards into Southern Asia with the ocean to the south reaching across to the northern margin of Gondwana.

The Laurentian platform became considerably extended through uplift and the incorporation of the broad Caledonian fold-chains, so that by the early Devonian Laurentia—" Eria " —had attained to a maximum, embracing most of North America and Greenland and fully two-thirds of Europe, and stretched eastwards right to the Ural trough. This growth was profoundly to influence its physiographical and climatic history as exemplified in the huge development of continental deposits with remarkable thickness of the "Old Red Sandstone" type.

Such a facies indeed appeared at the close of the Silurian in the peculiar Downtonian and Dittonian phases with their ferruginous red mudstones and clayey sandstones with fish remains and occasional marine mollusca—regarded by W. W. King (34) as waste deposited in shallow seas—represented along a belt running through southern Wales, west Midlands, Scotland (Lesmahagow and Forfar), Norway (Oslo and Trondhjem) and Spitzbergen but not Greenland. Equally instructive are those occurrences along the southern border of " Eria " that link the marine and continental phases of the Lower Devonian, for example, in Gaspé, Artois, Brabant and the Ardennes, and perhaps best seen in Devon and Cornwall.

Upon this extensive continental surface, in part peneplained, in part highly diversified, was laid down the " Old Red Sandstone " in depressions following largely the Caledonian trend-lines, that became accentuated under the succeeding Acadian deformation. In the British Isles the various basins have been named " Lake " Caledonia, L. Orcadie, L. Cheviot, L. Lorne, L. Fanad, etc., and were seemingly parted from one another by rugged country and only became interconnected when their watersheds were overtopped by sediments. In certain areas their floors were highly irregular. Characteristic are the vast accumulations ranging commonly from 2000 to 5000 m. in thickness, with well over 7000 m. in Spitzbergen. In certain regions volcanic flows contributed largely thereto.

The Lower Old Red Sandstone occupies a zone extending from Gaspé through western Newfoundland (where paralic

phases are present),[1] Ireland, England, central Scotland and
Scandinavia to Spitzbergen, with the Middle division generally
situated a little further to the north-west. The Upper division
was, however, the most widely spread, with a strong unconformity at its base, due largely to the intra-Devonian Acadian
disturbance. Thus it is represented in the Appalachian-
Acadian region, eastern Quebec, southern and south-western
Ireland, in many areas between Devon and the Shetland Isles,
eastern and northern Greenland, southern part of Ellesmere
Land, Spitzbergen, Bear Island, Timan, Novaya Zemlya (with
abundant volcanics) and Baltica, the last-named on the margin
of the Devono-Carboniferous sea of Russia. Instructive, too,
is the development in Siberia,[2] far to the east of the Ural trough,
in the Minusinsk region around the head of the Yennesei River
of the Old Red Sandstone facies with strong basal unconformity
and marine calcareous intercalations in its middle part—towards
the north-west—associated with acid and basic volcanics and
tuffs.

Well known to all is the peculiar lithology of the Old Red
Sandstone—the great masses of dominantly red, brownish or
chocolate-coloured coarse sandstones, often cross-bedded and
ripple-marked, the pebble-beds and coarse, poorly stratified
conglomerates or breccias, especially towards the base of each
series—sometimes of huge thickness—the subordinate deep-
coloured shales and clays associated with " cornstones " (and
gypsum in eastern Greenland) and the considerable development
of lavas and tuffs. The scanty life—mainly fishes and primitive
plants—is remarkably uniform in character, strata from even
widely separated regions containing common or closely allied
species.

Whether a purely terrestrial origin under an arid climate, as
is commonly supposed, or one of accumulation in shallow seas
or in lagoon-like depressions be adopted, there can be no doubt
of the general persistence of those peculiar physiographical conditions throughout the Devonian, and indeed in certain places
from the closing Silurian right up into the early Carboniferous
(see Chapter XIII). It is, furthermore, striking that away from
the old shore-line that ran from Acadia through Great Britain
to Spitzbergen no undoubted marine phases occur, not even on
the north-eastern or south-western sides respectively of those
several countries that are at present parted from one another
by ocean. This suggests that during the Devonian the intervening sections, now covered by sea, were not gulfs, but land,
that is to say, Laurentia or Eria was a *continuous land-mass*.
Now, since the areas of Old Red Sandstone fall within a region
some 45 by 28 degrees, one has under current views to postulate

[1] Schuchert and Dunbar (34).   [2] Backlund (35).

a wonderful uniformity of physiographical conditions over that truly enormous territory. On the other hand, under the Displacement Hypothesis the area shrinks somewhat, though still possessing a remarkable size.

Attention must be drawn to the surprising analogy shown with the equally extensive Siluro-Devonian sandstone facies of Western Gondwana (Chapter IV). The latter was, however, accumulated on a subsiding peneplained region under a moist and cool climate, whereas the Old Red Sandstone was formed at a slightly later date on a more diversified country back from the shore and under a drier and presumably warmer environment, save in eastern North America, where cool conditions have definitely been determined [1] (Chapter XIII).

ACADIAN OROGENY

The principal movements occurred during and just after the mid-Devonian—the " Shickshockian disturbance " of Schuchert —in Quebec, New Brunswick, Nova Scotia and perhaps central Newfoundland, along axes parallel to those of the Taconian, Fig. 18. In the British Isles, from Kerry to the Orkneys, the Upper Old Red Sandstone rests upon tilted strata including the Lower division, well seen, for instance, along the Highland border in Scotland, the boundary faults of which (like that of the Great Glen and other north-easterly directed fractures— such as some in the Orkneys) being of this age. The disturbance is traceable in Scandinavia and Spitzbergen (Svalbard phase), in Brittany (Bretonic phase) and in the Lahn area of Germany.

Outstanding throughout Eria is the intense igneous activity connected with this diastrophism, signalled by lavas and tuffs, by masses of granite, diabase and ultrabasic rocks over the stretch from Georgia to Newfoundland (including the province of Quebec), the British Isles and probably western Norway.

CARBONIFEROUS SEDIMENTATION AND OROGENY

From the closing Devonian the physiographical evolution on both sides of the " Atlantic " became so intimately bound up with crustal movements that the description of the Carboniferous and Permian stratigraphies becomes essentially an account of the reaction of the earth's crust to the several orogenic waves that affected it throughout that lengthy period. The areas of such disturbance form a girdle—approximating to the equator of the time—following roughly a great circle from at least the Mexican border to Eastern Europe together with a widening region beyond trending eastwards through Central

[1] Willard (35).

Asia and China to reach the Pacific—as well as the zone of the
Urals conjugate thereto, and even a few outliers, such as that
of eastern Greenland.[1]

Into this wide branching belt, that can comprehensively be
called " *Hercynian*," enter the various tectonic elements known
as Appalachian, Armorican, Variscan, Uralide and Altaide,
proceeding from west to east.   As during the early Palæozoic,
the pressures can be visualised as caused essentially through the
mutual impact of Laurasia and Gondwana, but the general
direction of the push had in the meantime changed from N.W.'
to N.N.W. and N.   The pattern developed is, however, far from
even, being constituted by a series of interlinked wide arcs with
various deflections, recesses, etc.   Such irregularity is due to
the composite nature of the folding, the deviating influence of
older tectonic belts, the greater advance between the more
resistant blocks over the weaker zones, but, above all, to the
deformation by the subsequent Alpine diastrophism.   As
elaborated by van der Gracht, the successive phases are clearly
identifiable on both sides of the Atlantic, and such lack of contemporaneity as is found can be accounted for by the great
distances involved and by the unevenness and spasmodic nature
of the advances to be expected in the several orogenic sectors.
In the western half of Laurasia the overturning and overthrusting
was to the *north-west* or *north*, save over limited sections and
through secondary causes, but in the eastern half, that is to say,
in Asia, to the *south* generally.

That no N.–S. ocean comparable to the North Atlantic was
in existence during the Carboniferous and Permian is shown,
first, by the lack of any general facial change from terrestrial to
marine types as individual formations are traced towards the
present ocean margins, and, secondly, by the very close relationships of the floras of North America and Europe.

During the late Devonian Laurasia reached its greatest
extent, but was nevertheless composed of the two masses of
Laurentia and Angara separated by the persistent Ural geosyncline that linked the equatorial ocean with the Cordilleran
by way of the Arctic.   The early Carboniferous saw spreading
of these waters over those margins generally—northern Greenland, New Siberia, etc.—though just failing to reach eastern
Greenland, Spitzbergen and Bear Island.   On the contrary, the
Appalachian trough became cut off entirely from the " Atlantic "
region, and throughout the Carboniferous its history is one of
intermittent upheaval and erosion on its south-eastern side
(Appalachia) and silting-up of the shallow sea over the foreland
to the north-west, the water becoming deeper in the Central and
Cordilleran regions.   The south-eastern border of Appalachia

[1] Koch (35), 139·44.

remains, as before, unknown, but in the north-east the coast-line ran through New York, New Brunswick, the south-western corner of Newfoundland and thence through southernmost Ireland, southern England and Brabant to the Baltic, while farther to the south parts of France and Germany made a large island.

Early Carboniferous clastics border these coasts, those of Acadia including much gypsum and some dolomite with an unusual fauna, while the same peculiar facies distinguishes Spitzbergen. Noteworthy are the coals of Culm age in Appalachia, Acadia, northern Ireland, Scotland and Spitzbergen. Negative movements led to the deposition of terrigenous muds off the mainland and around the aforesaid island, but of widespread shales and thick limestones in the deeper seas during Viséan times.

The close of the Lower Carboniferous was marked out by the upthrusting of the first of the great fold-ranges, largely within the geosyncline itself, dividing lengthwise the clear sea and thus providing a source for the abundant coarse detritus shed off on either side to form the estuarine and deltaic Upper Carboniferous rocks. In North America the big unconformity at the base of the Pennsylvanian over much of the central and eastern states, and the clastic nature of that division, mark out the *Wichita Phase* of this orogeny with its double pulse in the central region, traceable to the north-east (Tenesseian) into the piedmont region and possibly traversing a buried portion of Appalachia.

In Europe the equivalent *Suedetic Phase* folded and upheaved a wide zone in northern Spain, Brittany and thence through the Vosges to the Suedetes, forming the inner arc of the Variscan folding. In the narrowed foredeep situated to the north of those chains, extending from Acadia to northern Germany, the Millstone Grit was laid down, largely deposited by rivers draining Caledonia and Scandinavia, coals being patchily developed therein. Of high significance is the presence in the Glasgow area of the American brachiopod *Meekella*, unknown elsewhere in Western Europe, though common in Russia and Asia.

Within the Ural geosyncline folding took place during this phase, the pressure being directed eastwards [1] with deep-seated basic and ultrabasic intrusions, following which the sea transgressed over northern Siberia, Bear Island and Spitzbergen, depositing with unconformity the Upper Carboniferous and Permo-Carboniferous limestones with Russian faunas.

During the Upper Carboniferous the developing of conditions favourable to the laying-down of coal seams was patently deter-

[1] Bubnoff (26), 103.

mined and controlled by the rising of lengthy anticlines and the deposition of the waste from those emergent ridges within the adjacent deepening synclines. It is indeed striking that all the great coalfields of Laurasia dating from this epoch—from Texas on the west to China on the east—should have been intimately connected with the Hercynian deformation. Cross-warping, largely due to the resistance offered by ancient blocks, played an important part in producing the requisite lakes and swamps, and stratigraphical overlaps and local unconformities are hence common. In the Saar coalfield, for example, the Coal Measures are strongly discordant on folded Devonian, whereas to the north in the Belgian coalfield the Devonian and Carboniferous are conformable.

In the central part of North America came the *Arbuckle Phase* during the Upper Pennsylvanian with push from the south and south-west, a movement that extended farther to the west, but is not detectable in southern and central Appalachia, though, as in the case of the Wichita, it may well have affected the region more to the south-east. From the work of Billings (29) this disturbance would appear to have been developed in the Boston basin and perhaps in Nova Scotia.

As the *Asturian Phase* it is strongly expressed in Spain, southern Ireland, southern England, Brittany and thence along the outer Variscan arc through Belgium to Westphalia. There Stephanian and older strata have been driven northwards over the foreland region with much crumpling, accompanied in the Franco-Belgian sector by powerful overthrusts having big displacements and by extensive plutonic intrusion. The front thereof runs in a north-westerly direction through southern Wales, westwards through Wicklow in Ireland, where it is deflected southwards by the Leinster " Caledonian " granitic mass, and thence south-westwards through Cork and Kerry with a few minor ripples in the (northern) foredeep. At this time came the *Second Ural Phase* with tendency towards overfolding to the east.[1]

The relatively weak *Saalian* and *Pfalzian Phases* marked out the end of the Lower and Upper Permian respectively in the central Variscan zone, and are recorded by Fleury[2] on the south-western side of the Portuguese Meseta accompanied by down-faulting against that block. The Pennsylvanian is involved in south-western Newfoundland (between Cape Aguille and Stormy Point, near Cape Ray and east of Deer Lake), Nova Scotia, Maine and east of the Green Mountain, and in places some Permo-Carboniferous, as in the Boston basin, where in a curve convex to the north-west the beds have been thrown into folds and moved forward upon low-angled thrusts with travel

---

[1] Bubnoff (26), 103.   [2] Fleury (24), 505.

to the north-west of at least some kilometres,[1] though, according to Woodworth (23), with local overturning to the south-east.

From Pennsylvania onwards Carboniferous and older strata have been pushed north-westwards towards the foreland with crystalline and metamorphic rocks exposed on the east, the deformation becoming progressively more intense in the central Appalachians, where folds are commonly inverted with related thrusts, bringing forward for long distances great wedges of Cambro-Ordovician beds. In the southern Appalachians the folding is more open, though thrusting is quite as strong and spread over a greater width, the overriding Cambro-Ordovician being otherwise but little disturbed.

There are reasons for believing that the southern Appalachian system is continued south-westwards beneath the covering of Cretaceous and then through Arkansas and Texas into Mexico in an S-shaped curve crossing the slightly older Wichita system, which trends W.N.W. into New Mexico, both systems showing several phases of intense folding and overthrusting towards the foreland [2] (Fig. 18).

While the majority of geologists ascribe the Appalachian crisis to Permian or even Permo-Triassic times, it is only proper to point out that the youngest strata involved are the Pennsylvanian in the south-west and the presumed Lower Permian in the north-east. Seeing that this orogeny was by no means simple and manifestly of long duration, it could well have started back in the late Carboniferous and therefore have corresponded more generally with the Asturian and only partly with the Saalian Phase of Europe. Favouring that view is the lithological break marked by the clastic " Waynesborough " widespread in West Virginia, Ohio, etc., and taken as forming the base of the Permian. The " Appalachian " disturbance is hence viewed as mainly of latest Carboniferous age with some renewal during the succeeding Permian.

Waterschoot van der Gracht's able and illuminating reviews of the evolutionary history of North America and Europe during these times should be carefully studied.

[1] Billings (29), 104.   [2] van der Gracht (31 ; 35).

CHAPTER VIII

LAURASIA—SECTION 2

Carboniferous-Permian Geography. Hercynian-Appalachian Orogenies and their Import. Carboniferous-Permian " Red Beds " Facies and its Life. Permian. Late Palæozoic Floras. Triassic. Jurassic and Cretaceous. North American Pacific Border during the Mesozoic. Laramide Orogeny.

CARBONIFEROUS-PERMIAN GEOGRAPHY

A CONCRETE picture of Laurasia at about this time should prove useful. As during previous epochs, not only was it separated from Gondwana by the inter-continental southern ocean or primitive Tethys, but it was divided into the two portions of Laurentia and Asia (Pal-Asia) by the persistent Cordilleran-Arctic-Ural trough. Such simplicity of pattern became interfered with by the initiation along its southern side of the broad system of Hercynian foldings, which it is important to observe was not truly marginal but *developed inside the continent and somewhat oblique to its southern shore.* The latter ran more or less from Texas to Indo-China, the belt of disturbance from Mexico to China which, as will be seen from Chapter XIII, was roughly parallel to the equator of the period.

Radical changes in geography were brought about by the penetration of the bounding seas into the synclinal depressions so developed, namely, from the south (Mediterranean-Himalayan), centre (Ural) and extreme east and west (Pacific). Thus during the Lower Carboniferous several wide troughs were given off eastwards through Central Asia and China, which ultimately reached to the Pacific. Indeed, during this epoch Laurasia must have become subdivided by those fluctuating epeiric seas into at least *six sub-continents*, which during the early Mesozoic were linked together once more. The later Carboniferous saw a general shrinkage of these unstable synclines, the waterways within the lengthy stretch from central North America to eastern China becoming almost wholly filled up with paralic or terrestrial coal-bearing strata. Deepening proceeded during the early or Middle Permian, but towards the close of that epoch the seas were driven out to the continental margin, though without destroying the Ural-Arctic geosyncline.

This particular trough was throughout characterised by its Boreal faunas that mingled towards the south with the contrasted and probably more rapidly evolving ones of the tropical regions. Thus the southern and south-eastern sectors of Laurasia were distinguished in the early Carboniferous by corals, in the later Carboniferous by foraminiferal (Fusulina) limestones, in the Lower Permian by the Productus limestones and in the Middle Permian by the Lyttonia fauna. These incidentally prove useful criteria of former climatic zoning.

THE HERCYNIAN-APPALACHIAN OROGENIES AND THEIR IMPORT

The significance of this stupendous zone of crustal squeezing lies essentially in its spatial relationships to the several preceding orogenies, with which it possesses a vital element in common, namely, a *causal thrust directed towards the north-west and north*. As set forth in Chapter XVI, the several phases thereof affected a belt running fully half-way around the globe, quite narrow in Europe, but expanding prodigiously in North America and Asia. Its rear is well seen only in Europe, but its front is more or less sharply defined throughout (Fig. 18).

Conspicuously over its full length this region of crustal collapse is marked out by the emplacement of enormous quantities of granite, introduced according to F. E. Suess [1] after each orogenic spasm and always on the side from which the pressure came, with the suggestion that these many elongated parallel or sub-parallel bodies, closely following the strike of the beds, tend to unite in depth. To them is due the importance of the Hercynian as a *metallogenic period*, evinced in the numerous deposits of tin, tungsten, uranium and, above all, of pyrite in Saxony, Cornwall, central France, Meseta of Iberia, probably Nova Scotia and Quebec, and seemingly in much of the Appalachian region as well.

Staub [2] has clearly depicted its peculiar Ω or "omega-shaped" course in Europe, due to the northerly advance of the crustal waves between "jaws" made by the Hungarian and Franco-Spanish blocks until opposed by the Brabant-Anglian massif, which impressed itself on the northern arc and thereby gave rise to the sharp inflections on both north-east and north-west—the Variscan and Armorican bends. Relatively narrow on the east the belt broadens over central France with a wide "inter-fold" region, the rear making a bend towards the Mediterranean with occasional overturning to the south on that

[1] Bucher (33), 278.   [2] Staub (28), Fig. 39.

side.   Entering northern Spain with S.S.W.–S. trend, it curves
south-eastwards towards the Mediterranean, doubles back
south-westwards along the (later) Betic line and reaches the west
coast of Portugal with W.–N.W. strike.   Its southern limit must
lie far to the south, since Brichant has recorded equivalent folding directed W.S.W. in north-east Morocco.   In Cantabria the
overturning is to the east, in southern Portugal to the south-
west,[1] which suggests that the intervening Meseta has been
squeezed from both sides and that a fan-structure is represented
therein.

The rough S-shaped plan in Western Europe, deduced from
fragmentary exposures through the younger covering, cannot
wholly be primary, but must be interpreted, following Staub,[2]
as largely due to extensive distortion produced by the Tertiary
compression in the Pyrenean region on the one side and the
Betic on the other, as can convincingly be demonstrated with
models.

The outstanding problem, that of the *Northern Hercynian
Front*, requires consideration next.   Proceeding from Germany
westwards, the front is found to pass the southern or advance
member of the Taconian orogeny in the Ardennes and thereafter
to meet the south-eastern margin of the main Caledonian folding.   In southern Wales the latter becomes deflected from its
normal south-westerly into a west-south-westerly direction,
identical relations appearing in Brittany, where the Caledonides
trending in the north a little south of west are crossed in the south
by the Armorican folds, overthrusts and nappes striking W.N.W.
(Kerforne (24)).   In a curve concave to the south the front runs
through southern Ireland and with W.S.W. trend passes out to
sea in Kerry, crossing obliquely—and incidentally being modified by—the older Caledonian plications, which last have their
own front situated 150 to 200 km. distant in northern Ireland,
with the Taconian thrust-zone—corresponding to that of north-
west Scotland—hidden presumedly beneath the northern ocean.

While in America the Caledonian and Acadian foldings
affected central Newfoundland, Gaspé and probably northern
Maine, the Appalachian crumplings lay principally to the
south-east of them.   In Newfoundland little disturbed Carboniferous is present near the centre of the island, so that the
narrow belt of post-Pennsylvanian folding and faulting trending
north-eastwards from near Cape Ray at its south-western corner
to White Bay [3] represents presumably an advance member, like
that known in central Ireland.   The true front can therefore be
taken as passing along the north-western coast of Cape Breton
Island, turning westwards between Nova Scotia and Prince

---

[1] Suess (06), II, 126.   [2] Staub (28), Fig. 39.
                [3] Schuchert and Dunbar (34), 108-13.

Edward Island to Joggins, and thence south-westwards down
the coast of New Brunswick and through the centres of Maine,
Massachusetts and Connecticut, the Taconian-Acadian Green
Mountain massif having doubtless been responsible for the
strong southerly deflection of the arc in that section.   In eastern
New York, however, the younger foldings overcame that
obstacle, overran at a wide angle the older corrugations and
poured out westwards into the foredeep in waves of diminishing
amplitude over central Pennsylvania and Maryland, to renew
their normal south-westerly course in western Virginia so far as
Alabama, and then generally westwards to northern Mexico.

South of New York the " Hercynian " front steps clear of
the Taconian and thereafter maintains such a lead. As E. B.
Bailey [1] has so graphically put it, the *crossing of the several
older palæozoic orogenies, begun in southern England, is completed in Pennsylvania*, a view receiving strong support from
van der Gracht and approved of by F. E. Suess.[2]   The brilliant
syntheses of Bertrand and E. Suess are re-established, though
in a new guise !   Important now is the inference that Appalachia could have been integral with the masses of the Bahamas
on the one hand and of Iberia and Morocco on the other, and
that the ocean connecting Mexico and the palæozoic " Mediterranean " lay further to the south and crossed the Sahara in an
easterly direction onwards into Southern Asia.

Posted in imagination at the point of virgation in New
Jersey and looking north-eastwards one would see three, probably four, great tectonic fronts, each caused by pressures
directed towards Canada—the oldest on the left, the youngest
to the right—losing themselves in the ocean.   Across the waters,
similar in all essentials—as though 3500 km. of sea did not
intervene—one would sight their continuations diverging more
and more widely through Europe (Fig. 18).   *Spaced as they are
today, these two sets of radiating fronts are not congruent,
though, if America be brought some 1700-2000 km. nearer
Europe, a remarkable " fit " would reveal itself*, which becomes
even more striking when the subsequent distortion of Iberia is
allowed for and that block rotated clockwise, as has been done
by Staub, van der Gracht and Baker.   Newfoundland is thereupon brought next to and in structural agreement with Iberia
and the British Isles, Greenland opposite Scandinavia and
Spitzbergen, with the Arctic basin " closed-up."   Then the
numerous stratigraphical, palæontological, climatic and other
details, too numerous to be set down in these pages, acquire a
new meaning and unite in revealing a picture harmonious in
its design, with which the post-Permian history is furthermore
in full accord.   In addition, the " Atlantic type " of coast-line,

[1] Bailey (29), 75.   [2] Suess (36), 827.

M

first discriminated by Suess, finds its physiographical explanation.

This crossing of all the palæozoic fold-systems in the North Atlantic region is, precisely like that already described for the South Atlantic (Chapter V), held to form one of the weightiest arguments in favour of the Displacement Hypothesis, though conveniently ignored or strongly disputed by the orthodox school.

The Hercynian equivalents of Eastern Europe and of Asia— the " Uralides," " Altaides," etc.—and their place in the general scheme of Orogenesis are dealt with in Chapter XVI.

CARBONIFEROUS-PERMIAN " RED BEDS " FACIES AND ITS LIFE

Arresting is the distribution in space and time of that sedimentary facies which is held to indicate in general dry or semi-arid conditions, and with it, as Case (19) has so well pointed out, relics of the amphibian (tetrapod) faunas.

The idea enunciated by Grabau—and already used in the case of Gondwana—that the progressive elevation of the Her-cynian ranges, by *cutting off moisture from the great troughs to the north, was the prime cause of the corresponding aridity*, is without question the correct explanation.

Desiccation shows itself first in the Lower Carboniferous in the red sandstones with gypsum bands of east Lothian, Scotland, and also in the Windsor Series of Nova Scotia (with its salt), but becomes wider spread during the Upper Carboniferous, as manifested by strong red coloration of the sandstones and marls, the so-called " Spirorbis limestones " of the Staffordian and Radstockian of England and Scotland, the Pennsylvanian of Prince Edward Island, Rhode Island, Pennsylvania, western Virginia, etc. (Pottsville, Connemaugh and Mononga-hela series), and, in the opposite direction, in Czechoslovakia in Central Europe. Such can indeed be correlated with the Suedetic and Asturian phases.

During the Permian, which is indeed difficult to differentiate from the Carboniferous because of the persistence of such conditions, the spread becomes greater and extends through northern Spain, France, Germany and Czechoslovakia into Russia on the one hand, and through Acadia, Appalachia and into Texas on the other—almost continuously in an E.–W. zone running parallel to the fold-ranges—and so far north as Scotland and northern Ireland. Noteworthy are certain similar deposits of Carboniferous and Permian age in eastern Greenland within an area affected by equivalent movements.[1]

[1] See Koch (35), 62-5, 70, 139-45.

In eastern North America the stratigraphical rise of " red beds " conditions in a westerly direction was long ago pointed out by D. White, while a similar change northwards and eastwards has been established for Europe.[1] The climate in both these areas changed progressively from mild, humid and equable to " unreliable " and relatively arid.

In the case of the vertebrates intermigration was presumably not easy for these amphibians, since such would have depended upon the suitability of their environment, but, where that was favourable, a profusion of life resulted that followed up systematically the spreading of red beds conditions, and was inferredly dependent thereon. Striking accordingly in Laurentia is the community of Permian amphibians, the descendants of homogeneous Carboniferous forms, for instance, in Ohio, Ireland and Bohemia (Nopsca; Case),[2] the earliest recorded being from the Lower Carboniferous of Scotland. Unfortunately in America no land vertebrates have been preserved to bridge the gap between the Permian and Triassic.

PERMIAN

During the later Palæozoic Laurentia remained a sea-girt mass linked intermittently to Asia by the evolving latitudinal Hercynian anticlines. In the foredeep just north of the latter in America and Western Europe sedimentation was dominantly continental. Such is clearly seen in the West Virginia-Pennsylvania basin with its marine facies to the west, and along the arc extending from Rhode Island through Boston into southern New Brunswick, Prince Edward Island and Nova Scotia. Conspicuous in this eastern section is the conglomeratic nature and red coloration of much of the strata. The Appalachian foldings were still in progress, and lofty ranges may have reared their heads, sufficient at least to have borne piedmont glaciers, as indicated by the celebrated Roxbury conglomerate and the Squantum tillite of the Boston area (Sayles (14)).

The Permian is absent from Newfoundland, western and southern Greenland, western and southern Ireland, northwestern Spain and Scandinavia, and doubtless that extensive and continuous region was then being eroded ; indeed, in the British Isles the scattered outliers of such continental Permian were deposited unconformably upon earlier systems. The Armorican influence is seen in the arch separating the Permian of Somerset and Devon from that of the Midlands, in the " Howgill arch " striking E.–W. through Westmorland and in the conjugate Pennine and Malvernian warps and faults. In the Scottish Lowlands sinking went on between the old boundary

[1] Case (26), 199.   [2] Nopsca (34), 77; Case (26), 194.

faults with simultaneous rising of the abutments followed by
E.–W. faulting. The " brockrams " of the Midlands show that
movements between fault-blocks were in progress. Sinking
during the Lower Permian—Saalian Phase—accompanied by
fairly widespread basic eruptions, brought in as a branch from
the Ural trough the Zechstein Sea over northern Germany,
Belgium, eastern and central England and locally in Ireland
(Armagh and Down) with its deposits of thick dolomitic limestones, red sandstones and marls and " evaporites " that characterise this basin or series of basins, that were more or less
isolated from the open ocean.

Appalachia and the Atlas region, which are believed to have
been connected, suffered, on the other hand, elevation and
erosion.

Finally at the close of the Permian, during the Pfalzian
Phase, the central plateau of France and the Saar region were
upheaved along an E.–W. axis, the Zechstein Sea displaced
towards the east and its connection established with the Alpine
" Tethyan " geosyncline, from which an arm reached westwards
over to the Pyrenean area.

The first half of this epoch, rather like that of the later
Carboniferous, was marked out by strong volcanism over
Western Europe—England, France, Germany and Alps—
represented by masses of quartz-porphyries, porphyrites, basalts,
etc., in sheets, plugs and dykes, together with breccias and tuffs.
It is worth noting that during the succeeding Triassic such
volcanicity became shifted systematically *southwards and
westwards* and so came to affect eastern North America.

LATE PALÆOZOIC FLORAS

A wonderful uniformity of vegetation marked the Lower
Carboniferous, not only of the Northern Hemisphere, but of the
entire globe, well brought out by the distribution of the *Rhaco-
pteris-Aneimites* flora, significant being its presence in Kashmir,
China and Queensland. In Laurentia a rather abrupt floral
change occurred thereafter, evinced in the extensive elimination of species, one which may have been more or less
general.

From the Upper Carboniferous onwards strong floral
differentiation took place, that reached its zenith during the
later Permian, and led to the evolution of well-defined and
distinct floras with convergence thereafter, though world-wide
uniformity was not attained until the close of the Rhætic. The
boundary between the Carboniferous and Permian has by agreement been drawn by palæobotanists—wherever such is possible
—by the incoming of the genus *Callipteris*, usually *C. conferta*,

that species having been identified from localities so far removed as Texas and Shansi in China (Sze).

Three Floral Provinces are represented, which, however, overlap slightly in time and space (Fig. 39) :

(1) The Northern or Arcto-carbonic, comprising eastern North America, Europe and Southern Asia.

(2) The Angara or Kusnezk, embracing north-eastern Russia and Siberia (across to Vladivostock and south to Lat. 47°.

(3) The Sino-American or " Gigantopteris " (in part the Cathaysian of Halle), including South-eastern Asia and western North America (Texas and Oklahoma, but not Kansas).

These are distinct from one another and also from the Antarcto-carbonic or Glossopteris Province of the south, though the Angara Flora contains undoubted Gondwana elements. Future investigation will doubtless discover more points of community between these assemblages.

During the later Carboniferous (Stephanian) there was a remarkable uniformity of plant life on both sides of the North Atlantic, that is to say, in Laurentia, including such genera as *Sphenopteris, Alethopteris, Pecopteris, Odontopteris, Neuropteris, Sigillaria*, etc. D. White has remarked upon the high proportion of common or allied species in the Coal Measures of Missouri and the British Isles for example, and palæobotanists are now able to zone the Carboniferous of North America and Europe fairly well. The European genus *Lonchopteris* is not yet known from America, where an allied form is, however, present. White has shown that European elements were introduced into North America during the Stephanian. Europe was then cut off from Siberia by the Ural Sea, but by means of the Hercynian-Altaide anticlines Stephanian elements were able to migrate through the ocean-strewn regions into Asia, more especially in the south, for example, in the Suiyuan province of northern China (Sze), Malay States (Edwards) and Sumatra (Jongmans, Gothan, Posthumus), but some have even been found in Siberia.

The early Permian saw the extinction of representative Carboniferous types and the appearance of new genera, yet the host of common species on both sides of the North Atlantic testifies to the unity of Laurentia and of the evenness of ecological conditions therein. On the contrary, the curious genus *Gigantopteris* made its entry in Sumatra, and slightly later seemingly in Western America (Texas, Oklahoma), spreading through Korea and China so far north as Mukden (Halle). Though it failed apparently to reach Tibet, other members of this remarkable Chinese flora such as *Tingia* and *Protoblechnum* have recently been discovered by Bexell in the Nanshan region, where

it is overlapped by the Kusnezk Flora, a matter of high importance. The latter, which carries such forms as *Rhipidopsis*,
Callipteris, *Iniopteris*, *Pecopteris*, *Ginkophyllum*, etc., not only
occupies wide stretches in Siberia, but is found west of the Urals
in the Petchora region of northern Russia and even along the
Dwina River and its tributaries.

Gondwana elements, moreover, make their appearance in the
Kusnezk Flora, more particularly in the Dwina region, where
Gangamopteris and *Glossopteris* occur. For such Gondwana
affinities the Permian land-bridge must essentially have been
responsible, coupled, however, with the vital fact that both the
Kusnezk and the Antarcto-carbonic floras were of *temperate and
probably of cold-temperate characters*, whereas the Arcto-
carbonic and the Sino-American " Gigantopteris " floras were
either *sub-tropical or tropical* (Chapter XIII).

The outstanding feature of Permian floral distribution is
the fact, brought out in graphic form by Sahni (35), that the two
main regions comprising the Arcto-carbonic Province are parted
by the North Atlantic Ocean, and those comprising the Sino-
American Province by the North Pacific, whereas both impinge
upon the Angara one, a relationship which is not only in full
accord with, but constitutes a powerful criterion in favour of,
the Displacement Hypothesis.

TRIASSIC

Through persistence of the Arctic geosyncline Laurentia
continued as a unit until the later Triassic, when under the
general regression much of the Ural geosyncline became filled
up and the single continent of Laurasia was regenerated, that
now stretched from western North America eastwards to China.
The northern ocean spread from Alaska over north-eastern
Greenland, Spitzbergen, northern and north-eastern Siberia to
Alaska, the waters of which constituted Diener's " Boreal Sea,"
and harboured " northern " forms of life. In the south, on the
contrary, uplift had prolonged its margin, the land then sinking
fairly rapidly beneath the deepening Tethyan ocean that tended
to part Laurasia from Gondwana.

The writer is strongly of opinion that the current conception
of the Permo-Triassic " Tethys " as a regular trough of more or
less uniform width and depth—an idea gathered, it must be presumed, from Haug's palæographical maps—is erroneous. Closer
analysis suggests, on the contrary, at every stage a high degree
of irregularity in width and depth, as well as in the nature of the
materials deposited therein. The evidence points on the one
hand towards deeps of almost abyssal nature in which uncommon
radiolarian oozes were laid down, e.g. the Carpathian and

Albanian regions, and on the other towards land-bridges across it, e.g. in Persia. None of the existing oceans is exactly to be compared with the Tethys, though the Caribbean Sea comes nearest to it.

While the southern margin of Laurasia shifted considerably during the Triassic, it ran in a general way westwards from Indo-China through Tibet, Persia, Hungary, Switzerland, southern Spain, Morocco, to the south-east of Appalachia, through northern Mexico and California and northwards to Alaska, with Central America forming an island. Along this sub-tropical shore pulses of marine (neritic and benthic) life migrated, spread by presumably westerly currents, so that in the thick calcareous facies of Mexico and California the pelagic Alpine faunas are found. Migration between the Himalayan and Alpine sections was, however, only possible after the sinking in the early Triassic of the Anatolian " ridge."

The Appalachian-Hercynian ranges that stretched from Mexico to Silesia at least, although in process of erosion, were sufficiently high and wide to maintain semi-arid to arid conditions over the interior of the continent, which latter was being peneplained. In shallow warpings and intermontane troughs were laid down—almost everywhere with strong discordance—terrestrial, deltaic or lacustrine strata of the same general " red beds " type as during the preceding Permian. The physiographical parallel with the southern half of Gondwana during the Trias-Rhætic is surprisingly close (Fig. 41).

The North American red beds include in the eastern Cordilleran region Middle as well as Upper Triassic. To the latter alone belongs the Newark Series of the Appalachian-Acadian zone, deposited in troughs largely within the south-eastern or outer part of the now-reduced fold-ranges, partly in eroded strike-valleys, partly in rifts (gräben) in which vertical adjustments were in progress, but with a slightly more northerly trend than the grain of their framework. Throughout the stretch of over 2000 km., from southern Carolina to Nova Scotia, the deposits consist of " fanglomerates," conglomerates, arkoses, sandstones and varicoloured shales, with volcanic extrusives on several horizons. Fresh-water conditions dominated, while in the south-west coals were even formed, recalling the corresponding seams of Sweden and Germany (Lettenkohle).

Throughout the later Triassic Appalachia—and with it Newfoundland seemingly—was subjected to tension, resulting in strike-faulting that culminated in the " Palisade disturbance," under which the Newark Series became broken into fault-blocks, which were tilted and injected by sills and dykes of dolerite. This was towards the close of the Triassic, though a little *earlier* than the extensive volcanicity in Western Gondwana at the

beginning of the Jurassic. The corresponding marine phases
on the Pacific side will be dealt with later.

In Europe, on the contrary, sedimentation stood in close
relationship to the marine transgressions from the east and south-
east, and was therefore more complete. Thus along the western
side of the shallow sea of Central Europe and thence in an
S-shaped curve through eastern Spain, red Bunter (Lower
Triassic) beds were laid down characterised by salt, gypsum and
anhydrite. The sea made temporary incursions westwards
depositing dolomite, but failed to reach Portugal, Ireland or
England. Of interest are the pebbles of the " Grès Armoricain "
derived from Brittany in the Bunter beds of Devon.

Subsidence led to the Muschelkalk (Middle Triassic) facies
over west and central Germany and central and eastern Spain,
and developed the Germanic facies of the Keuper (Upper
Triassic), i.e. red, oxidised, false-bedded sandstones, clays and
marls, dolomite and some gypsum and rock-salt, over a broad
belt running through Morocco, Algiers, south-eastern and
northern Spain, much of England, central plateau of France,
Denmark and western and central Germany. It is noteworthy
that still farther back from the shifting shore-line the Keuper
is, just as in Nova Scotia, of fresh-water type with plant remains,
for example, in western and southern Portugal, fringing the
Spanish Meseta, in north-eastern Ireland, Scotland and southern
Sweden (Schonen). This strongly supports our hypothesis that
the broad region extending from Portugal through Ireland to
Scandinavia, which was land throughout the Triassic, represents
*the subsequently torn-off margin of the North American con-
tinent.*

It is admitted that the volcanic phase of Acadia is not clearly
displayed in the Keuper of Western Europe, though a possible
Triassic age is suggested for some of the basic dykes and sills
cutting the Coal Measures in the English Midlands, while a
more careful study of certain of the so-called " Permian " sediments and volcanics of the British Isles may possibly discover
Triassic equivalents in them. Faulting is evinced in the " Malvernian System " in England. On the other hand there is a
wide development in Morocco, Spain and the Alpine region of
the so-called greenstones or " ophiolites," certain of which are
intrusive while others represent magma poured out on the sea
floor during this epoch, thereby recalling the " Panjal Traps "
of the Himalayan sector.

Noteworthy are stages of the Triassic deposited along the
eastern margin of Greenland and upon Spitzbergen, certain of
them in continental phases recalling those of the south, while
the marine faunas contain Himalayan elements, the whole
indicative of overfloodings of the Boreal Ocean.

Over practically all Europe, and eastern Greenland and Spitzbergen as well, the terrestrial Keuper passes up through the transitional Rhætic into marine Liassic as the consequence of the world-wide Jurassic inundation.

The plant life of the Triassic is scanty, but the much-needed revision of the North American assemblages should bring out more clearly their close relationship with those of Europe. Striking, however, is the uniformity of the Rhætic floras of eastern Greenland, Sweden, Germany and Switzerland, which include a remarkably high proportion of common species.

Comparing the vertebrate life of North America and Europe, the Rhynchosaurians have curiously not yet been found in America, though allied forms and even common genera of phytosaurians, stegocephalians (amphibians) and sauropods (dinosaurs), are represented in both countries, for example, *Metoposaurus, Stagonolepis and Thecodontosaurus*, while the Middle Triassic of Spitzbergen carries the European amphibian *Trematosaurus*.

Directing attention to South-eastern Asia at this epoch, one finds the continental margin penetrated according to Fromaget (29) by various gulfs taking off from both the Tethys and the Pacific. The strong movements that took place at the close of the Triassic seem to have produced wide shiftings of the strand-line with complicated linkings and unlinkings between Indo-China (Laurasia) and Indo-Australia (Gondwana), disclosed by the floral assemblages, as already mentioned in Chapter VI.

JURASSIC AND CRETACEOUS

Hitherto the crust of Laurasia had repeatedly been subjected to compression directed more or less north-south, to which its lesser width in that direction has obviously been due. For some time thereafter, save in the Donetz region of Russia, the movements affecting Laurasia were essentially epeirogenic in their character. Furthermore, important depressions indicative of crustal tension were initiated athwart the continent, the rhythmic development of which stood in causal relationship to the alternate opening and closing of the Cordilleran geosyncline of North America on the one side and of the Ural furrow on the other. It was along such zones of crustal sagging directed roughly north-south that rupture of the mass subsequently took place in the North Atlantic region, thereby paralleling the behaviour of Gondwana at about the same time.

Not only did the Rhætic-Liassic sea of Central Europe—an expansion from the Tethys—transgress over southern Sweden, Denmark, British Isles, France, Spain and Morocco, but it even penetrated northwards along the western side of Portugal into what can be viewed as the primitive basin of the North Atlantic ("Poseidon"). Making towards it during the mid-Jurassic came the waters of the Boreal Sea in their expansion over eastern Greenland, Spitzbergen, Franz Josef Land, King Charles Land, etc., which spread southwards down the coast of Norway (Lofoten Island) and even lapped the north of Scotland. The Ural trough was furthermore reopened, to remain a seaway between the Arctic and Mediterranean (with mingling of faunas) during the Upper Jurassic and Lower Cretaceous, after which regression became pretty general.

The Tethys of the Jurassic did not impinge on eastern North America, which must still have been directly connected with Western Europe, as shown by the common sauropod genera *Stegosaurus* and *Camptosaurus* together with allied forms, a condition which persisted into the Wealden, though curiously *Titanosaurus*—which seemingly entered Europe from Asia— is absent from North America. During the later Cretaceous North America became separated from Europe and their faunas proceeded to evolve on different lines, while Trachodonts and Ceratopsians entered the former from Mongolia—*via* Greenland according to Nopsca (34), but with greater probability through Alaska as supposed by Osborn, though so northerly a route has been doubted by Gregory (30).

The Cretaceous saw the progressive spreading of the Tethys —outside Appalachia—over Alabama and Georgia, and ultimately to the south-eastern coast of Massachusetts, and its penetration respectively between Labrador and western Greenland—effected in the Senonian—eastern Greenland and Spitzbergen, Scandinavia and Scotland, Scotland and Ireland, England and France, with wide overlapping upon the Iberian block, while the Ural Sea spread far over northern and northeastern Siberia, parts of Alaska, the McKenzie Valley, etc., though driven out later. A closer analysis of the stratigraphies and faunas of these various areas would doubtless disclose the precise order of fragmentation in the North Atlantic and Arctic territories. As previously mentioned, still greater inundation occurred over the Mediterranean region, North Africa, Arabia and Persia.

A brief review of the Asiatic section during these epochs is illuminating. Movements, following roughly the old Altaide lines, took place during the Jurassic in the Donetz basin of Russia and produced interior troughs in Mongolia, around the Ordos "platform" of China and in Indo-China.[1] In these, resting on crystallines and palæozoics, were laid down the younger divisions of the continental coal-bearing Angara Series.

[1] Grabau (28), 307.

At the close of the Jurassic came folding corresponding with the late Cimmerian disturbance of the Caucasus, with abundant eruption of porphyries and with development of fold-chains in the Tsinglingshan, Yenshan and around the Ordos platform, turning north-eastwards in the eastern Siberian ranges towards Verkhoiansk and the Arctic Ocean. Of high significance are the contemporary chains marginal to the continent running north-eastwards from Indo-China through south-eastern China and Korea to Sakhalin. On their Pacific side there developed all through Cretaceous time the lengthy Hongkong-Nippon-Anadyr-Alaskan geosyncline.[1] The vast interior of Asia remained land, though in various basins in Mongolia were laid down Cretaceous beds celebrated for their dinosaur remains. On the south, as already detailed, Asia was separated from Gondwana by the Tethyan geosyncline throughout this long period, save apparently during the Upper Jurassic in the Afghanistan region, by which route the specialised Asiatic flora presumably entered India (Grabau), and during the Upper Cretaceous betwen India and Indo-China (Fig. 42).

Geologically and structurally Eastern Asia is almost the " mirror reflection " of western North America, a fact which leads us to view Asia as having evolved on identical lines, that is to say, through a relatively youthful crustal creeping in the south-easterly direction, i.e. towards the Pacific. That Asia possessed a tendency to draw away from Europe is shown by the persistence of the Ural syncline—indicative of E.–W. tension —from the Middle Jurassic to the Upper Cretaceous save for a brief interval. In Tertiary times such continental flowage was accentuated through the mass of India pressing into its flank, as discussed later.

THE NORTH AMERICAN PACIFIC BORDER
DURING THE MESOZOIC

So close stage by stage are the geological and structural analogies of the western sides of North and South America that the tectonic histories of those masses cannot but have been similar (see Chapter V). North America is hence viewed as having experienced successive displacements west-south-westwards towards the Pacific, each of which involved the piling-up of frontal ranges with consequent overfolding or overthrusting towards the interior " foreland." Of such pulses there were at least five major ones from the Jurassic onwards.

During the Triassic, as previously mentioned, the western end of the Tethys crossed Mexico and joined the Pacific, which transgressed over California, Nevada and Montana, and in the

[1] Grabau (28), 456, Fig. 581.

north from Vancouver to Alaska in argillaceous and calcareous phases, but with an enormous amount of erupted matter, particularly in the north. Mediterranean forms made their appearance in California. Almost identical conditions persisted during the Jurassic with continued eruption of pyroxene andesites and basalts, but with a temporary shallow incursion from the northwest into the Great Plains region.

Within the Upper, though not latest, Jurassic [1] came the great disturbance, called the "Cordilleran" by Smith and "Nevadian" by Blackwelder (or Nevadan), in which the mesozoic and older systems were folded and faulted along N.N.W. axes within a belt extending from California to Alaska that attains a maximum width of 800 km. along the 42nd parallel, though generally narrowing away therefrom. According to ordinary concepts the pressures came from the west by the crushing of the ocean floor against the continent : under our hypothesis through the westerly drift of the continent against the oceanic block, in a series of pulses.

Plutonic intrusion occurred on a vast scale throughout this region, the most conspicuous mass being the "Coast Range Batholith," 2000 km. long, probably the greatest single intrusive body known. Granites predominate, but basic and even ultrabasic types are also represented. A broad complex series of marginal ranges came into being—Sierra Nevada, Cascades and Coast Ranges (British Columbia)—parting the low-lying interior region from the western ocean.

This barrier was considerably eroded, and the coarse detritals therefrom deposited with an Indo-Pacific fauna in the Lower Cretaceous "Comanchean" or "Shastan" sea on the Pacific side, over the folded Jurassic, batholithic bodies and older formations, e.g. Sierra Nevada. Through renewed movement and uplift during the Cordilleran "Intermontane" disturbance, which incidentally extended north-westwards to Alaska [2] and beyond, the Upper Cretaceous "Chico" was laid down disconformably or unconformably upon the Comanchean, still on the Pacific side, both of those divisions of the Cretaceous attaining an immense thickness in places.

In the related interior-depression the waters from the Tethys invaded Mexico, Texas and Arizona during the Comanchean, depositing limestones and marls with a Mediterranean fauna and in strong lithological and faunal contrast to the equivalent strata of the Pacific side. Marine conditions spread later to Wyoming with a fresh-water facies—"Kootenay"—beyond, in the trough that extended north-north-westwards into Canada. Following the mid-Cretaceous disturbance a through channel was established between Texas and the Arctic, which towards the

[1] Hinds (34), 191.   [2] Mertie (30).

close of that epoch was replaced by lakes and deltas. Throughout this great furrow extremely thick non-marine clastics (with
coals) flanked the rising western ranges, e.g. in Wyoming and
Alberta, thinning out eastwards over the Great Plains, through
which this extensive " foredeep "˙ became ultimately silted-up
(Fig. 42).

THE LARAMIDE OROGENY

The close of the Mesozoic was marked by intense compression along a belt extending from far beyond Alaska to Central
America and thence eastwards into the Antillian region. The
territory affected *lay mostly inside and east of the Nevadian
folding*, coinciding with the Rockies in Canada and Montana,
curving southwards east of the Great Basin, then east-south-
eastwards around the Colorado Plateau and south-eastwards
through Mexico with an inner branch through Wyoming and
Colorado. With such a lengthy zone, involving various
elements, it is not surprising to find that the major pushes did
not synchronise everywhere, occurring in some places during
the late Cretaceous, in others only during the Eocene.

The tectonic pattern embraces many arcs linked together,
sometimes *en échelon*. In a rough way compression started on
the west and was propagated eastwards. Characteristic are the
gigantic and lengthy overthrusts—in the opinion of Taylor and
according to our hypothesis really " underthrusts "—usually
rising towards the north-east at low angles, along which masses
of mesozoics and palæozoics were pushed north-eastwards over
tilted or flat strata, some so young as earliest Eocene (Palæocene),
e.g. Virgin Mountains, Wasatch, Hearst Mountain, in the
Bannock and Lewis overthrusts and those in the Clarke, Purcell
and Selkirk Ranges of Canada. The measured horizontal displacements of from 20 to 40 km. may perhaps be exceeded in
places. It is estimated that lofty chains, rising indeed to thousands of metres, thereby came into existence. Today for long
distances those much-dissected, overriding sheets build the
Rockies and Cordillera and overlook the eastern undisturbed
Tertiary lowlands.

The Laramide movement affected not only Mexico but the
Antillian region to the east as well as eastern Siberia. Throughout it was accompanied by volcanic eruptions of " Pacific type "
on a considerable scale. It corresponded precisely with the
First or *Andean Phase* in South America.

CHAPTER IX

THE TERTIARY HISTORY OF THE LANDS

Introduction. European-African Sector. Asiatic-Indian. East Asiatic. East Indian. Australasian. Antarctic. South American. North American. West Indian and Central American.

INTRODUCTION

THE state of the earth during the Tertiary was one of marked unrest punctuated by powerful diastrophic and volcanic spasms.

Such multiple orogeny furthermore differed fundamentally from those of preceding epochs in that—

(1) It was not altogether inter-continental in the strict sense of the term.

(2) The fracturing of Laurasia and the drifting of its parts caused that continent now to assume a more active rôle.

(3) Over lengthy stretches the crustal pressures were developed between conflicting blocks that were moving along paths not directly opposed to one another, but at varying angles thereto, leading to unbalanced forces or couples.

(4) Between any two diastrophic phases the relative positions and directions of movement of such conflicting blocks became altered, often considerably.

(5) Rotation of the blocks tended to occur, accompanied by shear and by distortion of the zones of crumpling with the production of arcuate, vortex and échelon structures.

(6) These fold-belts were injected by extensive ultrabasic and basic plutonic bodies, more especially during the late Cretaceous and Eocene, taking their rise from immense depths.

(7) Disjunctive basins were developed in regions of strong tension or torsion, e.g. Arctic Ocean and Caribbean Sea.

(8) Vast eruptions of " plateau basalts " took place in such regions of stretching.

(9) " Isthmian land-bridges " were produced between pairs of continents, and in most cases destroyed subsequently ; while

(10) Sheltered behind its " fold-rampart " each block became subjected in the main to epeirogenic disturbance only, with the local development of fracture-patterns.

174

· So complex have been these movements as to demand a separate volume for their proper treatment ; only the merest outline can, however, be presented here.

The outstanding development during this era was, of course, the great double " Fold-girdle " or Mobile-zone around the earth with its extensive mountain ranges towering to the skies and its associated lines of active volcanoes situated along such zone of extreme crustal instability.

If one allows for the continuations beneath the present oceans, this fold-girdle consists essentially of *two closed orogenic rings* which surround the dispersed portions of Laurasia and Gondwana respectively, and which have been pressed together in mid-section, i.e. between the East and West Indies—the " twin-zone " of the earth as Hobbs has called it—outside of which lies the immense Pacific Ocean. Such is well depicted in Holmes' diagrams (Figs. 19 and 20), which fail, however, to show the continuations across the Atlantic. Were this double system to be transferred from the surface of an artificial globe and rolled out flat, it would form a rough figure 8, within the loops of which would fall in striking fashion all the lands, save a number of small oceanic islands. Wherever visible, the structural pattern would consist of regular marginal foldings, commonly overturned towards the land, thrown up, so it is maintained, by the outward-spreading of the continental fragments and the crumpling of their leading edges ; only such unilateral character has been interfered with in the " twin-zone " by the clashing together of the two great rings with the production along that sector of a twofold or bilateral disposition. In particular areas there is a fair degree of symmetry with overturnings in opposite directions towards the respective " forelands " with perhaps an interfold or median region as · well, thus conforming with Kober's special scheme of orogeny. Such is manifested, for example, in the section between the Betic Cordillera and the Atlas.

Where one block has overridden the other, as is frequent, the interfold region has naturally been squeezed together and destroyed and a strong asymmetry developed with thrust-slices and nappes, though the fundamental bilateral character can usually still be made out. Such is well seen in the Alps and Himalayas. Away from the twin zone the plication is with rare exceptions, as long ago emphasised by Suess, unilateral, Kober's scheme not applying.

Details of the complicated tectonics within this double-ring are still wanting for many parts of the earth, but, wherever the history has been worked out, at least *three distinct* orogenic phases can be distinguished, and would seem to be practically world-wide, namely, the *First* during the late Cretaceous or

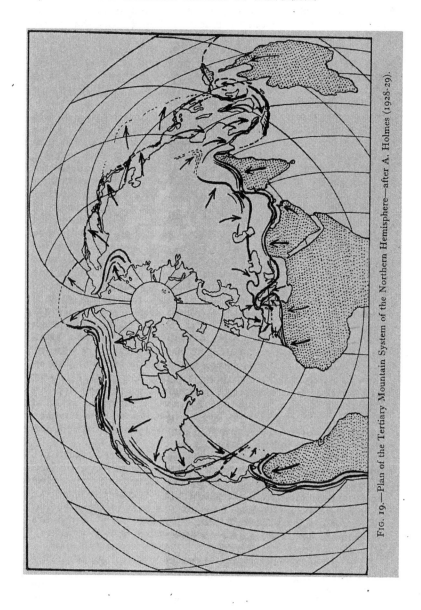

Fig. 19.—Plan of the Tertiary Mountain System of the Northern Hemisphere—after A. Holmes (1928-29).

early Eocene, the *Second* during the late Oligocene or early Miocene, and the *Third* during the late Pliocene or early Pleistocene.   In places a phase may have consisted of more than

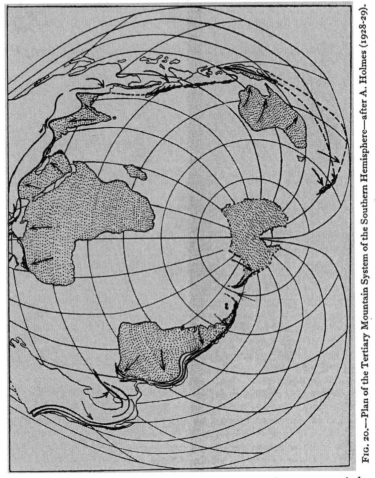

FIG. 20.—Plan of the Tertiary Mountain System of the Southern Hemisphere—after A. Holmes (1928-29).

one pulse.   Each orogenic spasm was normally accompanied or shortly followed by igneous activity.

It is considered that this diastrophism was the culmination of crustal pushes initiated during the closing Jurassic and well under way during the Cretaceous, following upon dispersive movements by the continents.   It is highly suggestive that each of the Tertiary phases began and ended earliest in the Mediter-

N

ranean region, also that the tangential movements reached their
zenith in the twin-zone during the Third phase, whereas elsewhere the latter was normally marked out by strong elevation
rather than by plication. Such would accord well with the
hypothesis of an intense squeezing of Eurasia by Africa plus
India with the pressures becoming less farther out to east and
west.

For convenient description the Fold-girdle will be divided
into the following sectors :

(*a*) European-African   (*b*) Asiatic-Indian
(*c*) East Asiatic   (*d*) East Indian
(*e*) Australasian   (*f*) Antarctic
(*g*) South American   (*h*) North American
(*i*) West Indian and Central American.

EUROPEAN-AFRICAN

For our purpose it becomes unnecessary to discuss at length
this excessively complicated region for which the reader must
consult the authoritative syntheses by Heim, Suess, Kossmat,
Argand, Collet, Termier, Kober, Staub, Bubnoff, Heritsch and
others. While there are several schools of thought, opinion has,
on the whole, favoured the view of the overriding of Europe by
Africa, as ably voiced by Argand, Collet and Staub. In 1910,
however, F. B. Taylor had pointed out that the part played by
Europe was not, as had been supposed by Suess, a merely
passive one, since the opening of the basin of the Arctic would
have enabled Europe to move southwards and so come into
active conflict with the approaching larger mass of Africa. This
concept of *two aggressive bodies*, the greater of which halted
and then forced back the lesser, was graphically developed by
Argand and Staub, with the second of whom the writer finds
himself in remarkably close agreement.

Before its repeated compression—which really started quite
far back in time—the Tethyan geosyncline from which the
Alpine chains arose must have had a width of well over 1000
km. in places and formed a more or less E.–W. trough connecting at both ends with the Pacific and apparently of abyssal depth
in certain localities, e.g. Alps and Barbados. Its warm waters
separated Eurasia from Gondwana. The Rudistid limestones
of the Cretaceous were replaced by the Nummulitic limestones
of the Tertiary seas, and it is instructive to observe that Voitesti
(28) in a study of the Upper Eocene nummulites has found the
species *N. gizehensis*—characteristic of Palestine, Egypt, Tunis
and Morocco—in Greece and the northern Adriatic, a region
which, from other lines of evidence, must, together with Sicily,
have lain farther to the south, presumably forming the tip of

North Africa. Incidentally, a remarkable fauna of sirenians and proboscidians became evolved in Egypt during the Middle Eocene, which spread to Italy.

The sequence of events may be outlined as follows :

(1) Near the end of the Cretaceous, Eurasia and Africa began moving actively towards one another, geanticlinal systems being initiated along their respective E.–W. trending and opposed borders, as shown diagrammatically in Fig. 21 A, i.e. (i) the Betic-Majorcan-Corsican-Alpine-Carpathian-Balkan-Caucasian and (ii) the Atlas-Sicilian-Appenine-Dinaride-Hellenic foldings —*b*, *m*, *c*, *a*, *cr*, *bl* and *at*, *s*, *ap*, *d*, *h* respectively.

(2) During the latter part of the Eocene the intervening Tethys became narrower, and Western Europe, weakened by the moving off of North America, now became distorted, with the opening of the Bay of Biscay, the anti-clockwise rotation of the Iberian peninsula about the Riff (*r*) as a hinge, and the consequent squeezing together of the Pyrenean sector (*p*)—which by the way was developed outside the Alpine geosyncline— resulting in intense overfolding and the formation of nappes and thrusts directed southwards in the western but northwards in the eastern portions, which clearly denotes a *twisting* of the mass.

(3) Advantage was taken of such weakened territory during the general northward advance of the Oligocene, when the two geanticlinal systems were not only pressed closer together but pushed north-westwards in an arcuate front into the space between Spain and Bohemia.

(4) In the Italian sector the southern system—which incidentally had grown in a part of the trough with peculiar sedimentary facies—became doubled back on itself like an " S " with overfolding to the north in its mid-portion (Fig. 21 B).

(5) The limbs became pressed together during the Pliocene with piling-up of chains, intense overfolding towards the north along the entire Betic-Caucasian front, and backfolding in the Atlas over the Sahara and in the Dinaride-Hellenic sections.

(6) As the drift of Africa was slowed down, part of its momentum became transferred to Europe, reversing the motion of the latter and causing it to retreat northwards.

(7) The disjunctive basin of the Mediterranean proceeded to open in its rear, breaking the connections between the Betic Cordillera and Corsica and between Tunis and Sicily. Spain, however, remained attached to Africa and so experienced some rotation, while the drawing-away of the Appenines from the Dinarides caused the opening of the Adriatic (Fig. 21 C).

Such movement is precisely the reverse of that postulated by Argand,[1] which view has been severely criticised by Bucher.[2] The grouping noted by Bubnoff of the young intrusive granites

[1] Argand (24), 357-59, Figs. 22-7.   [2] Bucher (33), 257-59.

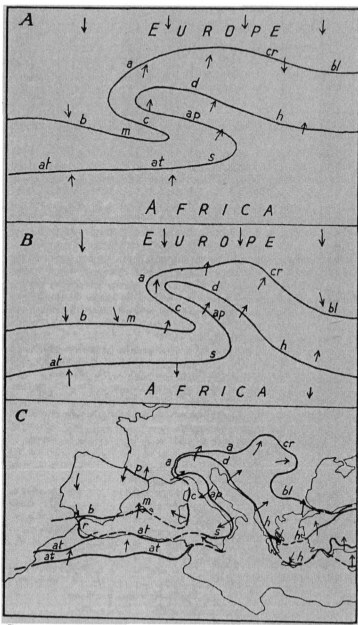

FIG. 21.—Showing three stages in the hypothetical evolution of the Alpine Foldings in the Mediterranean Region. The arrows denote the directions of inferred crustal movement.

in the southern part of the Alps around the head of the Adriatic is favourable to the scheme proposed.   The torsional deformation of the crust is signalised by the spiral arrangement of anticlockwise character shown by the Alpine ranges, long recognised by Suess,[1] which, twisting, may have been responsible for the rifting of the Rhine Valley and for the recent trough-faulting in the Netherlands.

The problem of the westerly extension of the Alpine (Betic and Atlas) foldings into the Atlantic will be touched upon later.

Discussion of the various movements in Asia Minor and Persia is superfluous since in that sector Laurasia and Gondwana, after becoming pressed together, remained so united.   It will merely be pointed out that the curving ranges running through Kurdistan have been overthrust upon the southern foreland of Syria and Mesopotamia from the N.N.W.   Their convexity is marked, as Suess has indicated, by the mighty volcanoes of Armenia and northern Persia, which are located upon lines of shearing.   The several tectonic phases in Persia correspond closely with those in the Himalayas.

ASIATIC-INDIAN

The Alpine zone of compression occupies a width of between four and ten degrees in the stretch between Asia Minor and Tibet.   Thereafter it expands enormously, to contract again beyond, its northern limit running through Kiev, Aral Sea, south of Balkash Sea, south and east of Lake Baikal, to the Aldan River and Okhotsk, while its southern one turns abruptly southwards from northern Burma in an immense even curve through the Andaman and Nicobar Islands into the Malay Archipelago.   Situated between these two diverging fronts lies the block of Indo-China, which appears to have largely escaped deformation, though it experienced fracturing and warping.   In general, the intensity of disturbance increased from north to south.   Throughout one can observe the superior momentum and plicative tendency of Eurasia.   In the west the steady opposition of Arabia—backed by Africa—determined the evenly running Iranian chains ; in the east serious obstruction was lacking and Asia escaped in numerous folds ; in the centre India made a valiant and successful attempt to force back its larger opponent, and their conflict created the most stupendous mountain system of the world.

The greater part of this Iranian-Himalayan-Indo-Chinese sector occupies territory that fell at one time or another within the Tethyan geosyncline, but the fold-region to the north lay outside that trough.   It has already been indicated that the

[1] Suess (04), I, 271.

latter seemingly became interrupted during the later Cretaceous in the sector between Tibet and Assam.   Such can, indeed, be regarded as brought about by an initial creeping-away from Africa of the Indian mass to the north-east, which not only ridged-up the floor of the trough in that quarter, but enabled those waters to be by-passed around Southern India and Madagascar.

This *First Himalayan* or *Karakorum Orogeny* [1] at the conclusion of the Cretaceous—corresponding to the First Alpine Phase—affected the sector trending through Baluchistan and New Guinea, the foldings being associated with abundant intrusions of granite and serpentine.

Sinking during the Eocene led to the re-establishment of the Tethys between peninsular India, Tibet and China and to the deposition of limestones and associated strata, and at later stages of thick "flysch," estuarine and fluviatile sediments recalling those of the Upper Cretaceous of that region.   Even away from the axis of the furrow vast areas were inundated and so became covered by Middle Eocene Nummulitic limestones, e.g. Sind, Cutch, Baluchistan, southern Persia and Arabia, Socotra, northern Somaliland, the western side of Madagascar, Burma and the East Indies.   Tension in the rear of the travelling Indian mass is displayed by the vast outpourings of the Deccan "Traps," which may indeed have covered originally a million square kilometres and to a maximum thickness of over 3000 m. The presence of many dykes usually striking E.N.E. is suggestive of fissure eruptions.   Their age is either very late Cretaceous or, more probably, early Eocene (Sahni).

The *Second* or *Sirmurian Phase*, that culminated in the Middle Miocene with further igneous intrusion, acted from the north as well as the south and drove out the contracted intercontinental sea, whereupon India became an integral part of Asia.   In the numerous interfold basins of Turkistan, Tibet and Mongolia, and between the rising Himalaya and the Indian peninsula—also in Sind and in Burma—terrestrial orogenic deposits were accumulated, known in India as the Siwaliks (Mio-Pliocene) with a remarkable vertebrate fauna.

Finally came the *Third* or *Siwalikian Phase*, between Upper Pliocene and Middle Pleistocene, when Peninsular India was thrust far beneath the southward-advancing Asiatic mass, which became raised thereby to form the Pamir and Tibetan plateaux with the Himalayan chains overriding the Indo-gangetic lowlands.   The generally low-angled surface of dislocation, which is traceable from Hazara to Assam at least, is known as the "Main Boundary Fault" though in places it is compound, and along it strata ranging in age from Archæan to Miocene (Muree)

[1] De Terra (36).

have been thrust, often in inverted order, over the Siwaliks, which are not to be found to the north of that dislocation, while nappe-structure is represented in places.[1] This gigantic zone of crustal fracturing follows the foot-hills of the Himalaya and movement must still be in progress along it, as shown by the serious earthquakes that have from time to time shaken the territory in question.

The numerous and often highly divergent opinions regarding the tectonics of this immense and complicated Indo-Asiatic region have been attractively summarised by D. J. Mushketov (36), to whose paper the reader must be referred, while the evolution of the structural pattern has been ably set forth by J. S. Lee in a highly illuminating essay (29). As in Central Europe, so also in Central Asia, only on a far larger scale, important older E.–W. trending Altaide (Hercynian) elements became involved in this Alpine compression. Argand holds the extreme viewpoint that the latter was not appreciably influenced by such Altaide remnants, while others deny this. F. E. Suess and J. W. Gregory take up the reasonable attitude that the older fragments acted as more or less passive cores, upon which the younger corrugations had to mould themselves. Much of the structure is certainly " germanotype."

There has been a general tendency to regard India as a passive block, but a careful analysis scarcely bears this out. For peculiar reasons—due certainly to a lesser altitude, and possibly to a greater thickness and velocity—this much smaller triangular mass was in its north-easterly drift able to penetrate deeply into and beneath its gigantic opponent before being arrested, the piled-up crustal folds on its two corners—in Kashmir and Burma—resembling the waves thrown off by a submarine. Comparison can also be made with a tablecloth stabbed obliquely by a skewer and rucked up into V-shaped crinkles in consequence. Appropriate, therefore, is the name " Jhelum wedge " given to the Kashmir salient. Wadia's[2] survey of that area shows that the curving major thrust-planes and the accompanying foldings run continuously round from the western to the eastern side of the " wedge," dipping to the north. They seem, for the most part, to be due to underthrusting of India and not to overthrusting of Afghanistan and Tibet directed separately from N.W. and N.E., in which case appreciable differences in phase could well be expected on the two sides of the salient. Taylor's[3] analysis of the problem is thoroughly convincing.

That India pushed its way far northwards under the Tethyan cover of Tibet is proved, first, by the facies of the strata trans-

---

[1] The reviews by West (34a; 37) are worth studying.
[2] Wadia (31), 215.   [3] Taylor (28), 160, Fig. 26.

ported southwards in cores or slices ; secondly, by the strong deflections around the two salients so far to the north as the Trans - Alai and Pamir and the Namkiu respectively; and, thirdly, by the presence in Ferghana of great overthrusting to the north and by the inversion over a wide area of the axial planes of the Asiatic foldings, which now dip south instead of north as is the rule throughout Tibet and Western China ; the hidden influence dies out farther towards the interior. Again, as Holmes has pointed out, the high stand of Tibet would imply an abnormal thickening of the crust (sial) over that large region. All this supports Argand's view that India underlies much of Tibet, which means that the block so involved must be far larger than present-day India.

The excessive loading of the northern side of the block could have been responsible for the tilting-up of the southern end of the peninsula, to which allusion is made on p. 258, as well as for the sinking of the vast Indo-gangetic "foredeep" with its enormously thick orogenic sediments that are being augmented by the existing rivers. Wadia has produced good reasons to show that the former drainage along this hollow was directed towards the W.N.W.

Had Asia been unopposed, it would doubtless have given rise to a single arcuate loop between Syria and Korea. The intrusion of India meant, however, interference and the production of a *triple* arcuate series with each loop convex to the south, the lateral ones having been formed exactly opposite the gaps developed between the parting masses of Africa, India and Australia. It is significant that a small anti-clockwise vortex should have been created at the Kashmir salient and a larger clockwise one to the north of the rigid block of Assam at the opposite corner, each just where the crustal streaming would have been greatest.

EAST ASIATIC

In analysing this sector it is unnecessary to go further back than the Triassic, when the Verkhoiansk geosyncline in Siberia led from the " Northern " ocean southwards between the Lena River and the Sea of Okhotsk, past Vladivostock and east of Korea to Kioto in Japan, constituting what has been called the " Ussuri Gulf." [1] It was bounded by the mainland on the west, the Cathaysian mass on the south and that of Kamtchatka on the east, and harboured a boreal fauna.

This furrow persisted into the Jurassic, lengthening and developing branches trending south - westwards, namely, the synclines of Yenshan passing through Peking, the Tslingling and the East Cathaysian, traversing the coastal portion of

[1] Grabau (28), 75.

China. In these extensions the sedimentation was essentially continental with plant-beds and coal-seams, a facies styled the younger Angara Series, but in the main Verkhoiansk section marine intercalations are present. In subsidiary related furrows within the continent in Mongolia, around Irkutsk and Ferghana, similar terrestrial strata were laid down. In certain areas deposition went on more or less continuously into the Cretaceous, but elsewhere, particularly in the east, the synclines were filled and compressed by the close of the Jurassic, when folding took place, with the overturning—whenever such occurred—directed mostly towards the interior. Two great series of fold-ranges thus came into being—the inner Verkhoiansk-Yenshan arc that deviated westwards into Mongolia and the outer marginal Asiatic chains that trended north-eastwards from Tonking along the edge of China through southern Korea and Ussuri and beyond, fronting the foredeep that was now developing on its south-eastern side—the Hongkong-Nippon-Anadyr geosyncline, which was being prolonged eastwards into Alaska.[1]

This new trough became filled with Cretaceous beds, of which much was terrestrial, for example, in the Anadyr Peninsula, while a similar facies was deposited in related tectonic basins inland, e.g. in Manchuria (eastern Gobi and Yenshan) and China (Shantung, Shensi, Szechuan, etc.), with plants, fishes and fresh-water mollusca ; those in Mongolia contained numerous dinosaurs. The eastern extremity of Siberia remained joined to Alaska right through the Jurassic, but during the Lower Cretaceous the Boreal Sea merged with the Pacific on the American side, bringing in Russian elements, after which the circum-Pacific up-foldings reunited those lands, for the first half of the Tertiary at least.

The previously described zone of the First Himalayan orogeny of Tibet and southern China is traceable eastwards and north-eastwards either within or just without the above Cretaceous arc, its inner margin running nearly northwards through Korea, between Vladivostock and Nikolaevsk (Sikota-Alin) to Okhotsk—with some compression at right angles thereto in the Verkhoiansk arc—and thereafter eastwards to the Anadyr Peninsula, while the outer arcs embraced Formosa, Japan and Sakhalin with a branch from Hokkaido through the Kurile Islands and Kamtchatka (which may be following an older line). The Aleutian Islands are seemingly set on another branch, which enters Alaska on the south-west, and, after making a strong inflection, turns down the coast of North America. In Korea, however, and probably elsewhere, the dominant phase of folding was late Miocene, but the disturbance was not very intense.

[1] Grabau (28), 456.

Throughout these northern regions of Asia and America the Tertiary strata are mainly non-marine and commonly carry coals, while volcanics abound, particularly andesites. Grabau [1] has postulated the existence during the late Mesozoic of land on the eastern side of Kamtchatka, which incidentally would have favoured the migration of the dinosaurs from Asia to America, a view that would meet Gregory's [2] objection that their remains have not yet been found more than a little north of the 50th parallel. Be that as it may, there can be no doubt of the unity down to relatively late times of Siberia and Alaska, as shown by the stratigraphical and tectonic evidence ; indeed during the Miocene elephants entered into America from Asia. H. B. Baker [3] has pointed out the close agreement in outline between the opposed coasts of Behring Strait, which supports the idea of the origin of the latter through rifting, the Asiatic side having been relatively displaced some seven degrees towards the south. The depth in the strait does not normally exceed 50 m.

Much has been written about the interpretation of the Eastern Asiatic arcs, namely, the Aleutian, Kamtchatkan, Kurile, Sakhalin - Japanese, Korean - Formosan, Formosan-Philippine and Malayan. As Wegener has observed, they are all comparable in size, regular, linked together *en échelon* and convex to the Pacific ; each shuts off a large portion of sea and fronts an oceanic deep, while the concave side bears a row of volcanoes. To Suess we owe the conception of the development of successive arcuate and asymmetrical fold-waves migrating outwards from the more stable " Amphitheatre of Irkutsk," which led to the progressive expansion of Asia towards the Pacific. While this hypothesis has since had to be appreciably modified, its fundamental ideas have been brilliantly confirmed by subsequent investigations, thereby supporting the views expressed by Taylor, Argand, Holmes, Staub, Tokuda, Lee and others concerning the ocean-ward flow of Asia, in opposition to those of Hobbs (23).

Of high importance is the échelon structure of certain of the lines, which Tokuda has shown must have been produced by pressure coming from the interior of the arc, a view confirmed by Lee, while Fujiwhara (34) has demonstrated the presence along the Japanese volcanic lines of vortices. The overlapping *en échelon*, which is general, displays that property known as " packing " whereby a particular *length* of fold-zone has been forced to occupy a shorter distance.[4] Already Wegener [5] had insisted that the linked arcs of this region were indications of a crustal shortening to an amount from 11,000 km. down to 9100

[1] Grabau (28), 506.   [2] Gregory (30), xcvii.
[3] Baker (32), 91, Fig. 30.   [4] Bucher (33), 205-8.
          [5] Wegener (29), 201.

km. at least as the result of squeezing directed respectively from the north-eastern and south-western sides of Asia.

Significant are the oceanic fossæ that immediately front the convex sides of the arcs—foredeeps subsiding in *advance of* the outward-moving geanticlines, and incidentally tracts of marked crustal instability. Developed in their rear through tension are the backdeeps made by the partially enclosed Okhotsk, Japan, Yellow, East and South China Seas—a " geosyncline in the making "—and particularly the great rift valley (with its basaltic effusions) that divides Korea nearly meridionally (Kobayashi).

EAST INDIAN

Including Further India with the oriental Archipelago, the region in question is nearly circular and some 4000 km. across. Tertiary strata, both fresh-water and marine (in which nummulitic limestones are conspicuous), together with volcanics, build up much thereof, usually resting discordantly on far older rocks that appear in anticlinal arches—portions of the post-Jurassic platform of Cathaysia or ancient Indo-China.

As seen in plan, the Tertiary foldings appear to hang between Burma and Formosa in several great loops with surprising regularity save in and immediately to the east of Celebes. As Baker has remarked, the festoon geography reaches its culmination here. Almost centrally placed is Borneo, a mass torn, it would seem, as suggested by Baker, from Cochin China with which it shows stratigraphical and structural agreements though now parted by the South China Sea that sinks towards the north-east to − 4000 m. Borneo forms, however, together with the Java shelf and adjacent lands of Java, Sumatra and Malay States a relatively stable unit, the seas parting them having the ruling depth of only − 60 m. A second such block is made by the southern part of New Guinea, the Aru Islands, Sahul (Arafura) shelf—and probably Timor—which really form the edge of the Australian mass. Though considerably folded, these two blocks must be fairly rigid now, as shown by the scarcity within them of earthquake epicentres.

According to Brouwer (20), the intervening mobile region embraces four strongly curved tectonic elements : (1) the zone of overthrusting (Timor-Ceram row of islands), (2) the marginal zone of plication without overthrusts (Sulu–Misool–western New Guinea and probably the Aru Islands), (3) the inner zone of young active volcanoes and (4) the zone between (1) and (2) of older volcanic rocks (north coast of Timor, Wettar, Ambon, south-western Ceram and Amblau).

Under our hypothesis the diverging paths taken by India and Australasia opened widely the long-existent gulf between them

(p. 97), and the crustal waves induced in South-eastern Asia moved out but little opposed into the Indian and Pacific Oceans, incorporating or overwhelming the weaker marginal crumplings of Gondwana. Their outer and active front runs in a regular curve from Assam through the Andaman, Nicobar, Sunda, Tenimber, Ceram and Philippine Islands to Formosa, but does not enter into Halmaheira or New Guinea. Inside this front, where the enlarging arc has been put under *tension*, lies the belt marked out by active volcanoes. Overthrusting has mainly been towards the outside of the arcs, while big lateral displacements have occurred between certain pairs of islands, e.g. along the Sunda and Manipa Straits. The orogenic phases are synchronous with those of Asia and India, the main diastrophism being Miocene.

Kuenen[1] has pointed out that the larger ocean basins situated within the above-mentioned front have flat and more or less horizontal floors and so contrast with those just outside, which have narrow shapes, are linked together and vary considerably in depth. Significantly the former are regarded by him as sunken blocks bounded by flexures or fractures. Wing Easton views the Sunda Shelf as a submerged peneplain.

Wegener has made the pregnant observation that the majority of the Dutch geologists working in the East Indies are either supporters of the " Drift Hypothesis " or favourably inclined thereto ; one can cite such eminent names as Molengraaff, Wanner, Wing Easton, Brouwer, van Vuuren, Escher and Smit Sibinga. Yet such is not surprising when one considers the unique tectonic structures with which those observers have been concerned, all but inexplicable under current ideas and mechanically impossible under the often-quoted " contraction hypothesis " as stressed by Smit Sibinga. Speaking eloquently of crustal flowage, for example, is the island girdle (with Celebes) hemming in the deep Banda Sea, the structure constituting an obvious anti-clockwise spiral or vortex with its centre sinking to − 5500 m. (Fig. 22).

The problem is essentially a hydrodynamical one, since the Banda vortex could only have originated through the actual streaming of the wrinkling crust past some rigid obstruction, which in this case was manifestly the New Guinea mass that terminates in Halmaheira and is viewed as having moved, together with Australia, north-eastwards towards the Pacific deep, while the East Indian festoons were enlarging southwards. Interference would have taken place and the torn-out hole of the Banda Sea could thereby have resulted. The obstructing western end of New Guinea would have played in relation to

[1] Kuenen's memoir (35) is not yet available to the writer and the above is taken from the review by Schuchert (36).

the East Indian crustal stream the part of an oar dipped into a flowing river, for an experiment—with the same relative setting—will show a vortex developing with anti-clockwise rotation just beyond the tip of the blade. The published accounts of this remarkable region of crustal eddying appear to accord well with such a supposition. The opposition experienced by the western end of New Guinea is evinced by strong overfolding and overthrusting directed towards the south. The neighbouring almost land-locked Celebes Sea sinking to −6150 m. can conceivably represent an additional vortex.

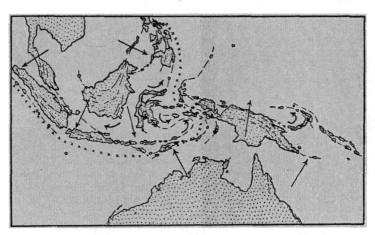

Fig. 22.—Showing Crustal Swirls in the East Indies. The broken lines mark Tectonic Axes; crosses, the Zone of Negative Gravity Anomalies; curved arrows, Crustal Vortices; straight arrows, General Directions of Crustal Flowage.

Remarkably, too, at the eastern end of New Guinea lies another anti-clockwise spiral in the shape of the Bismarck Archipelago (New Britain, etc.),[1] as though formed by the more rapid streaming of the Solomon–New Hebrides–New Caledonia arc north-eastwards past the slower-drifting mass of New Guinea. Significantly the overfolding and overthrusting in the adjoining territory of Papua are directed north by east.

Returning to the main Sunda arc, we find that in its advance towards the Indian Ocean it developed a parallel deep that descends to over −6000 m. off Java, which furrow the lofty island chains overlook. Of paramount importance is the fact that the outer edge of these island arcs for some 7500 km.—from Sumatra to the Philippines—corresponds fairly well with the unique narrow belt of strong negative gravity anomalies

[1] Wegener (24), 68.

determined by Dr Vening Meinesz, a peculiarity which will receive fuller attention in Chapter XII.

It is instructive to note that the belt, which is presumed to form the approximate limit between the conflicting blocks belonging to Laurasia and Gondwana, agrees closely with " Wallace's Line," which marks the boundary between the two strongly contrasted and unrelated life provinces of today— the Oriental and the Australasian. The forming of this deep oceanic fossa has apparently secured the Australian region from invasion by the large Asiatic carnivores.

Outstanding is the discovery at the extreme north-western part of Australia, in the Cape Range, of an important system of warping of post-Miocene—probably Pliocene—age and perhaps recent movement as well, which is striking north-eastwards,[1] and hence in the direction of Timor as indicated in Fig. 22. This affords proof of the northerly creep of Australia.

AUSTRALASIAN

After the Cretaceous the part played by Australia was, like that of Antarctica, a purely passive one. The rising, curving geanticline made by Timor, New Guinea, Solomon Islands, New Caledonia and New Zealand formed an advance-guard against the assault by the fold-waves of the East Indies, and Australia came to rest under the protection afforded by New Guinea.

Every one seems to have regarded this great island as an integral and northern part of the Australian mass, rigidly anchored thereto, a view fostered by the shallowness of the parting seas. Its present position is nevertheless anomalous in that its pre-Tertiary grain is *strongly discordant* to that of near-by Australia, as has been pointed out by David and others, though without realisation of its implications. No one appears to have given serious thought to the possibility that *New Guinea might actually be a slice of eastern Queensland that has become displaced to the north with anti-clockwise rotation*, and so be homologous with Madagascar.

Were it to be replaced, as shown in Fig. 7, making full allowance for the wide shelf off the Townsville-Rockhampton coast, the following relations would become apparent :

(1) If the geologically youthful lowlands are ignored, the southern side of the mountainous interior will show a surprisingly close parallel with the opposed mainland.

(2) It will come to form part of the region involved in the Samfrau geosyncline and prolong that structure well beyond Cape Yorke.

[1] Condit (36).

(3) Its trend-lines will agree closely with those of the mainland, whether affecting the oldest or youngest formations present.

(4) Its wide axial belt of presumably pre-palæozoic gneisses, schists, slates, etc., will be comparable with similar rocks that within the Australian coastal region appear only in the Queensland portion.

(5) Admittedly its little known, strongly folded Jurassic and Cretaceous display no clear resemblances to those of Queensland, but these strata could well be the marine facies, that was developed along the *outer* side of the Samfrau trough, of the continental equivalents known on the mainland.

(6) According to David, its backbone appears to be a horst made by a highly elevated and dissected post-Cretaceous peneplain, and therefore structurally and physiographically similar to the margin of Eastern Australia.

When Papua is better known geologically, a more intimate comparison will become possible, but for the present there would seem to be on record nothing inconsistent with the hypothesis here put forward.

In our reconstruction there fall upon a nearly continuous arc—concentric with the mainland,—New Guinea, Solomons, New Hebrides, New Caledonia and New Zealand, orogenic fragments which during the Tertiary were intermittently submerged, elevated, folded, flooded with andesitic or basaltic lavas and injected with peridotitic rocks during their outward drift toward the Pacific, with overfolding and overthrusting normally directed away from that ocean, as well displayed by New Caledonia. Significant, and in close correspondence with the marked structural bend near Brisbane, is the virgation in New Zealand (N.I.), where the N.W.–S.E. folds meet the N.E.–S.W. ones at a wide angle, the locus, in fact, of several diverging tectonic lines. In the southern part of South Island and in Stewart Island the longitudinally trending folds curve away to the south-south-east and hence in the direction of Western Antarctica.

The larger islands of Melanesia, situated to the west of a line drawn from Yap to the Tonga Islands, in many cases consist of, or else contain cores of, ancient continental sediments or plutonics, while several others lying towards the east are either known or suspected to have bases of such rocks. There is therefore an increasing tendency to regard those far-flung islands as the elevated parts of a continuous continent—an extension of Australia. Such a view was stoutly advocated by Gregory,[1] while other persons have even favoured its prolongation in the past right across the Pacific to South America. In contradic-

[1] Gregory (30), cxxiii.

tion these numerous terrestrial relics are viewed under our hypothesis simply, yet more in keeping with isostatic principles, as
fragments of the eastern edge of Australia, that is to say, of the
curved segment extending from New Guinea to New Zealand
and thence—as shown by the islands to the south and east of the
latter—to Antarctica.

Long ago Dana recognised the linear arrangement of these
various clusters of oceanic islands, while Suess [1] grouped them
along several arcs—the " Oceanides "—confocal with Australia
and fronting the Pacific and its abysses, which were visualised
as their foredeeps. By other workers, such as Phillips, Marshall,
David, Schuchert and more particularly by Staub,[2] these broken
and sunken tectonic chains have been differently aligned.
Henderson regards the regular submarine ridge — with its
island clusters — running from New Zealand to the Tonga
Islands as the upthrust edge of a crustal block overlooking from
the west the Kermadec and Tonga deeps.

The interpretation advocated here is as follows :

The intermittent drifting of the crumpling New Guinea-New
Zealand segment led to the deposition of thick marine tertiaries
on its flanks and in its troughs during rest periods, and to their
deformation under compression against the Pacific floor during
advances, with back-folding towards the Coral and Tasman Seas
that opened in the rear. The *First Orogenic Phase* was about
Lower Eocene, the *Second* about Middle Miocene (marked by
considerable elevation), the *Third* during Upper Pliocene (with
further uplift and intense strike-faulting, particularly in New
Guinea).

The radial motion, as a whole, caused the end-portions to
draw apart, rupturing the arc-like folds, whereupon the sections
dispersed, developing the Oceanides with convexities facing
the Pacific. The wide depression created between them and
Australia is notably all but free from islands and can be regarded
as a *gigantic rift*. The pendent nature of the festoons from
New Guinea and New Zealand respectively is clear, but, whereas
they are still in attachment with the latter, they have broken
away from their northern anchorage and have advanced in that
quarter beyond New Guinea. Since on the opposite end of the
latter the East Indies was, as we have already deduced, simultaneously moving south-westwards, New Guinea would have
been acted upon by a dynamic couple and rotated anti-clockwise
with a crustal spiral developing at either end, whence its present
tectonic discordance to Australia (Fig. 22). Well out in the
ocean, however, lies the island host of Polynesia formed of
volcanic matter or coral limestone, oriented on W N.W –E S.E.

---

[1] Suess (09), IV, chap. ix.
[2] Staub (28), 92-6.

lines and doubtless perched upon *advance-waves* belonging to the Australian system.

Conspicuous, furthermore, are the widespread volcanic activity at various dates, the evidence of shifting of volcanic centres, the number of active vents and their regular alignment on arcs or presumed cross-fractures, just as in the case of the East Asiatic and East Indian festoons. The nature of the ocean floor will be discussed in Chapter X.

Sheltered behind this island-strewn quadrant Australia remained almost undisturbed throughout the Eocene and Oligocene, and consequently became "reduced to one of the most perfect peneplains imaginable" (David), upon which a reddish-brown lateritic crust became widely developed. During the Oligocene the lag of the mainland led to widening and deepening of the Coral and Tasman Seas, to a limited amount of folding along the south-eastern coast of Queensland, to fracturing off its eastern margin and to outpourings of (the Older) basalts. The Miocene brought a gentle back-tilting from north-east to south-west, accompanied by appreciable subsidence in the south, which was continued into the Pliocene with renewed volcanicity in the east, involving the eruption of rhyolites and alkaline types followed by (the Newer) basalts.

The Kosciusko epoch (Andrews) of the late Pliocene produced universal uplift ; the early Tertiary peneplain was raised with warping to considerable heights in the east and in Tasmania —to over 2000 m. in places—throwing the drainages inland and south-westwards from the eastern divide, while concomitantly the ocean-side sank deeply in the Great Barrier Region. The evidence is in harmony with the hypothesis of strong down-faulting along the eastern edge of Australia (Chapter XII), which action was propagated southwards to Antarctica and is displayed spectacularly in the tremendous scarp bounding South Victoria Land.

The widespread fault-pattern in this part of the globe is consistent with the view of tension in a latitudinal direction associated with counter-pressures exerted from the north, thus producing a torsion in the Australian mass.[1] Illustrative thereof are the heavy down-faulting and block-faulting on both the eastern and western margins, faulting in the south—St. Vincent and Spencer Gulf, Bass Strait, the horst of Tasmania— and extensive interior warpings—Lake Eyre, Lake Torrens— all of which is in agreement with and supports the Displacement Hypothesis. The central depression of Australia is actually homologous with that of the Paraná valley in South America and the Kalahari in South Africa.

New Guinea and New Zealand form extensions of the

[1] For the Queensland coast see Steers (32), 294, Fig. 60.

Australian continental shelf and hence belong geographically to Australia. Furthermore, 300 km. south of Tasmania there rises an extensive submarine ridge from −4000 m. to −1100 m., possibly a horst, indicative of a former connection between Tasmania and Antarctica.

ANTARCTIC

Peeping out from under its carapace of ice, East Antarctica forms a surprisingly regular oval 4500 km. in extreme length, its convex eastern side following the polar circle quite closely, the straighter western one bulging opposite the pole, bounded along the Ross Sea by a tremendous fracture-system (Fig. 35), and not improbably similarly faulted along the Weddell Sea. It is essentially a high-standing shield of crystallines capped by flat-lying Gondwana strata and mesozoic dolerite sills. Prince Luitpold Land fronting the Weddell Sea was only seen and not examined by Filchner, but its morphology and the morainal blocks of conglomeratic arkose and sandstone suggest the western edge of the Gondwana-capped plateau of South Victoria Land. West Antarctica consists of two widely separated masses, the block of King Edward VII Land, Rockefeller Mountains and Edsel Ford Mountains to the east of the Ross Sea, and the isolated exposures beyond, and the Antarctic Archipelago of North Graham Land to the west of the Weddell Sea (Fig. 7).

The problem of the structural relations of these major ice-swathed masses is not yet settled, and although Byrd has not been able to find evidence for a strait between King Edward VII Land and the polar plateau, there are reasons for presuming that the Ross graben continues, though perhaps with weaker development, through to the Weddell Sea. South Graham Land is now known to form part of the continent.

It is the logical outcome of our hypothesis that the Samfrau geosyncline, and consequently the Gondwanide foldings, should have passed through the Weddell region and to the west of the pole, the most probable situation of that belt being along or adjoining the Edsel Ford Ranges—to the north of King Edward VII Land—and thence towards Eastern Australia.

Furthermore, the drifting away from Africa at the close of the Mesozoic of the South American-Antarctic-Australasian mass must have produced marginal folds within the Antarctic section, just as in the Patagonian and New Zealand portions. In evidence, therefore, the Tertiary Cordillera or "Antarctandes" curves southwards for 1000 km. with an average width of 120 km. through North and South Graham Land into Hearst Land, where its apparent continuation in the Eternity Ranges, about

160 km. wide with peaks rising to 3300 m. high, has been seen from the air by Lincoln Ellsworth,[1] striking towards the foot of the Weddell Sea with a high plateau beyond having isolated mountains at intervals.

It can be presumed to run next to, or else to merge with, the hypothetical Gondwanide belt thereafter, conceivably in the vicinity of Sentinel Peak (7500 m.), Lat. $77\frac{1}{2}°$ S., Long. 86° W., since to the north of the latter lies a plateau suggestive of horizontal (Gondwana) beds. Suggestively, too, the unexplored Edsel Ford Ranges, which are in a comparable position, make an arc sightly convex to the north,[2] which strikes somewhat north of west between Lat. 74° and 77° S. and Long. 130° and 145° W., and these are fancied to be built of folded strata. It is significant that, while the Rockefeller Mountains to the west-south-west consist of pink granite, the much nearer King Edward VII Ranges contain grano-diorites, and though the latter do not accord altogether with the Andean and Graham Land varieties, neither do the corresponding intrusions of New Zealand situated 3000 km. distant, that are admitted by all to belong to the circum-Pacific volcanic girdle. The islands to the south-east and east of New Zealand with their folded rocks of continental types and sub-alkaline eruptives obviously belong to this great Tertiary fold-system.

Returning to the better-known section, that of North Graham Land, its eastern side and islands constitute a " foreland " made of practically undisturbed Upper Cretaceous and marine Tertiaries of Indo-Pacific affinities containing many Patagonian forms, while the curved western Cordillera spreading southwards into South Graham Land discloses Andean characters by its folded sediments, grano-diorites and andesitic lavas and tuffs.

Long ago Suess[3] visualised the Andean system as proceeding in a great loop eastwards from Terra del Fuego through South Georgia, doubling back to the South Sandwich and South Orkney Islands and then traversing Graham Land and Antarctica. Indeed, from the close resemblance in pattern to that of the West Indies, he proposed calling this curved chain the " Southern Antilles," and explained it as due to the advance of the Andean folds into the Atlantic for a distance of about 2400 km. Many have nevertheless doubted such apparent simplicity of pattern, for example Nordenskjöld and Kühn, while some, for instance Staub, have thought that the Andean loop is smaller, not extending further east than Long. 50° W. The actual arrangement is probably complex.

The andesitic and basaltic lavas described by Backström and Tyrrell (31) from the South Sandwich Islands certainly suggest

[1] Ellsworth (36), 9.   [2] Gould (35).   [3] Suess (09), IV, 489.

Andean affinities, a point in favour of Suess' interpretation. On the other hand, South Georgia and the South Orkneys, with their strongly folded Ordovician and younger strata and the abundant igneous rocks of alkaline characters in the former, show no analogy with the Andean province (Ferguson). These islands have accordingly been regarded by Gregory [1] as fragments of a South Atlantic land. Windhausen has suggested affinities with the basement of Patagonia.

Our interpretation views the near-by Falkland Islands as having drifted from some position off the south-western Cape (p. 120). South Georgia and the South Orkneys could in similar fashion have been derived from the north-east and inferentially from the Samfrau geosyncline. Indeed, the author favours the view that the true Andean loop lies *inside of*, that is, on the concave side of, that made by the South Georgia–South Sandwich–South Orkney Islands, and that the latter represents a disjointed outer portion of the Gondwanides, which has become involved locally in the Andean deformation. The calcic composition of the South Sandwich lavas would then be due to a slight overlapping of the two orogenic belts. Alternatively, if the South Sandwich arc mark the true apex of the main Andean looping, then the South Orkneys and South Georgia, along with the Falklands, would have to be visualised as alien elements thrust south-westwards into the Tertiary system. It is somewhat suggestive that the axes of the foldings should trend W.N.W. in all three cases.

The latest bathymetrical map, that by Pratje [2] (Fig. 27), shows the crescent ridge of the South Sandwich Islands followed to the east by a narrow deep that sinks to below −8000 m. with concentric rises and troughs beyond, suggesting crustal waves. An analogous hollow exceeding −6000 m. lies just north-east of the Falklands.

All these data accord well with our hypothesis of the breaking away from Africa during the Cretaceous of South America, Antarctica and Australia, which masses during their subsequent drift remained connected down to a relatively late period by a narrow and readily deformed link, which became bent and stretched. The " South Antillian " loop is analogous to that of the Banda Sea, the rapid westerly drift of the South American peninsula past the slowed-up or stationary Antarctic block having developed on the lee side of the latter a vortex which descends to over −5000 m. and of anti-clockwise character.

Strikingly brought out in Pratje's map—though better still on a globe—is the way in which the major undulations of the Atlantic floor run in broad zones concentric with the African coast, namely, (*a*) the inner row of deeps, (*b*) the mid-Atlantic

<hr />

[1] Gregory (29), cx.   [2] Pratje (28), Fig. 2.

ridge that is prolonged eastwards into the Indian Ocean, and
(c) the outer row of deeps fronting South America, Falklands
and Antarctica and reaching close against the convex South
Antillian arc (Fig. 27). A pattern like this over so gigantic a
region argues for a single controlling cause, and finds its proper
answer only in *Continental Sliding*.

SOUTH AMERICAN

Much has still to be learned about the Tertiary history of
this great continent, more particularly of its northern and north-
western portions, while it is not easy from a study of the literature
to apportion correctly the several tectonic phases. The evidence
available is, however, consistent with the supposition of a mass
drifting spasmodically westwards with some clockwise rotation.
In places the width of the active zone, in its present compressed
state, exceeds 500 km.

The *First* or *Andean Orogenic Phase* (Steinmann) at the
close of the Cretaceous produced folding throughout the Andean
region and led to the emplacement of many of the great batholithic bodies of dominantly dioritic character between Cape
Horn and Colombia, and to the eruption of vast masses of
andesites and tuffs.

Gregory [1] has drawn attention to the important observations
that the north-westerly trending post-Cretaceous foldings in the
western part of central and northern Peru pass obliquely out to
sea, that near Guayaquil their strike becomes nearly E.–W. and
that the present ocean *cuts across the grain* of the country in the
great bulge of the continent. Inland the chains proceed northwards, while beyond in Colombia the trend changes to N.N.E.,
that is to say, parallel to the ocean. Steinmann would indeed
prolong to the Galapagos Islands this north-westerly directed
system of Peru, a viewpoint that has much in its favour.

During the Eocene-Oligocene the Andean chains were much
reduced by erosion in Venezuela, Peru, Patagonia and probably
generally, the waste therefrom being shed to the east and west,
forming in Peru the Rimac formation on the west and the Puca
on the east, and in Argentina giving rise to the famous mammal-
bearing beds of Patagonia. Sinking of the " foreland " took
place, bringing in the Atlantic over the eastern side of Argentina
during the Oligocene, wherein was accumulated the Patagonian
" molasse."

The *Second* or *Incaic Phase* (Steinmann) during the second
half of the Miocene drove out this sea, crumpled up the region
on the west and led to strong block-faulting inside that line,
with gigantic eruptions on the west of andesites, basalts,

[1] Gregory (30), civ.

trachytes and markedly in the section from northern Chile to Peru of rhyolites (Brüggen). Igneous intrusion on an enormous scale took place—grano-diorite, monzonite, granite, porphyry, etc.—coupled with intense mineralisation. Again the neighbouring erosion caused the overloading of the " foredeep," enabling the sea to flood the eastern edge of Patagonia and to penetrate for some distance northwards between the Brazilian shield and the Andean ranges, while littoral sediments were deposited off the western and northern coasts, for example, in the petroliferous formations of Peru and Venezuela.

The *Third Phase* during the Pliocene seems to have been strongest in the south, where compression and overthrusting to the east occurred. To the north and north-west the movements were mostly in the vertical sense, involving strong uplift of considerably eroded regions—to the extent of at least a mile in Bolivia, according to Berry. A host of volcanoes came into being, chiefly on the west, arising from this platform. In Venezuela the Pliocene was a period of erosion rather than diastrophism (Liddle).

The *Fourth Phase* was provided by the Pleistocene with great upheaval of the Andean and adjacent coastal regions, local folding and the building-up of giant volcanic cones. Strong climatic changes took place, incidentally leading to the widespread loessic deposits known as the " Pampeano " with their marine intercalations in the Paraná region.

Instructive is the repetitive nature of the Andean movements from the Cretaceous onwards, each complete cycle representing, under our hypothesis, the westerly urge of the continent, the piling-up of frontal folds, the slowing-up of the drifting block, the introducing of igneous matter into the zone of crumpling, the eroding of the latter and the depositing of the waste therefrom on either side, the sagging of the fore- or backdeep and the invading of the former by sea, the amount of advance being presumably some hundreds of kilometres at a time. At the conclusion all was set ready for another cycle and for a further stage of drift. Speaking generally, from the Cretaceous onwards, outward or forward migration of the marginal folding and of the zone of volcanicity took place, causing a *net gain in land-area* despite the compression of the advancing edge.

There is a strong suggestion that the Amazon River formerly discharged westwards, but that its flow became reversed through the rise of the Andean chains, a view supported by its unique " fossil delta " far inland that faces the wrong way, as described by Sherlock (34).

One must not omit to draw attention to the backward bending of the Andean plications in Venezuela and Terra del Fuego until an E.–W. trend was acquired in Trinidad and at Cape Horn with the thrusting directed inland in each case. Furthermore, the sub-parallel folds on passing northwards through Colombia diverge regularly on the one hand eastwards into Venezuela, and on the other north-westwards into Panama in a great virgation.

The analogy thereby displayed with the spreading waves thrown off from a blunt-nosed ship is accentuated by the regular curvature in the section extending from Venezuela to the northern end of Chile beyond which there is a strong inflection, and the folds thereafter run due south, suggestive of the greater westerly travel of Patagonia as compared with Bolivia. The significance of the E.–W. curvature in the extreme north and south is discussed in the sequel.

NORTH AMERICAN

The history of this continent stage by stage repeats to a remarkable degree that of South America.

The *First Orogenic Phase* is the Laramide, described in Chapter VIII. Thereafter terrestrial Tertiaries were laid down at various dates in rugged inter-montane basins, and marine strata along the Pacific coast within a narrow and irregular belt, only patchily in places, as in British Columbia, though commonly great thicknesses were attained.

A great break normally characterises the Oligocene and early Miocene, with the deposition of the Middle Miocene unconformably upon older sediments and upon batholithic intrusions, e.g. in Idaho, southern California, Coast Ranges, south-western part of the Great Basin, Oregon and northwards. Vast quantities of lava were poured out, in places over areas of considerable relief, as in the case of the Columbia Basalts.

The *Second Orogenic Phase* is marked out by flexing and faulting of the Miocene on the inland side, by its folding and overthrusting in the Coast Ranges and south-western part of the Great Basin, and even by the intrusion of granite in Oregon. This was followed by great erosion and by considerable uplift during the Pliocene, well marked, for instance, in the Rockies and in British Columbia according to Peacock, and exactly paralleling the Andean region.

The *Third Phase* is Lower or Middle Pleistocene, and affected the Coastal Ranges of California—where thicknesses of up to 3000 m. of Pliocene and Pleistocene were folded—and parts of Arizona and north-western Mexico, though not Canadian territory. In certain localities on the Californian coast late Pleistocene is today actually being deformed. Recent work is enlightening us regarding the large number of erosion surfaces within this region, evolved between crustal warpings and intermittently upheaved under isostatic adjustment.

Reviewing the diastrophic history of western North America it is interesting to observe the systematic shifting of the locus of deformation with each spell of presumed westerly or southwesterly drift of the continent. The strip of crust involved in the arc of compression became appreciably thickened and was further reinforced by batholithic intrusion and got so strong, ultimately, that subsequent yielding had to take place on one side or other of the original welt. Thus the late palæozoic disturbance affected the interior or Plains region, the Nevadian orogeny the belt well beyond the latter and nearer the Pacific, the Laramide movement the neutral zone between the preceding lines and the Tertiary plication the coastal region, with regular ocean-ward migration of that mobile section under each successive pulse.[1] The orogenic history of South America is thus duplicated.

Torsion, due to a slight clockwise rotation, is well brought out by the arcuate trends of the tectonic zones, by the conspicuous lines of faulting of varying age, striking for the most part N.W. or N.N.W. with a complementary set trending N.E. or E.–W., while internal adjustments are indicated by the numerous volcanic centres and by repeated eruption on quite a large scale. The close relationship between igneous activity and metallisation scarcely needs to be stressed.

Considering the non-homogeneity of the advancing continental front, the presence therein of more resistant cores— sedimentary and plutonic—and of older lines of dislocation, the tendency of eroded ranges to rise isostatically and of basins of sedimentation to subside, and so on, one could reasonably anticipate the existence side by side at more or less the same time, of areas showing the contrasted effects of strong compression and of tension, for which accommodation had to be found in faulting, rifting and volcanicity. Nor have such adjustments ceased, as is shown by the slipping that is still occurring along fault-lines, with periodic great destruction of buildings and loss of life.

The sub-coastal portion of North America shows strong similarities with that of East Africa in the matter of highly elevated and warped peneplains and of important rift-valleys, from the floors of which have arisen volcanic cones, some of them still active, e.g. in Oregon and Washington. As convincingly set forth by Fuller and Waters (29), such horst- and graben-structure can only be interpreted as due to *tension*.

[1] For details see Schuchert (23).

WEST INDIAN AND CENTRAL AMERICAN

Following the synthesis by Suess [1] for this complex region, opinion for a long while favoured his view that the Andean foldings, after passing eastwards out to sea from Trinidad, swung northwards in the arc of the Lesser Antilles, westwards through the Greater Antilles, Yucatan and Mexico, and thence north-westwards along the margin of North America. Today it is generally recognised that the Lesser Antilles is only a volcanic linkage across the two curving major fold-belts of the Americas, which vanish eastwards into the Atlantic, but cannot be mere " free ends " and must accordingly be regarded as joining up with the corresponding double Alpine fold-system across the ocean. That is to say, the Atlas would unite with the Venezuelan—each backfolded over its continent—and the Betic Cordillera with the Antillian—the former backfolded upon the Spanish block, the latter bounding the persistent " foredeep " made by the Gulf of Mexico. Staub's [2] instructive presentation should be studied in this connection together with Woodring's Tectonic Map (1928), which has been reproduced by Bucher.[3]

The movements on the two sides of the Atlantic were in a general N.–S. direction, and the four main phases agree in their dating, as shown by fossils. Similar formations are involved too, namely, marine Cretaceous—normally of deep-water calcareous facies—and marine Tertiaries ranging from Eocene to Pliocene, some of bathyal character—with various and usually corresponding breaks. The faunal relationships are furthermore remarkably close.

Few parts of the earth serve better to display the superiority of the Displacement Hypothesis as an interpreter of Earth-evolution than the Antillian region. Accepting current ideas, it is not at all obvious why, to mention only a few such puzzles, the Greater Antilles should have acquired its particular arcuate form, why the Lesser Antilles should be convex to the east, why the Costa Rica-Panama Isthmus should have been formed just where it is and to have got so queer a shape, or why the Bartlett Deep should have originated. Orthodox palæogeographical maps, moreover, tend to obscure the ever-varying spatial relationships between the land-masses thus involved.

As Trask (36) has cogently remarked, for a sea viewed by Schuchert as unchanged in its position from Pennsylvanian to Recent, bordered throughout most of that time by land-masses, yet only 300 to 400 miles wide, it would indeed be remarkable that it should not have had its shape radically altered or have become silted up.

[1] Suess (04), I, 544·50.   [2] Staub (28), 102·10.   [3] Bucher (33), Fig. 2.

In his endeavour to explain the region on other lines Bailey Willis [1] propounds the conundrum, " Is the Caribbean a dynamic basin or a passive structure ? " which is not a fair distinction. By the term " dynamic " he would imply that the raising of the margins has been produced by the " uptrusion " of plutonic masses and the establishment of numerous encircling volcanic centres, which, however, merely leaves the question of the cause and localisation of such igneous activity unanswered.

Under our scheme, however, the perspective is vastly different ; the region becomes the ever-narrowing field of conflict between the converging continents of North and South America approaching from north-east and south-east respectively (though possibly at different speeds) with the throwing-up of frontal, diagonal and advance folds, their interference and deformation by renewed squeezing and shearing, and finally the development of a gigantic rift-valley system within the distorting framework of the Caribbean Sea. Indeed it can be affirmed that " Continental Drift " is painted in large and unmistakeable characters across this fascinating region.

Its complex geology, as set forth by Suess, Hill, Spencer, Vaughan, Woodring, Schuchert, Berkey and many others, has nevertheless to be rewritten from the new viewpoint, which, because of lack of space, can only be done in merest outline. The admirable review and the subsequent monograph by Schuchert (29 ; 36) must form the basis of our reconstructions of land and sea, though our interpretation (Figs. 23-6) differs radically from the maps of that eminent palæogeographer. For most of the Tertiary our restorations show the Antillian Islands, together with the Bahama platform, assembled more closely and united to Yucatan and Honduras as their geological structures would favour, but situated further to the north-east of South America.

This " condensed " area happens to coincide with the belt of Upper Permian folding that strikes from Guatemala to the Virgin Islands, bent somewhat by Pliocene deformation, though originally trending about E.–W. Such former parallelism with the Appalachian structures of the Mid-States region suggests not only a tectonic unity, but a closer geographical relationship, which in turn accords with the conclusion reached from other lines of evidence that the Gulf of Mexico developed only later through a south-westerly drift away from the North American mass of the Mexican-Antillian segment. The position of the Tertiary folding in the latter could well have been determined by the palæozoic grain, just as the initiation of the Betic structures in Spain was controlled by the Hercynian pattern.

It was pointed out in Chapter VIII that during the late

[1] Willis (32), 928.

Jurassic North America started to drift away from Europe, when mountains were upheaved on its Pacific border, curving through Mexico and the Antilles, in which quarter that "Nevadian" folding is found to fall off in intensity. The Tethys must therefore have expanded northwards between the Spanish and North American blocks and sent a gulf between Cuba and Florida that submerged eastern Mexico, the main sea stretching south of the Antillian land to the Pacific. The southern shore of the Tethys extended westwards from North Africa, all the northern part of South America being land. The two Americas were closer to Europe and to Africa respectively, and possessed, more or less, the same longitudinal position, though farther apart latitudinally than today.

The Cretaceous saw tension in, and enlargement of, the early Atlantic and the progressive submergence of the coastal portions of Venezuela and Colombia, of Central America, Antilles, the Gulf region and even at last of the Appalachian coast, though the Intermontane Disturbance maintained the great curving barrier between the Pacific and the Gulf region. Noteworthy, in the axial part of the trough, was the great development of limestones and calcareous and glauconitic sediments with a Tethyan "rudistid" fauna that gave way towards the north to colder water assemblages. Volcanicity was rife in the Antillian region.

At the close of the Cretaceous the drifting of North America to the south-west and of South America to the north-west led to the respective marginal plications of the *Laramide* and *First Andean Orogenies*, which movements were seemingly completed by the early Eocene in the Antillian region, namely, along (1) the line Honduras–Cuba–Haiti–Puerto Rico–Virgin Islands to the Betic chains with the Azores belonging, as Gentil has suggested, to the Spanish "meseta"; and (2) the line Venezuela–Trinidad to the Atlas by way of Grand Canary with the Cape Verde Islands corresponding to the "shield" in Mauritania. The Caribbean would hence be an inter-fold region or median area of the "Kober type" in which compression has not yet brought the two sides together.

The "intra-Atlantic" continuations of these two fronts and of their successors during the Tertiary need not, and inferentially did not, rise everywhere above sea-level, more probably forming sub-parallel island-chains, shoals, etc., within the Tethyan geosyncline. A through seaway between them from the Mediterranean to the Pacific could be postulated, that connected intermittently with the enlarging northern and southern Atlantic. Along such isthmian connections or island-chains terrestrial life could have spread with more or less freedom and marine shallow- or moderately deep-water faunas along their flanks. Stretching, due to the westerly urge of the Americas, could have temporarily depressed or broken such connections; pauses or transverse flowing of the sub-crust could have revived them, that is to say, down to the late Miocene, after which presumably the ocean would have become too wide, and the north–south compressive forces too feeble to maintain them.

Thus simply, yet effectively, can be explained the resemblances in the terrestrial or marine life at certain periods and the absence of such affinities at others, and that between these four land-masses not only latitudinally, but *cross-wise* as well. The close relationship between certain of the marine in-vertebrates of the West Indian and the Mediterranean forma-tions has been stressed by many persons, beginning with Guppy, as summarised by Gregory (29) and recently reviewed by Schuchert.[1] Gregory and Stefanini have recorded this for the echinoids and Vaughan for the reef-corals during the Upper Eocene and particularly the Oligocene, while Woodring (24) and Jaworski have noted the same for the mollusca and fora-minifera, such resemblances weakening during the later Miocene. It is generally agreed that for such organisms the *deep basin of the Atlantic would have proved as effective a barrier as land*. Similarly, the rising or sinking of the Mexico-Panama Isthmus is reflected in the absence or presence of Mediterranean elements in the Pacific faunas.

During the early Tertiary the curving Laramide and Andean fronts approached one another and branches were thrown off from them obliquely across the Caribbean inter-fold trough, structures corresponding to the lambda-folds of Lee and shown by him to develop through horizontal shear during compression, i.e. the several rays diverging north and north-east from the virgation of Colombia on the south and those trending south-east from the virgations of Haiti and Jamaica on the north. In Jamaica, Hispaniola and Puerto Rico the folds thus produced trend obliquely to the longer directions of those islands. The volcanic arc of the Lesser Antilles, with a probable sedimentary core, which was initiated during the Eocene—its northern end at least being land in the very late Cretaceous—originated perhaps upon such a shear-line running N.E.–S.W. from the one front to the other, thereby proceeding to shut off the Carib-bean Sea from the Atlantic.

A consequence of our dynamic scheme is that in the extreme west the floor of the Pacific would have been forced up in an approximately N.–S. swell or *advance-fold* in between the convex diverging fronts, and so have given rise to the nucleus of the Nicaragua-Panama Isthmus, the instructive history of which will be dealt with below.

[1] Schuchert (32), Parts II and III.

Plications synchronous with certain of those in the Alps occurred during the early Oligocene in particular areas, as in Jamaica, but, generally speaking, that epoch was a period of submergence, when all but the summits of the folded Cretaceous mountain chains were buried beneath deposits largely characterised by foraminiferal limestones and marls, the area concerned including Guatemala, though nearly all Mexico, Central America and the south-eastern United States was emergent.

The early Miocene saw continued depression on the whole, including the Costa Rica Isthmus for a short period, but was succeeded by the *Second Andean Phase* of the later Miocene, which revived the older lines and most of their branches, incidentally restoring the Costa Rica land-bridge. Échelon structures were developed, while in deeply depressed synclines there accumulated the well-known radiolarian marls of Barbados, Trinidad, Haiti and Cuba for which depths of anything down to a few thousands of metres have been presumed by various authorities. About this time, or else a little earlier, numerous masses of grano-diorite, diorite and other rocks were intruded, as, for example, in Cuba and Jamaica. During the Upper Miocene the important Tehuantepec " portal " came into being, allowing the waters of the Gulf of Mexico to join those of the Pacific until closed again during the Pliocene.

The *Third Phase* during the Pliocene was, as elsewhere, one of intense fracturing and block-faulting rather than of folding, accompanied by great volcanic activity.

Under our hypothesis North America through its greater drift overran South America, thereby putting the Antillian arc in tension and shear. The link of the Lesser Antilles was dragged westwards by its northern end and so developed a curve concave towards the west, and the Antillian segment was broken into various fragments, with Cuba shearing past the Bahamas block and acquiring a convexity at its western end. Freed from its eastern attachment, Central America was now able to move off westwards, thereby producing the Bartlett Deep and the depressions between Yucatan and Florida and to the north of Panama.

The distortion of the Caribbean framework is made apparent in our restorations of the Nicaragua-Panama Isthmus for various epochs (Figs. 23-6), which land-bridge can be viewed as having swung around a pivot made by Colombia, its length shortening with each south-westerly displacement of Honduras, while its altitude increased during advances towards the Pacific, but lessened during pauses. Thus can be explained in turn the limited opening of the Costa Rica " portal " during the Upper Eocene, the closure during the Oligocene, the reopening during the Lower and Middle Miocene as stretching took place (due to

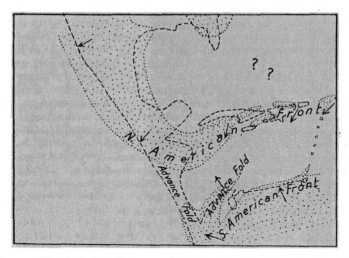

FIG. 23.—Hypothetical distribution of Land (stippled) and Sea (plain) in the region of the West Indies during the late Cretaceous.   The arrows indicate the inferred direction of crustal movement.

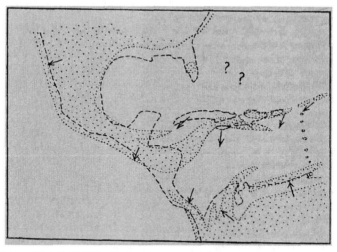

FIG. 24.—Hypothetical distribution of Land (stippled) and Sea (plain) in the region of the West Indies during the mid-Tertiary.

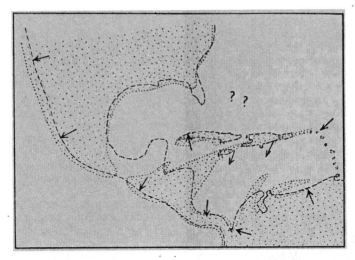

FIG. 25.—Hypothetical distribution of Land (stippled) and Sea (plain) in the region of the West Indies during the late Tertiary.

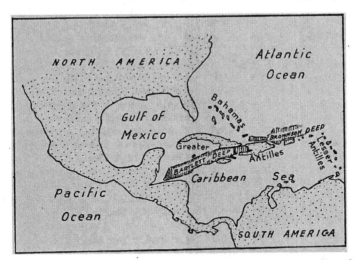

FIG. 26.—Present-day distribution of Land (stippled) and Sea (plain) in the region of the West Indies.

the superior westerly urge of Central America), the uplifting during the Upper Miocene, the developing of the double kink due to the mutual approach of Honduras and Colombia and, lastly, the tearing open of the Gulf of Darien in the rear of the westward-travelling and distorting barrier.

The Pleistocene was marked by some further deformation, but especially by block-faulting, the emergence of large areas and by subsequent and considerable submergence.

The mass of information about the Antillian region by investigators such as Woodring, Schuchert, Taber (23) and Hess and the recent *Gravity Surveys* furnish powerful support for this hypothesis, since at every turn intelligent expectation is found to accord with fact.

Thus the narrow Anegada Channel trending W.S.W., parting the Virgin Islands from St. Croix and Anguillia, and sinking rapidly to -4400 m., could mark the hypothetical line where the Antillian arc was snapped diagonally. Between Haiti and the Gulf of Honduras extends in a west-south-westerly direction the remarkable diamond-shaped Bartlett Deep, 1500 km. long by 150 km. wide, sinking at its deepest part to -7000 m. and clearly a disjunctive trough or " window " ; its submerged margins are presumed to be fault-scarps. The bottom is on the whole over 5000 m. deep and flattish, but broken into a number of tilt-blocks at its eastern end, while its exposed walls in western Haiti are bounded by high-angle faults that are presumed to be overthrusts. On its southern edge stand Jamaica and the Swan Islands, on its northern the Cayman Islands and Cuba, the straight southern edge of the latter rising abruptly in the Sierra Maestra to 2500 m., this coast-line showing strong east–west faulting according to Taber. An important criterion of rift-valleys, namely, the back-tilting of the sides (p. 250), is present, Taber having recorded on the southern coast of Cuba an inland dip which dies out to the north, while Matley has reported a westerly tilt in the Caymans. The shallower sea with its numerous " cays " between Nicaragua and Jamaica doubtless indicates the subsided connection between those masses. Hess (33) has ably shown that the Bartlett Deep may be ascribed to a horizontal shear, but, in accordance with orthodox ideas, makes the displacement one to the E.N.E. instead of W.S.W. as under our view. Probably, as Matley (26) has suggested, the Cayman-Cuba and the Honduras-Jamaica-Haiti features represent the separated edges of a great crustal fracture.

The disjointed Antillian arc is parted by a deep channel from the Bahama block, which Hess [1] has properly regarded as a sunken eroded mass, the valleys of which now descend to

[1] Hess (32), 30 ; (33), 45-7.

—4000 m. although they must have been *evolved sub-aerially*. In other parts of this region there has been great recent uplift, as in Jamaica, where coastal " benches " have been reported at various levels up to 600 m. The great instability of the West Indies is shown by frequent earthquakes as well as by periodic volcanicity.

The large *gravity anomalies*, which have been discovered by Vening Meinesz, Brown and Hess (33), are furthermore in complete agreement with our interpretation.[1] Thus negative anomalies, which are viewed as indicative of compression or of strong erosion (p. 265), mark out the northern length of the Antillian arc from the Virgin Islands to the western end of Cuba, the Bahama region and a part of Florida. The deviations are particularly large in the Brownson Deep, which descends to over —4000 m. to the north of Puerto Rico. Positive anomalies, regarded as characteristic of tension, of heavy sedimentation or of the ocean deeps, distinguish the Gulf of Mexico, Caribbean generally, the islands themselves and the open Atlantic. The positive values within the Bartlett Deep are found to reverse in sign over its sunken edges, and the arrangement is therefore strikingly the same as in the Red Sea, admittedly a rift-valley (p. 253).

The presumed easterly extension of the Antillian and Andean foldings within the Atlantic will be briefly referred to in Chapter X, where the history of that ocean is reviewed.

[1] Bowie's plan (35) of *averaging* the anomalies within each four-degree square is unscientific through lumping together adjacent values of opposite sign and not separating figures for land and sea, thus obscuring the true position. He finally averages out the entire region and reaches the amazing conclusion that it is almost isostatically balanced, thereby giving one the impression of a stability that is negatived by all the geological evidence.

CHAPTER X

THE OCEANS

Introduction. Pacific. Nature of Ocean Floor. Permanence of Basin. Evolution under Displacement Hypothesis. Atlantic and Arctic Oceans. Evolution of Atlantic-Arctic Trough. Tension and Volcanism. Movements during Recent Times. Indian Ocean. Origin of the Ocean. Land-bridges.

INTRODUCTION

VOLUMES have been written upon the fascinating theme of Oceanic Evolution. One school of thought favours the doctrine of the " Permanence of Ocean Basins "; the other supports interchanges of land and sea that involve the sinking of blocks of continental dimensions. Their respective merits need not be argued, since both are usually in open conflict with the Displacement Hypothesis. They regard seas as essentially due to movements of the crust in the vertical sense, and generally attempt to explain on more or less similar lines the fundamentally contrasted basins of the Atlantic and Pacific.

The Drift Hypothesis, on the contrary, has to view the great geosynclinal seaways of earlier geological history as branches, albeit important ones, of a universal ocean or " Panthalassa," which was transformed into the present oceans through the horizontal drifting of the crustal blocks towards a focus situated in the Pacific region. Much of the eustatic or world-wide oscillations of ocean level, currently styled " epeirogenic," can furthermore be interpreted as due to *changes in the shapes of the various basins* through such drifting rather than to the somewhat arbitrary up-and-down movements of the ocean floors usually postulated.

Were the waters to be spread out evenly over the earth they would cover the latter to the depth of about 2700 m.; today they are gathered together in basins that occupy 70 per cent of its area. Although soundings exceeding – 10,000 m. have been made, those portions below the – 6000 m. contour form only 1 per cent of the oceanic area, while the continental shelves— i.e. down to – 200 m.—constitute 5½ per cent. The grossly exaggerated profiles in vogue give a wholly incorrect impression of the relief of the floor. When plotted to scale, recognising the curvature of the earth, the oceanic slopes become slight, over wide regions imperceptible and nearly everywhere *convex upwards*. On a relief globe one metre in diameter the ocean basins would have to be sunk approximately one-fifth of a millimetre at the most.

THE PACIFIC

The limits of this vast ocean are ill-defined on the north and west because of the trespassing island-arcs, in strong contrast to its eastern shores. Its irregular shape recalls that of Australia—rotated through 45 degrees with its axis set N.W.–S.E. Lacking in symmetry it is 17,000 km. across from Mindanao to Panama and 15,000 km. from Behring Strait to Antarctica, and has twice the area, and more than twice the volume, of the Atlantic, its mean depth being about 4000 m. If divided along Long. 140° W., the two portions will show considerable differences.

The *eastern half* is characterised by few islands despite the generally shallower water. The portion down to about −4000 m. forms a relatively narrow border to North America, but expands prodigiously opposite Central and South America into the so-called Albatross plateau with surface sinking off Bolivia and Chile in a series of narrow fringing "deeps" to a maximum of −7635 m. The significance of this will be pointed out later.

The *western half*, though much deeper on the average than the other, is strewn with islands even far out from its bounding island-arcs, those nearest the centre of the basin being arranged in a belt trending N.W.–S.E. and consisting of coral, though flanked on north-east and south-west by others of volcanic origin. Fronting the island-arcs and chains lie roughly parallel and related narrow "backdeeps" which range in depth from −8000 to over −10,000 m.—veritable abysses! Remarkable is the chain of deeps aligned along a great circle trending north-eastwards from New Zealand to off Vancouver—possibly a rift-valley as suggested by Gregory;[1] dividing the ocean floor, it forms the antithesis of the mid-Atlantic Ridge. Athwart this feature are set the various sub-parallel island chains from the New Guinea–New Caledonian line in the south-west to the Hawaiian in the north-east, constituting ripples on the ocean floor. These form the "*Oceanides*" of Suess.

Kober[2] divides the basin into a northern part marked out by island chains striking N.W.–S.E. and a southern one free from islands. Both areas are regarded by him as having been

[1] Gregory (30), cxxvii.   [2] Kober (21), 243.

continents during the Mesozoic, and are supposed to be separated
by a tectonic belt trending N.W.–S.E.

NATURE OF THE OCEAN FLOOR

Continental kinds of rocks build up the frame as well as
most of the larger islands in the west, but in the whole central
Pacific the islands are either coral or volcanic, and not a fragment of any granitic rock has been found in them, the area thus
occupied being stated to cover about one-fourth of the surface
of the earth. Among the lavas, acid types such as rhyolites and
trachytes are occasionally represented, but basaltic types are
dominant, the widely-spread olivine-rich kinds being appropriately known as " oceanite." The average density, based
on 72 examples scattered over the Pacific, is just over 3·0
(Washington),[1] such greatly exceeding the corresponding figure
for any other land or oceanic region.

Admitting the isostatic principle, the great depth of water
would imply a rather dense floor, a deduction in accord with
the generally positive nature of the gravity anomalies over the
basin. One finds it often implied or else stated that the Upper
Layer must be absent and that the floor must be composed of
basalt or, more vaguely, of Sima. Such is scarcely borne out
by seismographic records, which, on the contrary, could readily
accord with a " granitic " layer up to about 10 km. thick
(Stoneley, Jeffreys, Baird), though it might perhaps be thinner
in places. From among a conflict of statements emerges the
important conclusion that the surface " Love waves " travel
fastest beneath the Pacific. A denser medium or greater
elastic constants are accordingly suggested, though lower
temperatures as compared with the land at equivalent depths
would confer a higher rigidity, as Gregory has pointed out.
Gutenberg's views (36) on the subject are worthy of attention.

PERMANENCE OF THE BASIN

It would be unprofitable to discuss those speculations by
Osmond Fisher, G. H. Darwin, H. T. Pickering and others that
have attributed the primitive Pacific to the tearing-off of crustal
matter to form the Moon during the remote past, but the suggestion that that catastrophe could have happened so late as the
Tertiary as advocated by H. B. Baker (32), or even during the
Permian as fancied by H. Nissen (34), is contradicted by
stratigraphical evidence. While some have seen in it a basin
formed during the Earth's cooling, and persisting as such thereafter, others have regarded it as impermanent and have not

[1] Bucher (33), 45.

hesitated to throw across it in various directions and at different times land-bridges up to 17,000 km. in length, for example Frech, Haug, Gregory, Kober, von Ihering, von Huene and others, for which they have the backing of various biologists such as Huxley, Baur, Germain, Pilsbry, Cooke, etc.

Reviewing the great mass of data, Gregory (30) has eloquently argued for repeated and sweeping changes in the distribution of land and sea over most of this vast region from the Cambrian onwards, and for a trans-Pacific connection surviving until the early Tertiary. All the same, the Displacement Hypothesis can interpret otherwise, though quite as convincingly, most of his evidence, and can even explain away certain difficulties encountered by him. Furthermore, it effectively disposes of the stupendous mechanisms implied in those repeated continental connections across that vast and profound hollow.

Suess maintained that the Pacific had existed for a long time, and that it had acted as a rigid block or foreland to the encroaching fold-systems by which it became engirdled. He furthermore denied that the pressures responsible for those plications had been directed from the ocean side, though he does not seem to have realised the mechanical inconsistency of his orogenic scheme. The idea of such a superior rigidity is one widely held and is probably in the main correct, though of little real importance in our problem. Conceivably such was due to the lesser thickness of the blanket of heat-generating Sial.

While opinion has generally favoured the view that the Tertiary fold-systems framing the Pacific must have encroached on that basin, it was left to Pickering in 1907 to advance the revolutionary idea that such had arisen through the tearing-open of the Atlantic and Arctic on the opposite side of the earth and the drifting of the thus-severed lands towards this then much larger ocean, in which belief he has been closely followed by Taylor, Argand, Holmes, du Toit, Baker and Gutenberg.

EVOLUTION UNDER THE DISPLACEMENT HYPOTHESIS

The author conceives the Pacific during the Palæozoic as a single ocean or " panthalassa " with about twice its present area though about the same mean depth ; arms therefrom penetrated into the several geosynclines of the time. One can hardly doubt that the great continental fold-systems of the Palæozoic and Mesozoic, which are traceable right to its margins of the time, did not just end there, but continued toward its centre for unknown distances as peninsulas and island-chains, to be subsequently buckled by the younger foldings, though some may still be preserved in the present floor. Throughout history the part played by the basin itself has been rather a passive one, and the naming of this great body of water has been appropriate therefore.

Up to the mid-Mesozoic, marine faunal migration must be regarded as rather following the coastal shelves of Laurasia and Gondwana than crossing this enormous ocean, for which a mean depth of some 3000 m. may be conjectured. Thereafter, as the Atlantic, Arctic and Indian rifts widened, they became filled from, and at the expense of, the correspondingly contracting Pacific.

Clearly written in the rocks is the centripetal drift of the crustal blocks towards that great hollow, shrinking in size with each invasion of its margins. Striking, too, is the rhythmic nature of those incursions from the late Jurassic onwards, culminating in the three migrations of the Cretaceo-Eocene, mid-Tertiary and late Tertiary, almost in unison around the frame, coupled with rotation and some distortion of the masses as they converged upon the focus. Such is particularly well marked in the East and West Indies.

In their regular westerly and almost meridional advance the two Americas kept almost in step, though the northern mass moved further, possibly because of the lesser opposition in that quarter as compared with the wider expanse of shallower oceanic foreland in the south. In the west, on the contrary, the motions of the Asian and Australasian blocks with their extensive advance-folds were directed almost at right angles to each other; wherefore their respective fold-systems not only interfered with but *reinforced one another*, whence the extensive march into the ocean of the crustal waves, thereby leaving the parent continents far in the rear. The mobility of this island-strewn region is finely brought out on a map by the abundance of earthquake epicentres in the section from Kamtchatka round to Fiji.[1]

In the south, Antarctica was unable to keep up with Australia and South America and thus failed to press into the basin, which would explain the apparent stability of the South Pacific and the lack of islands therein.

While the strong asymmetry of the Pacific is due to differing degrees of mobility of the crust on east and west, the same prime agency has unquestionably operated right around its periphery. This is shown not only by its tectonic history, but by its volcanism and metallisation. Thus, in strong contrast to the enormous central *intra-Pacific* region with its basic lavas—largely of alkalic characters (basalts and nepheline-bearing rocks)—is the wide *circum-Pacific* fold-girdle marked off by eruptives of intermediate and calcic types—dominantly

[1] Daly (26), Fig. 8.

andesites. Impressive is the ring of active and extinct volcanoes that constitute the so-called " girdle of fire." E. C. Andrews (25) has emphasised the structural control of ore and oil deposits around the Pacific from the late Mesozoic onwards with the crowding of successive zones upon the central basin, and has stressed its unity of design.

Finally, while many matters still await explanation under the scheme of continental sliding adopted, it is imperative to note that the observed intensity of the " packing " of marine strata in the framework of the Pacific is inexplicable under current views through demanding contraction so great in its amount and so spasmodic in its character as to be most improbable.

THE ATLANTIC AND ARCTIC OCEANS

Form.—While an enormous amount has been written about the Atlantic, the Arctic is far from being well known. It is commonly overlooked that the latter is merely, as Supan has remarked, the continuation of the former, structurally as well as geographically. The combined oceans stretch over more than half the circumference of the earth, are antipodal to the Pacific, have just less than half the area of the latter and possess a mean depth of about 3500 m.

Their framework is much of it old and peneplained, not, on the whole, mountainous and without any true marginal ranges. Suess observed that the shores cut across the grain of the continents nearly everywhere and accordingly introduced the term " Atlantic type " for such a structural peculiarity. In 1907 W. H. Pickering pointed out that, although the width of the basin varied greatly, the opposed shore-lines, while much indented, showed throughout remarkable correspondences that recalled a zigzag fault-pattern. He suggested that, when the Moon separated and the Pacific was formed, the remaining crust was torn in two to make the Old and New Worlds. This is indeed the germ of the Displacement Hypothesis.

Floor of the Basin.—The principal feature of the Arctic section is the great Polar depression exceeding −3500 m. in places, and rudely D-shaped in plan, with a relatively narrow and straight shelf off the American archipelago and Greenland, and a broader curved one off Siberia usually less than −200 m. deep that reaches to Spitzbergen, with Barents Sea ranging between −300 and −400 m. (Fig. 28).

The rise that links Greenland to Scotland *via* Iceland and the Faroes is submerged by less than 600 m. of water, and forms the divide between the Arctic and Atlantic depressions. Through the medium of the narrow Reykjanaes ridge, trending south-

south-west from it, Iceland marks the beginning of the mid-Atlantic Rise, that remarkable feature, which can be followed southwards, save for one short break right on the Equator, to Bouvet Island, a distance of fully 16,000 km. Sharing in the changing trends of the associated coasts, it is nearly everywhere equidistant from them, more particularly if the measurements be made along axes running W. 15° N., while right in the south it turns eastwards between Africa and Antarctica, according to Pratje.

Its undulating crest commonly stands at between −2000 and −3000 m. and supports the volcanic islands of Iceland, Azores, Ascension, Tristan d'Acunha and Bouvet. The two flanking troughs are divisible into subsidiary deeps, descending to over −5000 m. and in a few places to over −6000 m., by rises set more or less obliquely to the central axis (Fig. 27).[1] In plan it bears some resemblance to a crooked, branching tree. The most striking of these lateral rises are the Walvis ridge off the Cape reaching to about 4000 m. above its floor, and the Rio Grande one off southern Brazil.

Origin of the Rise.—The interpretations put upon this prime feature have been many and varied. Haug, the first to say so, regards it as a young geanticline due to compression, in which view he is followed by Washington and Pratje ; Wegener and Windhausen, as the former floor of the rift when the Americas drifted off westwards ; Taylor and Daly, as a crustal strip left behind ; Molengraaff, as the " cicatrix " where the sima was laid bare when the continents moved away in opposite directions; Kober, as the sunken axis of a post-palæozoic fold-system ; Baker, as an arching due to tangential thrust in the sima when under the weight of the ocean the floor was forced to occupy a lower and shorter arc ; Gregory, as a central part of the fractured sunkland that gave the Atlantic ; Stille, as a rising epeirogenic structure ; van Bemmelen, as having been uplifted by lighter igneous differentiates. Staub is alone in denying any unity of plan to this lengthy feature, connecting its northern section with the trans-Atlantic Alpine folding, while leaving the southern half more or less unaccounted for.

Each and all of these many-sided explanations presents difficulties of greater or less degree. That normal compressive forces could have been responsible is scarcely favoured by the strong evidence for tension around the basin and, as Stille has pointed out, by the fact that in troughs any folding develops not in the centre but along one of the sides. Its apparent youthfulness, as emphasised by Gregory, Baker and Stille, is

---

[1] A photograph of the striking relief model of the South Atlantic basin based on the surveys of the *Meteor* will be found reproduced in Kober's *Die Orogentheorie* (Fig. 32).

adverse to the hypotheses of Taylor, Daly, Molengraaff and
Kober. Movement of the continents toward it on either side is
negatived by the lack of young fold-ranges on the western

FIG. 27.—The Bottom of the Atlantic Ocean (after Pratje and
Supan-Obst); scale 1:120,000,000.

border of Africa. Lastly, an identical feature is present in the
Indian Ocean running southwards from India to Antarctica
(Supan-Obst).
The question is hence raised whether such a medial ridge

might not develop normally within any large tension-basin as the consequence of sub-crustal adjustment, and that, too, whether one or both abutments moved. While stretching proceeded, the floor of the trough would be more or less even, but, as drifting ceased, adjustments would take place, so that warping would occur. Sedimentation—and consequently depression—would be greatest nearest the lands, whereas only oozes would accumulate over the central zone. Up-bulging would therefore tend to occur at some point in the mid-section. Once a faint arch had been established the conditions would be favourable for further disturbance and for volcanicity along its axis, thus accentuating the structure, and incidentally loading—and therefore sinking—the section on either side. Furthermore, such a structure need not be rigidly fixed, but could migrate horizontally, within certain limits at least.

In the case of the Atlantic the area drained is nearly four times that of the Pacific with twice its size, while it is suggestive that practically each of the deeps on either side of the central rise has one or more large rivers discharging waste upon its floor. Since the last Andean phase was characterised more by uplift than compression, the evolution of the mid-Atlantic Rise could have dated back to the Pliocene at least. If the explanation proposed be correct, the arch may still be slowly rising and moving westwards.

The volcanoes perched thereon from Iceland to Bouvet Island are situated mostly where lateral rises join the main axis and possibly stand on lines of fracture or shear, for, as Brouwer has remarked, some differential horizontal movement is doubtless in progress. The only spot where the central rise is broken right through is at the small but surprisingly deep Romanche sink, −7370 m., and it is significant that the detailed soundings recorded by Böhnecke reveal contours indicating an anti-clockwise vortex.[1] This spot lies, furthermore, just where, in conformity with the changing trends of the African and Brazilian coasts, a sharp bend takes place towards the west-north-west, this zone being also one of high seismicity. Such secondary movements could have brought up strips of crustal rocks to near or above sea-level, as in the case of Ascension Island (with granite fragments in its agglomerates) or St. Paul's Rocks.

The Tertiary Brazil-Guinea Land-bridge postulated by Bailey Willis will be discussed later.

It cannot be doubted that the Upper Layer of the crust, though thinner under the continents, is well developed beneath this ocean, as indicated, first, by the analysis of records of earthquake waves, as stressed by Gutenberg (36), and, secondly, by the absence of unduly large positive gravity anomalies.

[1] Pratje (28), Fig. 5.

From the viewpoint adopted the value of gravity over the central rise should not be very different from that found elsewhere in this ocean, which is the case along the line run by Meinesz from the Canaries to Haiti, though any large values could be explained as due to a veneer of basic matter capping the presumed lighter rocks of the rise. According to Daly the basaltic core of St. Helena is a heavy load on the earth's crust that in large part does not seem to be isostatically compensated, and the same may be true of Ascension Island.

The extraordinary pattern of the Atlantic-Arctic floor, i.e. its zigzag " stem " and its " branches " that reach out in various directions to the continental frame, is highly suggestive of E.–W. stretching. It shows close analogy with a sheet of the so-called " expanded metal " made by cutting parallel slots in a plate and then stretching it at right angles thereto so as to produce a trellis with diamond-shaped openings, only the Atlantic pattern is far less regular. Certain of these submerged branches are conceivably strips not completely torn away from the adjacent lands, as for instance the lengthy Walvis Rise, since between Walvis Bay and the Cape the dominant fold-system runs nearly parallel to the coast-line.

Staub [1] sees preserved upon the floor of the North Atlantic terrestrially evolved Alpine structures forming a series of strongly curved ranges—mostly convex to the south—with associated basins, all dependent upon Newfoundland–Venezuela on the one side and Spain–Morocco on the other—the manifestations of a sunken continental segment. The writer, on the contrary, interprets these rises and sinks as stretched, distorted and broken wrinklings of the crust evolved under the simultaneous drawing-away of the Americas from Eurafrica and the approaching of South America and Africa towards North America and Europe respectively. Under such combination of E.–W. tension, N.–S. compression and up-welling of the sima in between as the oceanic basin developed, complicated arcuate forms and incipient vortices could naturally be expected to result.

EVOLUTION OF THE ATLANTIC-ARCTIC TROUGH

It is generally conceded that this compound ocean could not have existed during the Palæozoic with anything like its present shape (Chapter VII). The occupation of the southern part of the North Atlantic region by a small ocean, called by Schuchert " Poseidon," is usually admitted and was accepted by Wegener [2] as the locus from which the rifting started. Possibly it consisted of two opposed gulfs leading out from the ancient Tethys,

[1] Staub (28), 128-31, Fig. 27.   [2] Wegener (24), Fig. 1.

which, during the Mesozoic, penetrated respectively north-north-eastwards between North America and Europe and south-eastwards between South America and Africa with progressive expansion during the Cretaceo-Tertiary.

The southern split proceeded southwards through Gondwana as far as Bouvet Island, where it met another great rift-system entering from the east, and by the close of the Cretaceous the Afro-Arabian mass had become detached from the rest of Gondwana. A second offshoot from the Tethys, propagated southwards across the Sahara and Nigeria to Angola—in which Mediterranean faunas were represented—closed up after the Eocene through having become short-circuited by the main westerly rift. An identical arrangement applies to the north also, where the mesozoic geosyncline of the Urals, taking off from the Tethys and running northwards into the Arctic, became short-circuited by the more direct rift that extended polewards from the North Atlantic.

At no time seemingly was the Arctic a stable region, for during nearly every epoch it was crossed by some synclinal depression in one direction or another, for the most part between the central margin of Siberia and either North America or Greenland, with wide over-spillings of the seas at certain dates and floodings by basalts during the Cretaceo-Tertiary. Grabau's palæographical maps (23-4 ; 28) bring this out well. Save in Spitzbergen practically all folding just within its confines was pre-Tertiary and most of it Palæozoic (Fig. 28).

By means of the two nearly straight lines *a–b–c* and *d–b–e* (Fig. 28), intersecting at an angle of about 75°, drawn from the Lofoten Islands to Amundsen Strait and from western Iceland to Cape Chelyuskin respectively, the present Arctic can be divided into *four contrasted quadrants*. The diamond-shaped North Atlantic trough then lies opposed to the D-shaped Arctic basin, and the elevated sector of Canada–Greenland to the slightly depressed Barents Sea " shelf." These two lines, which are traceable with surprising regularity across the sea floor, clearly mark out the courses of crustal fracturing and displacement, the latter mainly in a horizontal direction. Under our hypothesis the point *b* coincided formerly with *a*, and *e* with *b*. Further consideration brings out that such a particular pattern could have originated only in one way, namely, by the drifting of Greenland and America away from Scandinavia in a direction parallel to *a–b–c*, while *simultaneously* the Alaska - eastern Siberia sector swung outwards from *b* to *e* round a hinge situated in eastern Alaska. Stages in such sliding are shown schematically in Fig. 29. It will be observed that the distance *a–b–c* is only slightly greater than that between *c–e* when measured along the Siberian shelf, which discrepancy can be

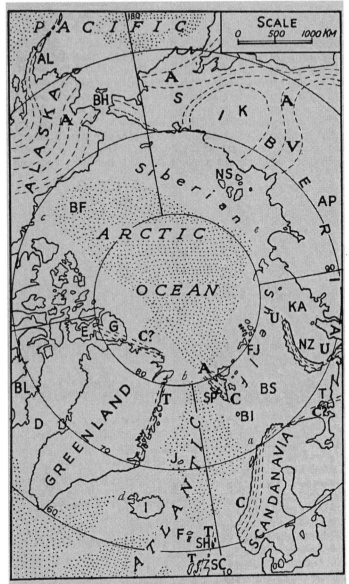

FIG. 28.—Map of the Arctic Region, showing Disjunctive Basins and Tectonic Zones. *Stippled*: Ocean exceeding 2000 metres in depth (based on the map by Nansen). *Tectonic zones*: *T*, Taconian; *C*, Caledonian; *U*, Ural; *V*, Verkhoiansk; *A*, Alpine.

AP, Angara Platform; *AL*, Aleutian Peninsula; *BF*, Beaufort Sea; *BH*, Behring Strait; *BI*, Bear Island; *BL*, Baffin Land; *BS*, Barents Sea; *D*, Davis Strait; *E*, Ellesmere Land; *F*, Faroe Islands; *FJ*, Franz Josef Land; *G*, Grant Land; *I*, Iceland; *J*, Jan Mayen Island; *K*, Kolyma platform; *KA*, Kara Sea; *NS*, New Siberia Islands; *NZ*, Novaya Zemlya; *SC*, Scotland; *SH*, Shetland Islands; *SP*, Spitzbergen; *T*, Timan.

ascribed to the minor rifting within the American archipelago
on the one side and to the compression within the Verkhoiansk
arc on the other.

The distance *b–e* being some 2000 km., a displacement along
it of such a magnitude ought obviously to stand revealed in
structural features within the continent. Such is indeed evinced
by the intense distortion of Asia, the vast width of the region
of Alpine folding, the peculiar V-shape of the structural lines
right through the centre of the continent from north to south,
the bending towards the east of Novaya Zemlya, the strong

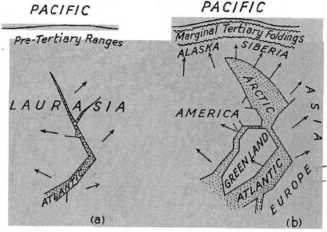

FIG. 29.—Two stages in the hypothetical evolution of the
Atlantic-Arctic Rift.

curving of the younger arcs in eastern Siberia and Alaska, the
lengthy belts of rifting next the Lena and Yennesei rivers, etc.,
all of which accords with such deduced opening of the disjunctive basin of the Arctic. When restoring the pre-Tertiary
Laurasia, a certain amount of remoulding is hence demanded,
a point which should be allowed for by critics.

The above is admittedly just the scheme so ably and
graphically formulated by Taylor [1] long ago, though with several
modifications, chiefly in a less uniformly radial flowage of the
crust than as indicated by him.

If, as under our restoration, the Arctic basin be closed up—
the edge of the American archipelago being brought against
that of the Siberian shelf—many remarkable structural relationships will find their interpretation. We then discover the Ural
geosyncline of the Mesozoic crossing the north of Siberia and

[1] Taylor (10), Fig. 7 ; (28), Fig. 29.

meeting at right angles the great Jurassic land-barrier of
Alaska and eastern Siberia which cuts it off from the Pacific
basin, along which it was forced to spread into British Columbia
on the one hand and the Okhotsk region on the other. We
observe its pulsating expansion during the Jurassic and its
boreal Russian faunas. We note its further enlargement during
the Lower Cretaceous with sinking of the Alaskan-Siberian
barrier until connection is established between the Pacific and
the Arctic trough, and American faunas appear in the latter.
During the Tertiary this barrier is restored and maintained.
Furthermore, each orogenic phase of the circum-Pacific fold-
ranges from the Jurassic onwards finds its expression in the
intermittent stretching of the Atlantic-Arctic basin and the
corresponding creep of North America and Eurasia in opposite
directions towards the Pacific. Not, however, until late times
was the Arctic rift able to penetrate this " compression-girdle "
of the Pacific. Striking, above all, is the *zigzag pattern* of the
original split that ran from Bouvet Island northwards to Alaska.

TENSION AND VOLCANISM

With such mighty displacements of the land-masses under
stretching it would indeed be surprising if in places some
rotation had not also occurred together with some local squeezing. Thus can be explained the minor disturbances in Angola
and Nigeria (related to the N.–S. branch from the Tethys), the
Isle of Wight, western part of Spitzbergen, etc.

The dominant feature, however, of this region, as so ably
expounded by J. W. Evans (25), is the widespread faulting that
follows a regular pattern and betrays not only simple tension
but strong torsion. That such should be the case is not at all
surprising, since the Alpine foldings were pushing their way
into Western Europe at the very time when this oceanic basin
must have been widening.

The physiographical reasons for considering most of its
diversified coasts as due to tension-faulting are subsequently
presented (p. 259). Such is in harmony with the views advanced
by Taylor [1] and Wegener [2] that Davis Strait, Baffin Bay and
certain of the waterways within the American archipelago have
essentially been due to the tearing away of North America from
Greenland. Support therefore is given by the complex block-
faulting recorded by Koch in north-western Greenland and
known also in eastern Greenland, Spitzbergen and elsewhere,
as commonly disclosed by fiords.

Volcanic matter has ascended freely along many of these
fractures, and the directions of the ensuing dyke-swarms, which

[1] Taylor (10), Fig. 4 ; (28), Fig. 29.   [2] Wegener (24), 58.

run independently of older structures, accord well with the course of the postulated rifting. The trend is from N.W. to N.N.W. in northern Ireland, northern and western Scotland—finely displayed in the swarms of Arran, Mull and Skye—and the Faroe Islands. In Iceland,[1] where " fissure-eruption " is still prevalent, those fissures as well as the abundant faults strike north-east, parallel to Denmark Strait and the Greenland coast ; in eastern Greenland and in Spitzbergen the trend is mostly north. A post-volcanic torsion is indicated by Gregory for the Outer Hebrides and by Peacock for the Faroes.

Impressive are the vast quantities of basic matter erupted at intervals from the Cretaceous onwards, particularly during the Tertiary, and continued to the present day, from points situated on the mid-Atlantic Rise, e.g. Iceland and Jan Mayen Island. Such plateau basalts characterise northern Ireland, western and northern Scotland, Faroes, Iceland (where they may exceed 4000 m. in thickness), Jan Mayen, eastern and western Greenland, Spitzbergen, Franz Josef Land, Novaya Zemlya, and from the mouth of the Yennesei up that river for an enormous distance. These lavas, of which the northern representatives have been styled the " Thulean basalts " by Washington, are of remarkably uniform chemical composition with alkaline tendencies.

It is commonly supposed that they are merely remnants of a huge lava-field that stretched during the Tertiary from the British Isles to Greenland and beyond, of which the bulk has foundered—a most unlikely thing if the principle of isostasy be admitted. Dispersal by Drifting would not only solve this difficulty, but demand a parent volcanic region having only a fraction of the size currently conceived. It is highly suggestive that the principal occurrences of such basalts—which are accompanied by some rhyolite—fall on a belt stretching northwards from the Irish Sea to eastern Greenland, at both ends of which the lavas, have been penetrated by masses of gabbro, granite and occasionally syenite.

Despite much hostile criticism, the broad association of *alkaline igneous rocks with regions of tension* would seem to hold. This so-called " Atlantic suite " is, however, practically confined to the centre and south of the North Atlantic, though spreading out widely in the South Atlantic and including portions of Brazil and Africa, ending at Gough's Island. The calcic lavas of the Caribbean and Lesser Antilles are, on the contrary, products of the transverse Alpine compression zone.

Brouwer (21) has called attention to the several centres of alkaline plutonics and effusives of Cretaceous and Tertiary age on both sides of the South Atlantic, while Beetz (34) has recorded many new occurrences in Angola and South-west Africa which

[1] Pjeturss (10).

confirm the idea of a regular volcanic belt following a zone of weakness stretching from Cameroons almost to the Cape, corresponding to that of eastern Brazil.[1] Strikingly, on the opposite side of Africa—from Abyssinia almost to Zululand— runs a third belt of alkaline rocks that is intimately connected with the Great Rift system, and which involves Madagascar as well.

The widely accepted criticisms by Washington (23) that the igneous suites down the two sides of the Atlantic show serious differences, greater indeed than their actual resemblances, has been disposed of in Chapter II.

MOVEMENTS DURING RECENT TIMES

In addition to the enormous early Pleistocene and Recent faults in Holland recorded by van der Gracht, crustal unrest is shown by the many earthquake epicentres falling upon the mid-Atlantic Rise, the heavy shocks recorded beneath the lands, e.g. Lisbon and Charleston, and the numerous minor quakes experienced on the eastern side of the United States and in Great Britain, some of which, in Scotland, originated upon older (Caledonian) fractures.

There is, moreover, evidence of actual westerly drift of the crust—Sabine Island on the east coast of Greenland and Jan Mayen Island—as revealed by astronomical observations (Chapter XV).

Coincident with a violent earthquake on November 18th, 1929, a dozen widely spaced telegraph cables were broken off Newfoundland. As fully set forth by Gregory (31), the breaks were arranged on nearly parallel lines and at various depths down to −5365 m., the evidence indicating a subsidence of the sea floor with intense disturbance of the bottom between two presumed faults along the sides of the so-called " Cabot Trench," the submarine continuation of the Cabot Strait and the St. Lawrence River.

This introduces the problem of the extraordinary submarine ravines—the extensions of existing rivers—off the coasts of North America (Bahamas, Hudson, Maine and Cabot canyons), Western Europe (" English Channel River," Garonne, Adour, Douro and Tagus), West Africa (Congo), Cape, Brazil, India and elsewhere. Although showing all the characteristics of sub-aerial river-erosion, they are traceable as winding and often steep-walled troughs sloping continuously down to depths of over −3000 m. in certain cases.

---

[1] Even Leme (29), an opponent of Drift, has admitted that the mesozoic lavas and the post-Cretaceous basalts and alkaline rocks on the two sides of the ocean correspond quite well.

Q

Since Hull advanced the view that an enormous depression of the lands must have occurred quite recently, there has been a reaction in favour of a certain amount of trough-faulting (Wegener, Gregory, Steers,[1] etc.), to which the Cabot disturbances would give colour. It can, however, be suggested that the strips of sea floor bordering the Atlantic coasts characterised by such gorges are merely marginal parts of the continents that have become depressed through down-warping and down-faulting during the stretching of the Atlantic rift-basin, and that they constitute unstable fringes in process of being torn off from the continental blocks as that basin continues to deepen and widen. These gorges would in short represent nicks sliced by the rivers through the fault-line scarps before and while the down-dragging in question took place.

Sedimentation just beyond the continental shelf is tending to sink the sloping floor, while erosion is causing the land to rise; hence the shelf region is developing on the whole an oceanward tilt, and occasional slumping could be expected in such positions. The evidence is particularly clear in South Africa. It is significant that these canyons, which should not be confused with true fiords, should be so finely developed on either side of an ocean that is regarded as disjunctive.

Since these lines were penned there has become available the instructive work on Angola by Jessen [2] in which, with the help of several convincing diagrams, the subaerial and submarine erosion-surfaces are correlated, a down-warping of the continental margin demonstrated and the drowned gorge of the Congo explained.

One cannot close without alluding to the mythical " Atlantis " with the recommendation that the reader should study that attractive volume by H. E. Forrest, *The Atlantean Continent* (2nd ed.), with its striking evidence of surprising geographical changes since the beginning of the Pleistocene epoch.

THE INDIAN OCEAN

Floor.—With slightly less than the area of the Atlantic this basin has a mean depth of about −4000 m. with maximum of −7450 m. in the Sunda or Java foredeep.

It is like the Atlantic through the presence of a medial rise that runs northwards from the Gauss Berg in Antarctica through Heard, Kerguelen, St. Paul, New Amsterdam, Chagos, Maldive and Laccadive Islands to the western side of India, which in similar fashion gives off branches to the Cape (*via* the Crozets and Agulhas Bank), Seychelles, Somaliland (*via* the Carlsberg ridge and Socotra) and possibly Tasmania. The

[1] See Steers (32), 53-7.   [2] Jessen (36), 325-58, Figs. 93-6.

floor is thus divisible into an eastern and a western portion, and those again into subsidiary deeps.

Madagascar stands on a deep shelf prolonged somewhat to the south, which is connected with the African mass, and has a crystalline basement. The curious curving Mascarene ridge to the east reaches above the sea in the granite of the Seychelles in the north, and the volcanic islands of Mauritius and Reunion in the south. The Carlsberg Rise, which divides the Arabian Sea into two portions, has been found by recent surveys [1] to be narrow at its north-western end but to expand towards the Chagos group with a subsidiary rise between it and the Seychelles, the depth of water above these features ranging from about – 2200 m. to – 3000 m. The depression on the east side sinks to close upon – 5000 m. ; that on the west to over – 5000 m. The Chagos-Maldive-Laccadive Rise ascends from over – 4500 m., its crest undulating considerably and bearing three clusters of coral islands parted from one another by wide channels from – 2000 to over – 3000 m. deep. Nowhere are the basement rocks seen.

ORIGIN OF THE OCEAN

Except where bounded by the curving Burma-Malay arc, its margin is made by fractured continental blocks. The hypothesis advanced to explain the profiles of the Atlantic can be applied to the central and southern parts of the Indian Ocean at least and to its medial rise. It is highly significant that the various volcanoes situated between Antarctica and the Equator have erupted lavas of dominantly basic composition but alkaline tendencies, recalling Atlantic and East African types ; those of the Gauss Berg include rare kinds characteristic of Kenya. Mounts Erebus and Terror in the Ross Sea are situated upon Tertiary faults within a gigantic rift-system reminiscent of East Africa (Fig. 35).

Evidence for crustal fracturing around and beneath the Arabian Sea is strong. The Gulf of Aden is universally admitted to be a rift valley, and recent work [2] has proved a series of parallel ridges on its floor in its central and northern parts that suggest the edges of fault-blocks. Reasons have already been submitted for regarding Madagascar as having been displaced towards the south. The plan of the four great topographical elements between East Africa and India forms roughly a sprawling W, the inner part of which has been deeply submerged. If India were to be pressed back against Africa, the four arms would fold together neatly and make a practically continuous elevated mass.

[1] Sewell (34 ; 34a).   [2] Sewell (34), 86.

All this goes to support the hypothesis that these elements represent sub-parallel slices of Gondwana that have become detached and rotated, without complete severance, when India moved away from Africa. Indeed, another great curving land segment, reaching from Somaliland to the Zambezi, has been actually severed by a chain of riftings from the rest of Africa and is on the point of following in the wake of the others.

Despite Bucher's [1] objections, the Laccadive-Chagos Rise with a length of some 2500 km. is interpreted, in agreement with Staub, as essentially a stretched and broken slice of " Indo-Africa." The rise from the Chagos Islands southwards to Antarctica is, on the contrary, regarded as a recent epeirogenic uplift, the counterpart in fact of the mid-Atlantic feature, and largely due to the relaxation of tension.

On the north-east the disjunctive basin of the Indian Ocean has been invaded by the marginal foldings of Asia, and that region not only differs structurally from the rest of the framework, but is characterised by young eruptive rocks of andesitic composition and typically circum-Pacific affinities.

LAND BRIDGES

The numerous alliances between the geology and former life of Africa, Madagascar and India have long been recognised, which indeed led Blanford, Suess and others to postulate more or less continuous land between these masses during the past. Such has since been narrowed down by Bailey Willis (32) [2] to a tortuous isthmus extending from the northern end of Madagascar to Seychelles, southwards almost to Mauritius and thence *via* the Chagos Rise to India, though such a route would involve a number of lengthy stretches covered by from 3000 to over 4000 m. of water today. It fails absolutely to account for the definitely established resemblances of the faunas of Madagascar with those of South America as against the lack of close affinities after the Lower Cretaceous with those of East Africa, not to mention various peculiarities in the associations with India and Australia (p. 124). It can be doubted whether anyone would have the courage to propose a second isthmian " ribbon " joining Madagascar to Patagonia, yet just skirting South Africa !

The Displacement Hypothesis is, on the contrary, competent to explain these and other puzzles of biological distribution, particularly in this great Austral region, in a simple and logical manner and without the violation of isostatic principles. *Current views of Continental Linking must therefore be firmly rejected.*

[1] Bucher (33), 96-8.   [2] Willis (32), Pl. 27.

CHAPTER XI

THE PARAMORPHIC ZONE AND ITS IMPORT

INTRODUCTION

CRUSTAL movement being closely dependent upon the nature and behaviour of the deeper-seated matter, the physical properties pertaining to the sub-crust become of paramount importance. In depth, increases must take place in temperature, pressure, density and basicity, while current opinion favours the conception of an outer zone (*sial*) resting upon a weaker sub-stratum (*sima*) in approximate isostatic equilibrium.

The majority of persons believe that, as the vertical stresses set up by loading or unloading exceed the limits of crustal strength, so the balance is restored through a lateral transfer of deep-seated matter. Down to the theoretical depth known as the "level of compensation"—or the "isopiestic layer" as Washington would term it—equivalent columns should be in approximate hydrostatic equilibrium and accordingly possess roughly similar masses.

In such a scheme the rise or fall of the surface would obviously be determined by the quantity of matter added to or subtracted from the base of a column. Thus, if one metre of rock of density s were removed from or added to the top of a column, the corresponding rise or fall of the surface of the latter would be s/S m., where S is the density of the material at the level of compensation. Assuming s as 2·7 and S as 3·5, the change in height produced would be $\pm$0·8 m. In the case of the ocean, the equivalent figure would be $\pm$0·3 m. This, however, presupposes, as is invariably done, that the physical state of the matter involved has not altered appreciably during this process. While we can well ignore the minute changes arising from elasticity, we cannot strictly assume that in the depths of the crust the physical conditions must have remained unchanged. On the contrary, cogent reasons will be advanced suggesting that, *in response to such variations in loading, related changes in the volume of the crustal column can and will result*, having such a magnitude as to affect materially the calculations got under current views.

Some doubts in such a direction have indeed been voiced by Barrell, T. C. Chamberlin and others, but the principle here involved does not seem to have been stressed before. It is nevertheless believed competent to explain some of the phenomena attending diastrophism and isostasy, such as the :

(*a*) Great maximal thickness of orogenic sediments ;
(*b*) Height of the lands and the depth of the oceans ;
(*c*) Hypsographic curve ;
(*d*) Crustal recovery after melting of ice-caps ;
(*e*) Immense height attainable by mountains ;
(*f*) Rift valleys ;
(*g*) Atlantic and Pacific types of coasts ;
(*h*) Central Atlantic ridge ;
(*i*) Peculiarities in the distribution of gravity anomalies ;
(*j*) Varying depths of the earth's interior shells as calculated from earthquake paths ; and
(*k*) Abnormalities in geothermal gradients.

Because of the above, and since it would seem to furnish a clue to the elusive mechanism of continental sliding, this principle is developed at some length here. Incidentally, it would appear to cut at the basis of many mathematical calculations concerning the Earth and its crust.

PARAMORPHIC ZONE

The concept of a very deep-seated—" infra-plutonic "— zone characterised by garnet and other dense minerals was clearly enunciated so far back as 1913 by L. L. Fermor (13 ; 14). Casual references thereto can be found in petrological literature, but the idea has nowhere been developed with all its implications, though it has been used by Holmes [1] to considerable advantage in several of his papers and was accepted by Evans.[2]

This is nothing less than the well-known and accepted principle of the formation under increasing pressure of *minerals of smaller specific volume* — Le Chatelier's Law — an action which is theoretically *reversible*. In its simplest form the hypothesis assumes that *within parts of the sub-crust variations in pressure will be attended by definite, related changes in the mineral composition of the rocks, the chemical composition of the latter remaining nevertheless unchanged*. Increases in

[1] Holmes (26 ; 28-29).   [2] Evans (25), cviii-cx.

pressure will promote molecular rearrangements and the developing of more, or of new, minerals having smaller volumes, while
decreases in pressure will have the opposite effect, so that the
crustal column will respectively shorten or lengthen.   The transformation postulated is indeed that known in mineralogy as
" paramorphism " and the region wherein it is considered to be
operative can hence be called the *Paramorphic Zone*, though
that particular term· is not entirely free from objection.
Alternatively, such can perhaps be styled the *Intermediate
Layer*, which has the advantage of being non-committal,
though it is strictly the deeper-seated part of the latter that is
regarded as being particularly involved.

Since the mass of the crustal column will remain practically
unchanged (though the position of its centre of gravity may be
shifted), the movement of its upper surface will become *additive
to* that due to lateral flow in the zone of compensation following
such loading or unloading.   To the normal isostatic displacement must therefore be added the paramorphic effect ; in other
words, the actual change in elevation of the surface of the
column, whether positive or negative, will be *greater than* that
calculated in the usual way.   Furthermore, a crust endowed
with such unique properties will be found to behave in a very
different fashion from that of current conception. Though the
mechanical analogy is not perfect, we can picture the lower part
of the crust with its paramorphic zone, situated between the
quite-normal upper crust and the basal sima, as comparable
with a sheet of sponge rubber sandwiched between two layers
of solid rubber.   The response of this compound mass to loading
and unloading would be immediate, perfect and considerable.

The essential difference between the above and Fermor's
view is that he visualises particularly the high temperature at
which such mineral rearrangements would take place within
the infra-plutonic shell, and stresses the liability of the " condensed " material to melt under relief of pressure.   The author,
on the contrary, while recognising that such might, and probably does, occur in particular cases, especially during orogenic
periods, feels inclined to regard the materials of the paramorphic
zone, although hot, as *normally crystalline and solid.*

THE CHARACTERS OF SMALL-VOLUME MINERALS

Such are the garnet, olivine and pyroxene groups, together
with kyanite, sillimanite, corundum, spinel, zircon, rutile,
ilmenite, magnetite, chromite, pyrrhotite and diamond. Holmes,[1]
extending the observations of W. H. Goodchild, has emphasised
the significance of the " formula-volumes " (= formula – weight

[1] Holmes (30), 58.

÷specific gravity) of various minerals, particularly in relation to isomorphism and polymorphism and to crystallisation under varying pressure. Thus the following have similar formula-volumes: pyrope (pyralspite of Winchell), almandine and spessartite; enstatite and hypersthene; diopside and hedenbergite (nearly); the members of the geikielite-ilmenite-crichtonite family; pleonaste and spinel; and picotite and magnetite.

In forsterite and the pyroxenes there has been a contraction over the sum of their constituent oxides in *their* contracted forms. The crystallisation as augite of a melt derived from a mixture of forsterite and anorthite would involve a contraction of $13\frac{1}{2}$ per cent. Entry into solid solution of soda and alumina in the jadeite molecule would produce a still larger shrinkage. Yet the stability of polymorphic minerals does not invariably follow density, as seen in kyanite, currently regarded as a stress mineral, though such a result is not in conflict with other evidence under our hypothesis.

CONDENSATION OF ROCK MATTER

Under pressure highly magnesian types would yield forsterite- or olivine-rocks of specific gravity about 3·2–3·35; if containing a little more silica, then enstatite would form; if some lime, then diopside, chrome-diopside and omphacite; if some soda, then a proportion of the jadeite molecule in solid solution with the pyroxene; if some potash, then phlogopite. Most basic matter contains alumina, which under normal circumstances is locked up in the plagioclase, hornblende or biotite. In depth those minerals would become dissociated and the alumina embodied in garnet, or, if still in excess, separated as sillimanite, kyanite, spinel or corundum. Red garnet (pyralspite of Winchell), being a solid solution, could form from materials of widely differing composition. Chromium and nickel would enter the femic minerals, while titanium, carbon, zirconium and sulphur would respectively crystallise as rutile, diamond, zircon and pyrrhotite, all highly dense and stable substances. Holmes [1] has depicted the transformation of gabbro, dolerite or basalt through transitional types, namely, uralitised gabbro or amphibolite, epidote-hornblende-schist, chlorite-epidote-schist and glaucophane-epidote-schist with shrinkage at each stage to give eclogite ultimately, with a total reduction in volume of 20 per cent. Two unpublished analyses of an amphibolite and an eclogite from a kimberlite pipe prove almost identical, while numerous other instances could be quoted. Joly and Poole found that on being melted, eclogite expanded almost one-fourth with a corresponding decrease in

[1] Holmes (30), 66.

density of 20 per cent as compared with the 4 to 8 per cent in the case of most other kinds of rocks.. Glimpses of the working of the paramorphic process are furnished by residual structures, sieve structures, pseudomorphs (such as garnet after chlorite), etc., particularly among the granulites and eclogites. All this strongly supports the hypothesis of *loading metamorphism*.

EXPANSION OF ROCK MATTER

The opposite action, that of *unloading or de-loading metamorphism*, is evinced by the reappearance of larger-volume minerals, the development of " reaction-rims " of hornblende, mica or minute picotites around garnets, the conversion of rutile and ilmenite into sphene and perofskite, and particularly by ex-solution phenomena, ranging from schillerisation through interlamination of felspar in pyroxene (in kyanite-granulite) to curious " myrmekites " and " eutectics," which resemble in some ways those arising in alloys. In the splitting-up of solid solutions, if the drop of temperature be too rapid for the new phase to develop along the boundaries of the grains, a part, or the whole of it, is constrained to separate out along crystallographic-, cleavage- or twinning-planes within the grains (van der Veen), as in the titaniferous magnetites. Striking in the diopside of certain of the kimberlite eclogites are the oriented rods or plates of garnet and prisms of sillimanite, and also the needles of rutile in the olivine of the associated peridotites. In more advanced phases there has been amphibolitisation of pyroxene, chloritisation of garnet, marginal alteration of kyanite and so on, much of which may have been due to the release of occluded gases.

As Harker [1] has remarked, the ebb of metamorphism has tended to leave minerals in turn in the reverse order to that in which they had developed. This is known as *retrograde metamorphism*. Such can only as a rule be studied in the rare samples brought up in the cores of anticlines, in much the same way as lumps in the process of plutonic assimilation have been preserved as " frozen-in " xenoliths along the contacts of batholithic bodies. A rapid fall in pressure or temperature might cause shattering of crystals, as is finely seen in the holocrystalline inclusions floated up from immense depths by the kimberlite magma. Indeed, it is probably because of their rapid elevation, under which the blocks have not had sufficient opportunity to revert to their low-pressure equivalents, that we are able to secure almost unaltered specimens of the directly inaccessible sub-crust.

From all such varied evidence, which could be amplified

[1] Harker (32), 353.

considerably, the *reversibility of the paramorphic cycle* can be established, though the response to pressure is apparently not immediate but shows a time-lag or hysteresis that must depend upon various factors, in particular the geothermal gradient. The postulated mineral transformation can in a way be compared with the annealing effect in metals and alloys, while at the same time recalling that curious property possessed by ice—regelation. Within the paramorphic zone deformation could be expected to take place without mechanical shear or fracture of the constituent grains, though an oriented or banded mineral structure could well result therefrom. In the higher and cooler zones the influence of shearing would become of greater importance and stress minerals would then be developed, such as kyanite, anthophyllite, glaucophane and mica.

We have some grounds for concluding, therefore, that unloading would tend to convert a part of such eclogite back into the diorite-gabbro-amphibolite facies.

THE RÔLE OF THE KIMBERLITE PIPES

These remarkable " chimneys " have provided a marvellous and comprehensive sampling of the crust from the very surface down to levels which are conjectured even to underlie the hypothetical paramorphic zone. One cannot over-estimate the value and significance, in problems relating to the interior of the earth, of the abundant inclusions which have thus been floated up, such as was first stressed by P. A. Wagner (1928). The wealth of new material described fully by A. F. Williams (1932) has not only confirmed Wagner's main contentions, but provided new information of outstanding importance in the interpretation of the constitution and reactions of the sub-crust. The pipes and associated " fissures " (dykes), of which hundreds are already known, and which appear to be of late Cretaceous age, are widely scattered over the southern half of Africa, being significantly all but confined to the interior region that has remained unaffected save by gentle vertical movements since the Silurian at least.

The proportion of foreign matter is so high in most kimberlite as to make it a breccia, for which the abnormally basic composition (and hence density) of the eruptive magma has obviously been responsible. The " frozen-in " samples can be roughly arranged according to their probable depths, largely, though not entirely, by their respective densities, namely :

(1) Sedimentaries or volcanics, referable to exposed formations ;

(2) Granite, gneiss, diabase, gabbro, hornblende-schist and amphibolite—sp. gr. 2·7–3·0 ;

(3) Garnetiferous amphibolite, pyroxene - granulite (often carrying garnet and sometimes kyanite) and eclogite—sp. gr. 2·9–3·6 ; and

(4) Pyroxenite and olivine-peridotite (often carrying garnet, chrome-diopside, ilmenite and phlogopite)—sp. gr. 2·9–3·3.

Numerous detailed descriptions, some accompanied by photo-micrographs and chemical analyses, have been given by Wagner, Williams and du Toit, to which reference should be made.

Group (2) is merely the universal Archæan basement (pierced by deep-seated intrusions of varying ages) passing downwards without any sharp break seemingly into Group (3), viewed as the transformed and condensed representative of materials in part like (2), resting, in turn, upon those of Group (4), which are dominantly olivinic and which, presumably, extend down to an immense depth with an ever-increasing basicity and density due partly to pressure and partly to higher proportions of Fe, Ti, Ni and Cr in the ferro-magnesian silicates, oxides and sulphides. With such deeper-seated zones we are not, however, concerned.

GRANULITE-ECLOGITE GROUP

A fine suite is available giving a full sequence from amphibolite through granulite to the various eclogitic types composed essentially of green monoclinic pyroxene and red garnet, though commonly carrying in lesser proportions one or more of the following minerals: brown and yellow mica, chrome-diopside, diopside (including some jadeite), corundum, kyanite, sillimanite, pleonaste, rutile, ilmenite, graphite and *diamond*. The garnet is a mixture rich in pyrope but sometimes containing some uvarovite. Certain garnet-corundum-pyroxene and, still rarer, zircon-pyroxene rocks also belong to this remarkable assemblage. Olivine is normally absent, though a few instances are known that are transitional to true peridotite.

The density is always high and indeed may just exceed 3·6, which is due to the abundant garnet, assisted by other small-volume minerals. The only light constituent is phlogopite, which is understandable, since this is the only substance into which the potassium can enter, but even here a second variety (optically abnormal) with a somewhat higher density may be present either surrounding, intergrown with or, less frequently, enclosed in the normal mica. While such inclusions are widespread geographically, they are not abundant, due first to their high density, when compared with the transporting magma, and, secondly, to their extraordinarily tough nature when contrasted with the brittle peridotites. Because of certain peculiarities Beck has proposed to call these types " *griquaite* " rather than eclogite. Chemically, both kinds agree in their relatively high silica (43-51 per cent), variable alumina (7-23 per cent), moderate lime (11-19 per cent) and magnesia (7-18 per cent), but small proportion of alkalies.

Despite A. F. Williams'[1] contention that the "griquaites" are "cognate inclusions" crystallising out of the kimberlite magma, and therefore of Cretaceous age, Holmes[2] and Paneth have secured definite proof from the helium ratios and other lines of evidence that they are not younger than *pre-Cambrian*. This upholds du Toit's view that these eclogites belong to the Earth's primitive shell and represent highly transformed and condensed material, having about the same chemical composition as the rocks presumed to form the lower part of Group (2). Types transitional to the undoubtedly metamorphic felspar-pyroxene-granulites, with or without garnet and kyanite, are actually represented. Wagner, following Eskola, provisionally subdivided these inclusions into igneous and metamorphic eclogites, but in an exhaustive review Williams has shown that the criteria used by the former are inconclusive.

It is only fair to point out that with materials of so peculiar a composition and crystallising at such enormous depths and temperatures, the current distinctions between "igneous" and "metamorphic" rocks could scarcely hold. Indeed, they could perhaps better be termed "*paramorphic rocks*" in the sense that their mineral nature has suffered radical changes. It should nevertheless not be assumed that these eclogites necessarily represent basic materials formerly near or at the Earth's surface that have been carried down into the depths and there transformed. Rather should they be regarded as representing the (normal) lower part of the primitive sial that has *experienced regeneration not once, but repeatedly*, during the revolutions of geological time. The rare examples found containing both olivine and kyanite suggest types transitional to the underlying Group (4).

PYROXENITE-PERIDOTITE GROUP

These range from almost pure pyroxene to pure olivine types, with a long series of intermediate varieties composed of nearly all possible combinations, in varying proportions, of enstatite (including bronzite) diopside, brilliant green and usually chromiferous diopside, phlogopite, garnet and ilmenite and, more rarely, corundum and graphite. The handsome garnetiferous kind known as lherzolite is rather common. The structure is holocrystalline allotriomorphic ; certain kinds show crystal areas several inches across, while in a single block great variations in mineral composition and texture are occasionally to be found.

[1] Williams (32), I, 316-36.   [2] Holmes and Paneth (36), 409.

A rude foliation or mineral banding is not uncommon, emphasised by phlogopite-rich varieties that tend to form streaks or layers in the mica-poor rock, while lenses of ilmenite may be present. Cracking and even shattering of the crystals is common, with displacement or rotation of the crystal-fragments in the case of the olivine, less frequently in the pyroxenes—probably a secondary phenomena due to rapid expansion or " decompression." Striking are certain eutectic intergrowths of enstatite and garnet, diopside and ilmenite and in ruder fashion garnet and ilmenite, while the pyroxenes may be strongly schillerised.

Chemical analyses are rather few, but show silica ranging from 45 to 49 per cent, magnesia from 34 to 41 per cent, oxides of iron from 6 to 8 per cent, the alumina, lime and alkalies being low. Serpentinisation has lowered their density to between 2·9 and 3·2, from perhaps an original value of between 3·1 and 3·4, which is slightly less than that of the eclogite group.

There can be little doubt that these highly magnesian types are also old sub-crustal rocks, despite a rather low helium content. Holmes has pointed out that such could be expected in view of the coarse-grained and commonly shattered nature of the inclusions and their heating in the kimberlite magma, a high antiquity being indeed favoured by other lines of evidence.

KIMBERLITE AND ASSOCIATED ROCKS

Because of their higher magnesia and lower silica content, it can be suggested that these late Cretaceous magmas, which have reached the surface through pipes or dykes, originated within the peridotite zone by a process of liquation and differentiation, the melting being accomplished by radioactive heating, towards which the potassium present would have contributed. The potassic, magnesic, calcic and sodic phases are respectively represented by mica-rich kimberlite, mica-poor kimberlite, melilite-basalt and nepheline-melilite basalt. This suite is characterised by porphyritic olivines, a little idiomorphic pyroxene, phlogopite (normal and abnormal), perofskite, ilmenite and chromite set in a highly basic and, in kimberlite serpentinised and often calcitised, glassy base. The kimberlite varieties contain abundant rock and mineral fragments derived from the eclogite and peridotite shells that have been perforated and must hence have taken their source at enormous depths.

Of significance is the fact that the radioactivity—and hence the rate of heat production—of eclogite is almost equal to that of peridotite, but is only about one-third to one-half that of kimberlite, amphibolite or plateau basalt, and one-twelfth to one-sixth that of granite generally.

CRUSTAL SHELLS

Following Bucher, the term " crust " has been defined as the outermost portion of the Earth, which, on the whole, possesses sufficient strength to offer resistance to deformation and to transmit long - continued stresses within certain limits. It furthermore rests upon a yielding sub-crust, sometimes referred to as the " asthenosphere " (Barrell). Such boundary cannot be permanent since changes of pressure or temperature would bring about a shifting of its position. *The thickness of the crust will therefore vary both in space and time.*

The crust is in turn composed of several approximately concentric shells, though opinions differ widely concerning the nature of the materials composing the deeper ones as well as their depth. Since samples therefrom are almost solely obtainable from the volcanic pipes, weight attaches to the inclusions derived from kimberlite, and hence to the fourfold grouping advocated above.

The *First* or *Upper shell* is really a double one, of which the outermost part is made by the stratified systems together with the oceans, both highly variable as to depth. Although missing over large areas, the sediments involved have been preserved in geosynclines to the thickness occasionally of 20 km., while the oceans exceed 5 km. in depth in places.

The innermost part thereof—the granite-diorite-gabbro-*shell*—can be taken to extend down to between 10 and 25 km., indicated through earthquake waves by the definite surface of discontinuity at between those levels.

The upper limit of the *Second*—the amphibolite-granulite-eclogite—*shell* or the Intermediate layer may in places be vague, but its lower boundary must be fairly sharp to judge from the marked differences in chemical and mineralogical composition of, and the scarcity of transitional types to, the underlying *Third*—the peridotite—*shell* or *Lower layer*.

That boundary can be correlated with the marked seismic discontinuity variously estimated at 30-35 km. (Jeffreys), 39-45 (Gutenberg), 42 (Tillotson), 50 (Matuzawa) and 60 (Mohorovičič). So wide a range is due partly to the methods of computation employed, and partly to actual differences in the depth of the break beneath the territory dealt with. For reasons connected with radioactivity Joly and Holmes favour the lesser values. There is good evidence to show that the upper two " acid " shells are thickest beneath the continents, but are thinner beneath the Atlantic, and more so beneath the Arctic oceans and reach a minimum beneath the Pacific, the floor of which must be composed largely, though not wholly, of basic matter. Deeper still, possibly at from 70 to 100 km. or more, and situated well within the peridotite shell, where temperatures and pressures are high and the ultrabasic matter loses its strength, lies the region of *isostatic compensation* wherein sub-crustal transfer of material takes place. Whether the matter there is crystalline or in the glassy state, as Daly [1] has supposed, does not concern us here.

This weaker phase forms the *Sub-crust* or the *Sima* of current usage, though, to follow Suess' [2] definition (*si*lica + *ma*gnesia), such should rightly include any overlying crystalline and still-resistant peridotite as well, the three outer shells because of their composition (*si*lica + *al*umina) constituting the *Sial* (or Sal).

VOLUME CHANGES IN THE PARAMORPHIC ZONE

The widely differing estimates for the thickness of the Earth's shells do not affect our problem materially. Selecting values on the small side, and assuming that the intermediate amphibolite-eclogite shell ranges on the average from 15 to 35 km. in depth, we obtain a thickness of 20 km. for the zone within which the paramorphic process could be operative. Paramorphism, due to loading or unloading, can be conceived as primarily affecting the *middle* and *upper* parts of the zone—the *lower* part having reached its maximum degree of "condensation"—and thus raising or lowering the boundary between it and the overlying acid shell, which in its basal part at least must be regarded as being made by potentially transformable rock. During the geological processes quite a large amount of material might thus become "condensed" or "expanded"—not, of course, to the theoretical maximum, but to a fraction thereof. The process of loading can be regarded as tending towards "*saturation*," that of unloading towards "*under-saturation*."

Let us assume that 1 km. of rock of density 2·65 is eroded from any area, then its effect on the paramorphic zone could be viewed, providing the response were perfect, as producing a change in the density of the latter proportional to the lightening of the mass due to such unloading. Assuming a mean density for the supporting shell of 3·3, the expansion would be 2·65/3·3 or 0·8 km. Now the additional rise of the surface due to isostatic adjustment through the sima (of density 3·5) would be 2·65/3·5 or 0·75 km. Therefore the absolute change in surface level would be 0·8 + 0·75 − 1·0 or 0·55 km., i.e. the area would actually stand *higher* after such erosion. Under loading the area would, on the contrary, have sunk to *below* its original level. This is shown graphically in Fig. 30. Assuming that a thickness

[1] Daly (33), 177, 180, 188.   [2] Suess (09), iv, 544.

of 15 km. of the Intermediate layer were so involved, the corresponding change in its mean density would amount to $\pm 5\cdot33$ per cent.

Admittedly the above is an extreme view, but important evidence regarding the real magnitude of this effect is indeed furnished by the study of thick sedimentary accumulations. Over and over again one finds instances of formations from one to several thousands of metres thick showing throughout their mass indisputable evidences of shallow-water (estuarine or marine) conditions, or what can alternatively be termed "constant-depth" phases. Current geology has not yet been able

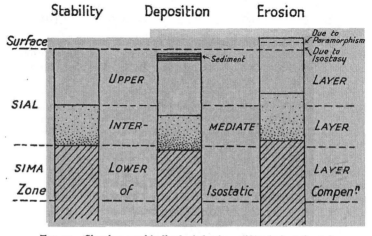

Fig. 30.—Showing graphically the behaviour of Equivalent Crustal Columns under Loading and Unloading.

to explain so remarkable an apparent balance between settlement and sedimentation, since, if the latter be regarded as the cause thereof, the amount of isostatic adjustment would be wholly insufficient for that purpose and the basin would soon become silted up. It is easy to show that for every 100 m. of sediment deposited not more than about 80 m. of sinking could be produced and, as Bucher[1] has done, that, if the sea were 100 m. deep to start with, not more than 500 m. of material could ultimately accumulate below water-level. In the face of this impasse a regional sinking through tension or other causes has invariably, though often gratuitously, been invoked. Actual stratigraphical sequences frequently show that sinking has roughly kept pace with sedimentation, irrespective of the ultimate thickness of the formation, wherever developed, which

[1] Bucher (33), 59.

strongly supports the hypothesis that such weighting of the crust has directly determined its corresponding depression. From the above it will be seen that a paramorphic response of only *one-fourth* or 25 per cent would just enable *subsidence to balance sedimentation*. If it were less, or else delayed, the basin would silt up : if more, then deeper-water deposits would form. There must naturally be some physical limit to this effect, just as to the response of a metal spring, but the above estimate suggests that its magnitude may well be of the same order as— perhaps not much less than—that produced by purely isostatic adjustment. Moreover, consideration of the phenomena of the uplift of continents under erosion, and particularly of the deformation of coastal peneplains, brings out the importance of the paramorphic effect in that connection.[1] Finally, the competent way, as set out later, in which various difficulties attending the dynamics of the crust can be solved or else eased by the application of this vital principle makes the author confident that in the paramorphic zone will be found the key to much that is obscure in the evolutionary history of the earth.

An hypothesis that can assist in explaining so many difficulties should have a fair amount of probability. It is maintained that the effect is by no means negligible, and must carefully be taken into account in all matters relating to the response of the crust under varying conditions. If with Daly and others greater depths are postulated for the base of this zone, then the reaction to pressure and the correction-factor to be applied would be correspondingly larger. On the other hand, the introduction of such a principle must lead to conclusions sometimes at variance with those held by the orthodox minded, and will incidentally demand no small modification in our outlook upon the problems of crustal movement, diastrophism, gravity distribution, seismic action, volcanicity, geothermic gradients and so on. R. D. Oldham has indeed made the notable suggestion that earthquakes might possibly be due to the sudden conversion of eclogite into its larger-volume basaltic equivalent.

In the gross the paramorphic zone can be pictured as showing a downward passage with alternation from gabbro and amphibolite through granulites of various kinds into dense eclogite, and the latter with fairly rapid transition into peridotite below. In detail, in analogy with the archæan complexes, it would display a heterogeneity due to bands of differing composition or of varying degrees of transformation cut by veins of acid and basic granulite and penetrated by tongues of basic and ultrabasic matter from the sima beneath. The arrangement would

---

[1] For example those brought out by Lawson in his analysis of the warping of the Sierra Nevada, California.

R

naturally be subject to secular change more particularly under orogenic deformation or under a raising of its temperature to the melting point.

THERMODYNAMICS OF THE PARAMORPHIC ZONE

It is generally admitted that in the higher grades of metamorphism the reactions are *endothermic*.[1] Under loading heat is absorbed, which would be furnished partly by the work done under gravity during the compression, and partly by the rise of the geotherms consequent upon such depression and also through the blanketing effect of the added covering.

With unloading the reactions are *exothermic*, but some of that heat would be absorbed in elevating the expanding crust. The term due to gravity is calculable,[2] but the heat of transformation of such materials is unfortunately unknown. The fact that in certain undisturbed regions the geothermic gradients are lower than the normal beneath areas that have been raised through prolonged erosion (South Africa, Brazil, parts of Canada and Iowa) but higher beneath basins of deposition (Gulf of Mexico), would suggest that the paramorphic heat-equivalent is not larger than the term due to gravity.

Up to now only vertical movement has been presumed, involving purely static paramorphism, but the above conclusions will demand modification when lateral movement takes place, since endothermic reactions will tend to become exothermic under shearing.[3]

Wagner has discussed the formation of basaltic magmas from the fusion of the intermediate gabbro-amphibolite-granulite horizon, while long ago Fermor deduced the consequences of the partial or complete melting of the eclogite shell under relief of pressure. Similarly, ultrabasic magmas would have been produced from the deep-seated peridotite shell with strong differentiation, as well seen in the kimberlite-melilite-basalt suite.

There is more than a hint that the paramorphic zone may have acted as a thermostatic as well as a thermodynamic regulator, but this problem must be left for geophysicists to investigate.

[1] Harker (32), 179.
[2] The energy required to raise a column of sial through a height of 1 km. would be equivalent to a lowering of the temperature throughout the mass of $9\frac{1}{2}°$ C.
[3] Harker (32), 180.

CHAPTER XII

APPLICATION OF THE PARAMORPHIC PRINCIPLE

Deltas. Problem of Ice-caps. Peneplains. Rift Valleys. Continental
Fault-line Coasts. Hypsometric Curve. " Roots " of Mountains.
Gravity Anomalies. Mobile Zones. Geothermal Gradient.

THE moment that paramorphic volume-changes in the deeper
part of the crust are admitted, even to a limited extent, the outlook upon crustal mechanics becomes revolutionised. The loading of a basin by sediment or by ocean brings about a related
condensation of matter in depth ; the unloading thereof a
corresponding expansion.

So far back as 1889, when setting forth his doctrine of
Isostasy, C. E. Dutton was driven into postulating increases and
decreases of density in depth, but without corresponding
changes in mass, although recognising the absence of any real
grounds for such a supposition. The agency is now provided
by " Paramorphism." Furthermore, instead of isostatic " compensation " is found a tendency towards " *over-compensation*,"
since the paramorphic effect becomes *added* to that due to
purely isostatic causes. Its application to certain major problems
must therefore be discussed, starting with the simpler and working up to the more complex cases.

DELTAS

Clear instances of steady loading of the crust can be found
in those great detrital fans built out by rivers such as the
Mississippi, Nile, Ganges or Yangtze. Few persons indeed have
doubted that a sinking of the sea floor must have taken place
pari passu with the deposition upon it of silt. Deltas constitute
a peculiar problem, however, since in certain cases the measured
gravity anomalies are *positive* instead of negative, such as would
be demanded by the very low density of the material laid down.
Over that of the Ganges they are slightly negative, according to
Burrard. For this and for other reasons Barrell took up the
extreme position that the existence of such broad accumulations,
so far from demonstrating isostatic mobility, proved instead a

high crustal strength. The geodesist has actually had to invoke a higher density within the crust beneath the delta as compensation for the lesser attraction due to the lighter veneer, whereas the geologist has hitherto been unable to find any support for such an hypothetical increase in sub-surface density.

The recognition of the paramorphic effect will harmonise these conflicting views and rehabilitate isostatic theory.

No better region can be discussed than the Gulf of Mexico, since the vertical movements therein would appear to have been

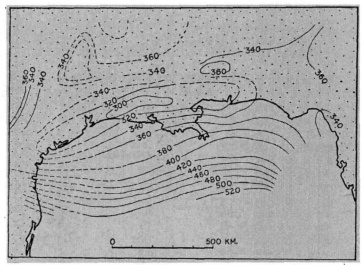

Fig. 31.—Map showing the Variation of Gravity over the Gulf of Mexico. (The values, in C.G.S. units referred to Orange, Texas, are reduced to sea level as if the land extended out across the Gulf.) Adapted from Barton and Hickey.

rather regular. The marine and estuarine deposits that have accumulated within it since middle Cretaceous times are estimated to attain a thickness along its northern shore of from 7 to 9 km., while loading is still in progress through the Mississippi delta. Just inland the Mississippi basin shared, though to a less extent, in such depression, but is today either stationary or is emergent, following erosion.

The results of the elaborate gravimetric surveys carried out over this region and brought together in Chart No. 7941 of the U.S.A. Hydrographic Office,[1] show dominant positive anomalies that increase regularly outwards into the Gulf. Our Fig. 31 is based on the careful analysis of the data by D. C. Barton

[1] " The Navy-Princeton Gravity Expedition to the West Indies in 1932," Washington, 1933.

and Miss M. Hickey (32). These observers were led to conclude that the peculiarities could best be interpreted by postulating a regular increase outwards in the density beneath the floor of the Gulf, such as might be produced by the replacement in that direction of lighter by heavier rocks—say of granite by basalt.

Such an assumption becomes superfluous under our hypothesis, since the immense load of sediments would automatically bring about condensation in the paramorphic zone. Farther out to sea sedimentation would manifestly be less, which would accord with the lower values for the anomalies found on approaching the coast of Cuba. Barton and Hickey have furthermore indicated a well-defined belt of minimum, though still positive, anomalies situated just inland, which according to the Hydrographic Chart would curve southwards through Florida. This they have had to regard as due to a buried syncline. It is, however, significant that that feature coincides with the hinge-line between the stable inland block and the down-warping region of marine deposition, for under the principle advocated here such a locus of inflection could result in a reduction in mean density beneath the faint monoclinal structure.

Again, the Bahamas must be regarded as a greatly eroded, triangular block drowned so recently that it has not yet become buried beneath sediments and is therefore unlike the region of the Great Plains. Suggestively, the anomalies over it are *negative*.

Of interest is the mention by R. M. Field of the prevalence of positive anomalies off coasts, with the opinion that such a relation might prove a general one. This could be anticipated, since waste is deposited mainly upon or just beyond the continental shelf, thereby loading the crust in such position. Indeed we can for such reason suspect the thickness of such off-shore sediments to be much greater than is currently supposed. Taken in conjunction with the simultaneous erosion of the adjacent coastal belt, an ocean-ward tilting of both coast and continental shelf could be anticipated. The south end of Africa with the submerged Agulhas Bank forms a fine example thereof.

THE PROBLEM OF ICE-CAPS

Intensive study of the Pleistocene history of Europe and North America has provided a wealth of information about the response of the crust to glaciation and deglaciation, the most up-to-date discussion of this problem being that by R. A. Daly in his *Changing World of the Ice Age*.[1]

The " basining " beneath the ice-cap is currently viewed as

[1] Daly (34), chap. iv.

due collectively to glacial scour, elastic compression of the crust and " plastic " escape of sub-crustal matter from beneath the loaded area.   Daly, however, points out that, if the last-named had taken place at a shallow depth, the ice-front would be bordered by an elevated belt, and in the absence of such a feature he visualises such transfer as happening much deeper down—perhaps at over 1000 km.—by which a fairly wide region around the ice-cap would have been raised evenly, though only slightly.   Upon the melting of the ice there would have been first the general elastic recovery and thereafter the plastic rebound or " up-punching " beneath the basin, outlined by a peripheral shear-zone in the crust, such uplift being of a spasmodic nature there.   Within the area now freed from ice the recovery would have been more even, though delayed, attaining its ultimate maximum at the centre of the glaciated region.

In the writer's opinion the true plastic effect must still be regarded as operating within the zone of compensation, but is deemed to be of small importance when compared with the paramorphic factor.   The fact that the crust is thick and rigid would prevent it from sharply responding at the ice-margin to the rapid falling-off in loading next the ice-edge, whereupon a belt characterised by shearing stresses would develop marginally, as Daly has indicated.

Off the ice-front any tendency towards up-bulging due to normal, sub-surface plastic flow would be counteracted by the tendency to dimple through spreading of the load over the fairly rigid crust (cantilever action), so that the marginal belt would usually be only slightly elevated.   Daly has calculated that the elastic deformation might have reached about 100 m. in Scandinavia and 160 m. in North America, and the " plastic recoil " about 500 m. and 800 m. respectively.   In the case of the existing ice-cap of Greenland the basining seems to be over 2000 m. (Fig. 36), Wager [1] indeed remarking that the depression is more than would be expected from isostatic theory.

For much of this plastic sagging the paramorphic principle could account, besides giving a better explanation for the topographical anomalies that have been recorded.   Assuming a thickness for the Intermediate Layer of 10 km., 500 m. of ice would correspond with a condensation of only 5 per cent in the lower part of that layer.

While the elastic recovery is rapid, the plastic recoil is not only delayed but spasmodic.   Such is essentially due to the resistance to bending offered by the relatively thick crust along the margin of the deformed area.   This lag would appear to be of the order of 5000 to 15,000 years, as calculated from the

[1] Wager (33), 149.

amount of the uplift and tilt of the raised beaches developed around the Pleistocene ice-cap as it melted away and the ground rose. Such movement is still in progress, showing that recovery is still uncompleted. In Scandinavia confirmation of this is provided by the change in sign of the gravity anomalies from negative over the interior to positive in the surrounding country as shown by Born.[1] Sub-crustal inflow must still be taking place therefore. Gutenberg [2] points out that the measurements made in Scandinavia and Canada are not inconsistent with isostatic theory, nor, it may be added, with the hypothesis here advocated.

The Paramorphic Principle can explain, first, the deep basining observed without any notable bulging outside of the depressed area, secondly, the inordinately large amount of such depression and, thirdly, the delay in recovery—as being due to the lag in molecular and thermal readjustments—while finally it eliminates the need for ultra-deep transfer of matter as postulated by Daly. The phenomena of glaciation thus provide an excellent measure of support for that deduced principle.

PENEPLAINS

The widespread production of plains of erosion during the past is one of the surprises of geology. Such planing-down has, furthermore, not been confined to the stable regions, where such an action would have been much easier, but quite as commonly has operated in the mobile zones although that has meant the wearing away of great ranges that have ever tended to renew themselves under isostatic uplift.

Assuming perfect freedom of response (which is, however, far from being the case) and a 10 per cent difference in density between the crust and the sub-crust, some 10,000 m. of rock would in theory have to be removed before a mass only 1000 m. high could be reduced to a plain! During the lengthy time required for the development of such an even surface orogenic movement must consequently have been lacking or trivial or else erosion must have been extraordinarily potent. On the other hand, rejuvenated or warped peneplains are quite abundant. Almost every important orogenic phase has been followed by rapid planation and that by upheaval and even by subsequent folding, though commonly along *new axes*.

Under any hypothesis whatsoever it is hard to account for the apparent crustal passivity indicated by peneplains, and at first sight the explanation would appear even more difficult should the paramorphic principle be conceded. Closer consideration will, however, show that the greater degree of renewal

[1] Wegener (29), Fig. 12.   [2] Gutenberg (33), 449.

of the mountain ranges due to the two combined causes, and the resulting larger amount of denudation thereby implied, would the more effectively relieve the compressive stresses in the crust and enable the latter to regain its former stability. Such would be brought about by (a) the emission of volcanic matter when compression had been relieved, (b) the absorption of thermal energy during the paramorphic expansion in depth demanded by the crustal uplift, and (c) the lowering of the geotherms with consequent cooling and thickening of the crust.

At the most the paramorphic process would merely have lengthened the total period required for planation; indeed, such would help to explain why the " perfect peneplain " seems never to have been developed.

Abnormal under current ideas is the conspicuous " bulge " of Africa, wherein the vast Central African peneplain projects well above the geoid surface (Heiskanen, Wayland, Holmes, Willis). Such super-elevation can, however, be ascribed to excessive and prolonged erosion and consequent expansion in the Intermediate Layer.

RIFT VALLEYS

(1) *General.*—Some of the most impressive features of the globe are its tectonic troughs known as " rift valleys " or " gräben," with which must also be genetically associated " tilt-blocks " and " horsts." They are sufficiently common as to occur in each of the continents and must therefore be viewed as a *product of the normal diastrophic process*. Such subsidences may have caused impounding of the drainage with the formation of lakes, e.g. those of Central Africa, or the sea may have entered, e.g. the Red Sea, or they may be completely submerged, e.g. the Bartlett Deep.

Widths between the boundary walls of up to 120 km. and differences in level between abutments and floor of between 2000 and 3000 m. have been recorded, though in the case of the great Bartlett Deep the maximum known relief is almost 9000 m. Their lengths are in proportion, running into hundreds or even into thousands of kilometres. Noteworthy is the way in which they cut across the " grain " of the country, sometimes obliquely to strong fold-ranges, though with local deviations and often with marked zigzags. Another striking feature is their usual association with volcanicity—obviously the consequence of such deep crustal fracture.

(2) *Rival Theories.*—There are two opposed schools of thought regarding their genesis :

The *tensional origin*, which is held by the majority of persons, having been strongly advocated by Suess, Gregory, Krenkel,

Obst, Evans, and Wegener, and which is, of course, the logical outcome of the displacement hypothesis.

The *compressional origin*, which is maintained by a minority, chief of whom are Uhlig, Kober, Wayland, Willis and Parsons. The ingenious attempt by Bailey Willis (36) to explain rifting in terms of deep-seated igneous intrusion and crystallisation leaves an impression of unreality. Were that view correct, rifting, if not the normal outcome of the petrological process, ought to be far more common instead of being a restricted tectonic manifestation. The association argued by him would seem to be largely consequential and not causal.

The opinion of the author is that much of the evidence used to deduce tangential compression has been wrongly interpreted or is of a secondary character and that true rift valleys are due strictly to *tangential crustal tension*. It is maintained that in such cases the crust has been torn apart right down to the Intermediate Layer at least, the bounding fractures flattening out progressively in depth, such as would follow from mathematical theory. Taber[1] calls these major structures " profound fault troughs." Krenkel's[2] term " *taphrogenesis* " or " *taphrogeny* " is accordingly favoured for such a type of crustal rupturing in contradistinction to the shallow kind of faulting to which normal fault-basins are essentially due.

The unique characteristic of rift valleys as compared with such fault-basins is the almost universal tilting-away of the margins from the depression (with the development of " consequent " exterior drainages), which has given rise to the firm-rooted idea that the original surface must first of all have been up-bowed along an axis and that the " keystone of the arch " then fell in. Admittedly such marginal back-tilting has not been satisfactorily explained under the tensional hypothesis, though Obst[3] has come nearest to doing so, whereas the phenomenon has been claimed by the other school as evidence for the up-thrusting of the walls over the edges of the subsiding floor, an action also described as " ramping."

On the latter assumption the sunken block ought not only to have retained some of the initial arch-structure, but to have had such warping *intensified* through the continued compression and the weighting of its edges by the overriding walls. One looks in vain for such structures among the normal rift valleys in the African continent, where level or inwardly tilted blocks, horsts, etc., are sometimes brought together in the most curious fashion. Furthermore, the existence of ramping at the northern end of the system, in the Dead Sea region, has been disputed by L. Picard. There are, however, cases, such as the Coastal Ranges of

[1] Taber (27), 591.   [2] Krenkel (25), 56, 240.
[3] Obst (13), 187.

California or the Harz Mountains, where a special type of flexing—called "fault-folding" by Stille—occurs, and even some overthrusting in addition to normal faulting. As Bucher remarks, this seems to have been produced by alternations of tangential tension and compression, which incidentally would not be incompatible with our hypothesis of drift. Bucher has also made the pointed observation that the zigzag pattern, so well developed in the Great Rift Valley, for instance, is characteristic of fracturing under tension and cannot be reproduced by compression alone. His discussion of the problem[1] is outstanding, and should be carefully read.

The writer maintains that in general : (a) the evidence cited for compression is deceptive, (b) there has been no true axial uparching prior to the rifting, (c) the observed tilting is a secondary action and only the consequence of a crustal readjustment, and (d) this phenomenon should more properly be termed "*pseudo-arching.*"

M. Weber[2] crystallises certain older views in emphasising that *compression along the length of the structure* has been an important factor in its development. One can make comparison with the vertical cracks produced inside a loaf of bread on pressing too heavily upon its top. It is striking to observe that each major rift-system of the Earth terminates either at a right angle or at a wide angle against fold-ranges of comparable age, intracontinental or marginal. This harmonises with the view that the *arresting of the moving block, by putting the latter under compression, has favoured its tendency to split-up lengthwise.* Particularly well does this apply to the stupendous Atlantic-Arctic " rift " that abuts against the Siberian-Alaskan fold-system (p. 222).

(3) *Influences of Isostasy and the Paramorphic Zone.*—Our picture of the rifting process is indicated diagrammatically in Fig. 32, with complications purposely omitted. In A is shown the initial sagging of the floor under tension, which in B is restored by inflow within the sub-crust and expansion in the Intermediate Layer due to lightening of the load—or " levity " as Willis calls it—the crust having been extended and now occupying a greater linear distance. In the case of some of the African rifts such extension has been reckoned at fully 5 per cent. This lightening must, however, affect not only the wedge-like margins of the down-sunken block, but the supporting edges of the adjoining walls between *a–b* and *c–d*, and these edges will in consequence tend to *rise* under the combined influences of isostatic and paramorphic expansion, as depicted in C, and so develop the backward tilting here called " pseudo-arching." Actually, instead of independent stages such as A, B, C, the adjustment

[1] Bucher (33), 325-47.   [2] Taber (27), 599.

would be more or less continuous, though doubtless spasmodic—
a " bursting tension " in the words of Suess.

In this scheme part of the vertical displacement along the
sides would be due to the *rising of the flanks* and not entirely to
the sinking of the floor. Such a view would agree with the
physiographical evidence, which, as Willis has well pointed out,

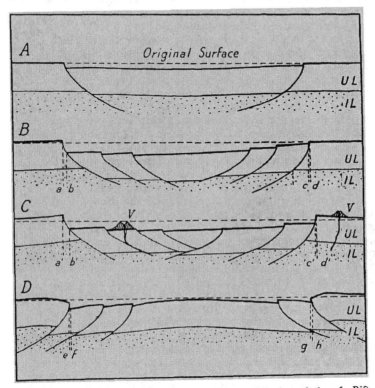

FIG. 32.—Cross sections showing diagrammatically stages in the evolution of a Rift
Valley: *A, B, C* under Tensional Hypothesis; *D* under Compressional Hypothesis; *U.L.* Upper Layer; *I.L.* Intermediate Layer; *V*, Volcano.

does not favour any extravagant amount of subsidence within
the rift itself.

Volcanoes would break out from fractures along the edges of,
or within the trough, thereby relieving pressures and facilitating
adjustments. A slight down-drag at the edges caused by settlement is frequently to be noticed.

If compression be the prime cause, then, as indicated in *D*,[1]

[1] This can be compared with Fig. 4 in Taber (27).

the increased load thrown upon the marginal sections *e–f* and *g–h* should induce sinking and hence the flattening of the initial back-tilt produced through ramping ; up-arching in the middle of the sunken block due to the compression could be expected, while with exhaustion of the lateral pressure the floor should proceed to bulge upwards in an attempt to regain its original level.   Such an arrangement is, however, *not* characteristic of rift valleys.   Furthermore, in such a case the gravity anomalies ought to be positive within the trough or along its margins at least, whereas they tend to be zero or negative there, which, on the contrary, is in precise accord with the tensional hypothesis.

It must be pointed out that the above arguments will only apply to rifts in which the width bears some close ratio to the depth of the fracturing, which presumedly must be deep.   It holds nevertheless for large "*tilt-blocks*" or large troughs bounded on one side by a monocline, which are merely asymmetrical gräben ; indeed, such structures are commonly associated with rift valleys, being often repeated in step-like fashion, and in Africa forming an integral part of the Great Rift System.

For the evolution of a rift we can picture : (*a*) the primary tensional furrow, which will or will not receive sediment, (*b*) the initial fracturing, simple or compound, (*c*) the dropping of the block or blocks as the sides recede, though not necessarily symmetrically, (*d*) the relief of tangential stresses by extension of the crust, (*e*) the lightening of the load per unit of length and the consequent " up-doming " in which the abutments share, giving a pseudo-arch, (*f*) the establishment outside of a grid-iron system of consequent drainages directed away from the edges, (*g*) the wearing-back of the scarps with lightening and further uplift, and (*h*) the melting in the Intermediate Layer and the pouring out of volcanic matter, mostly of alkaline tendency.

Under renewed tension the same process will be repeated, leading ultimately to complicated fault-structures and -patterns. Wedges pointing upwards and downwards would be produced, and in the adjustment between horsts and troughs local squeezing could readily occur.   The transfer and crystallisation of magma in depth will moreover induce unequal settlement.   Depending on local conditions the floor will either remain dry, be flooded with lavas or volcanic ejectamenta, be filled with fresh water or depressed beneath the ocean.

It is clear that this identical principle must apply to the *disjunctive ocean basins*, which are merely rift valleys of unusual width.   Thereby can be explained the *characteristic plateau-character and backward tilting of the margins of the bounding blocks*, as detailed below.

(4) *Gravity Anomalies.*—Krenkel's [1] analysis of the African

[1] Krenkel (25), 89-99.

rifts is most instructive. The " Erythrean Graben," or Red
Sea, was during the Cretaceous a broad syncline, which through
fracturing and sinking from the Lower Oligocene onwards became a trough of lesser width in which sediments—mostly
marine—were repeatedly deposited with marginal overlapping
upon tilted and broken Cretaceous and older Tertiary strata.
In the Lower Miocene the Mediterranean entered, and in the
Pliocene the Indian Ocean.   More recent upheaval of the lands
bordering the Red Sea is indicated by coastal terraces up to 200
and 300 m. in altitude, while there has been much faulting along
its margins.

Because of the heavy sedimentation within the trough the
gravity anomalies (Bouguer) are positive there,[1] changing to
zero at the margins and even becoming negative outside.   In
the inland rifts, which probably contain little silt, but in places
deep water, the anomalies (Bouguer) are, as shown by Kohl-
schütter's [2] observations, uniformly, and in places strongly,
negative.   Willis' review [3] is an important one, but should be
read with the above considerations in mind.

The criterion afforded by gravity is held to be sufficient,
despite Bullard's objection, to reject the hypothesis of compression with ramping.

(5) *Some Rift Systems.*—Pre-eminent, of course, is that of
Africa, so ably and convincingly described by Gregory, much
extended by further exploration and refreshingly reinterpreted
by Bailey Willis — the manifestation of a stupendous crustal
rifting, revealed by fault-troughs, tilt-blocks, warpings and widespread volcanicity (Fig. 33).   The main system of fracturing
runs from next the Taurus fold-ranges in Syria to across the
Zambezi, but minor faulting, and particularly warping, are
known to have affected a far wider region, extending westwards
well into the Congo basin and southwards into the Karroo,
attaining its maximum in the tropics.   Furthermore, the coast
of Kenya and of Somaliland is a faulted one.   This slightly
curved zone, 8000 km. long, displays a jagged series of rifts and
a mosaic of faults that could never have been produced under
compression alone.   Furthermore, wherever the fault-planes can
be seen, they are vertical or normal ; the few reversed faults
that have been reported are nearly all of pre-Rift age.   Torsional
influences are shown by the two well-developed N.E. and N.W.
trends set at right angles.   Noteworthy is the fact that the
territory traversed is mostly a super-elevated portion of the
geoid and also that it lies, as Gregory (21) has observed, antipodal to the Pacific basin.

For Nyasaland and adjoining territory Dixey has been able
to show that some of the major troughs were initiated so far back

[1] l.c. Fig. 12.   [2] l.c. Pl. XVII.   [3] Willis (36), 329-43.

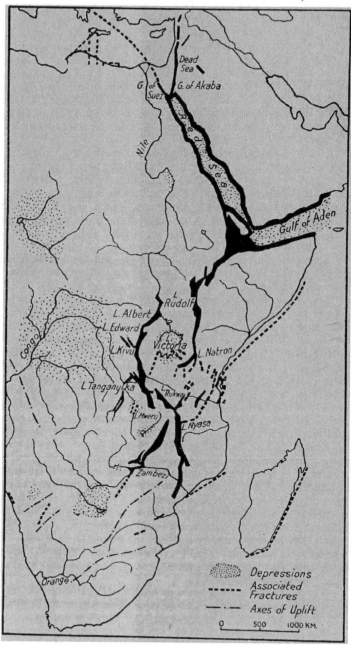

FIG. 33.—The African Rift Valley system (largely from J. W. Gregory).

as the Jurassic, were defined by erosion and faulting during the Cretaceous and were completed by heavy rifting during the later Tertiary. The same would seem to have been the case elsewhere. A state of east–west tension in Africa, that persisted from the mid-Mesozoic onwards, must accordingly be presumed —a conclusion in full accord with our hypothesis of continental evolution.

The hypothesis by Willis [1] of a vast cake of basic magma some 800 km. in diameter, situated beneath the Lake Victoria region giving outward radial pressures and constituting the *fons et origo* of this gigantic fracture-system, is highly speculative. Such a hidden body should stand revealed by the gravity anomalies, which are, on the contrary, low and often negative. Furthermore, it can scarcely explain the vast extent of external fracturing and warping of precisely the same kind as in the centre of Africa that stretches far to north and south, wherein volcanicity has been much weaker or is else wanting. On the contrary, we see Lake Victoria almost surrounded by rift valleys and we cannot but conclude that, deprived of much of its lateral support, that block dimpled at its centre and so enabled the lake—with its typical drowned topography—to come into being. A depression of only a few hundred metres could well account for it. Gregory's original interpretation is accordingly and wholeheartedly upheld.

Obvious, indeed, is the " *breaking-up of Africa.*" We observe the initial separation of Arabia ; we see Madagascar long-separated and displaced to the south-south-east from its original position in the bight of Tanganyika ; we perceive the curving sector from Somaliland to Beira that is about to follow suit ; we notice behind it the Lake Victoria block not quite severed ; and we detect in the hinterland, penetrating into the heart of the Continent, sundry fractures and extensive far-flung saggings of the peneplained surface. Can we doubt, then, that the partition of Africa from east to west is still in progress ?

With such colossal masses involved and with rifting taking place progressively, it would be remarkable if one block, through premature release or through differential movement, did not press upon another at some point or points, so paralleling the case of an uneven plane to a tension-fault. Thus can one account for the irregularly-directed reversed and thrust faults described by Parsons a short distance inland from Mombasa, for certain of the evidence recorded by Wayland and by Willis from the region of Lake Albert and by Willis from that of Lake Tanganyika and the post-miocene compressional disturbances reported by Fuchs from the Lake Rudolf region.

The much-studied Rhine graben, extending a little east of

[1] Willis (36), 72-97.

north for 300 km. from Basel to Mainz, agrees closely with that of Africa. Pseudo-arching is developed in the abutments made by the crystalline masses of the Vosges and Schwarzwald. The eastern end of the Jura foldings have not only closed the trough of Oligocene beds on the south, but poured northwards into it, thereby exerting a kind of " plunger action," which must have assisted in the forcing apart of the crust and the development of the graben. The arrangement of the main and subsidiary fractures is precisely the same as that obtained by Cloos in his experiments with clay under tension. The occasional thrusting, more particularly on the eastern side, is apparently due to the fact that the pulsations of Alpine pressure were directed not exactly from the south but from the south-south-east and hence at *less than a right angle* to the axis of the valley, thus giving a component across the latter. The abundant volcanic rocks are strikingly of alkaline characters.

In the western part of the United States differential movement during the later Tertiary is marked by considerable tensional faulting, some of it of the rift type. The Rio Grande valley of New Mexico traverses a pseudo-arch which terminates obliquely against the strong folds and overthrusts of the Mexican region. The western side of the Colorado Plateau and the Great Basin region display much block-faulting. Oregon exhibits a fine series of rift valleys and tilt-blocks with characteristic pseudo-arching, lavas and recent or active volcanoes that strongly recall those of Central Africa, the fault-mountains trending obliquely to the folds that have affected the Columbia basalts more to the north-west. Fuller and Waters (29 ; 31), investigating the cause of these phenomena, have given most convincing reasons for rejecting the compressional hypothesis.

CONTINENTAL FAULT-LINE COASTS

In dealing with the rift valleys the almost invariable, though slight upturning of their dislocated edges was discussed, and it was pointed out that such an inherent character was developed even where the floor of the graben or sunken block had become submerged by the ocean. This peculiarity will hence form a valuable criterion for the recognition in regions that are free from subsequent tilting or folding, of *fault-line scarps* resulting from tension-faulting.

Where appreciable time has elapsed since the fracturing, as in the cases about to be mentioned, the fault-face has been worn back, sometimes for long distances, into a somewhat ragged edge, to which the name "*fault-line scarp*" will apply, the true nature of which will, however, seldom be in doubt, although it may have retreated under erosion far back from the original fracture and may, therefore, be attended by no visible faulting of the strata in the immediate neighbourhood.

In the physiographical evolution of a land-mass not bounded by important faults the profile should roughly conform to that of a somewhat flattened dome, and develop a drainage tending to be approximately radial. Where, on the contrary, faulting has primarily been responsible for the outline, one will find an escarpment either close to or away back from the sea, while from the resulting plateau-edge the larger drainages will be directed *inland*, to reach the ocean only some distance away—

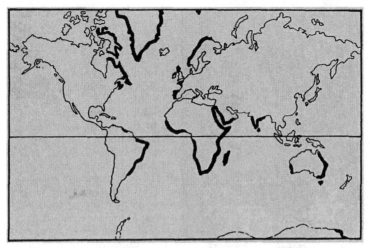

FIG. 34.—Showing Distribution of Continental Fault-Line Coasts.

in certain instances right on the opposite side of the continent. To do that the rivers may have had to pass through gaps in the escarpment or take a circuitous route to avoid the latter and have commonly developed cataracts. The majority of the great waterfalls of the world have been evolved in such positions and under such circumstances. While the significance of such a physiographical arrangement does not seem to have particularly impressed itself on geomorphologists, it is here viewed as a feature of fundamental importance. This extraordinary drainage scheme, though actually widespread, is not readily explicable under current views, although, on the contrary, the logical consequence of continental rupture. The writer first pointed this out in 1928, and, after fuller consideration, has come to regard it as a very powerful criterion in support of the Displacement Hypothesis, such geographical peculiarity being confined to those blocks which under that conception are regarded as having drifted apart (Fig. 34).

S

The following elements are then normally represented : the interior mature plateau with inland slope, the escarpment, the moderately or else considerably dissected coastal zone, the narrow continental shelf, and the steep slope to the ocean deep beyond.

These form conspicuous features in each of the portions of

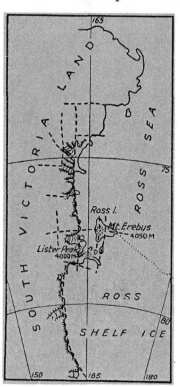

FIG. 35.—Fracture Pattern in South Victoria Land (after Priestley and David).

Gondwana, as finely exemplified in Africa and Arabia. In that continent we observe over enormous stretches of its marginal zone, more particularly in the southern half, situated at a distance back from the ocean of from 100 up to 1000 km., the so-called "Main Escarpment" inside of which lie curious interior basins traversed by lengthy rivers taking their rise quite near to the ocean, e.g. Zambezi, Limpopo, Orange, Congo, Niger, Senegal, Atbara and Blue Nile. The general altitude of the escarpment is close upon 1000 m., but over long sections may attain to 1500 m. or even more. This topographical feature involves many different geological formations and structures and is normally unconnected with visible faulting. That it is due to the very considerable retreat under head-stream erosion of marginal fault-scarps originated during the Tertiary is pointed to by all the evidence.

Right across the South Atlantic in Brazil is an identical arrangement with escarpment facing the ocean behind which the Paraná, Uruguay and São Francisco rivers take their rise and flow for extraordinary distances parallel to the coast, developing magnificent waterfalls and rapids before reaching the sea. In India the escarpment of the Western Ghats, rising to 2700 m., runs parallel to the shore and unbroken for 13 degrees ; the Eastern Ghats are not so persistent, regular or high, though still typical. The drainage system, seen in the

Narbada, Mahavadi and Godavari, conforms to expectation. Ceylon again shows two important peneplains, the higher standing at about 1800 m. Madagascar forms another example, being a crustal block tilted slightly to the west or north-west with straight, upraised and fractured eastern coast. Arabia and Western Australia are elevated peneplained masses also. Reference has already been made to the late Tertiary faulting off the eastern coast of Australia with upheaval of the eastern " divide " and diversion of the run-off south-westwards into Encounter Bay and Lake Eyre. Far to the south this is repeated in the massif of South Victoria Land capped by flat-lying Gondwana beds, with its ragged main escarpment jutting upwards many thousands of metres, bounded on the east by ice-eroded fault-line scarps overlooking the Ross Sea, but sinking on the west beneath the ice-cap (Fig. 35).[1]

Turning to Laurasia a similar physiography is extensively developed in the North Atlantic region. We observe the marked

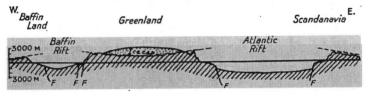

FIG. 36.—Cross section 3500 km. in length from Canada to Scandinavia, vertical scale greatly exaggerated, showing tilting of plateau-surface away from oceanic rifts. *F.F.*, Hypothetical Faults.

asymmetry of Scandinavia—the lofty watershed, the rapid fall of the Norwegian streams through intensely dissected country towards the Atlantic in contrast to the flatter sub-parallel Swedish drainages towards the Baltic and Gulf of Bothnia, remembering, as pointed out by Schuchert, that this land-mass must have stood even higher in the late Tertiary. In Scotland we note the same asymmetry—the westerly position of the main divide, the rapid drop to the Minch with a second easterly-tilted block represented in the Outer Hebrides. Beyond lies Iceland, a volcanic plateau with slight tilt to the north-west.

Again, is not the basin-like form of the ice-floor of Greenland, rimmed by lofty mountains, another example of crustal adjustment within a large block fractured on east and west and, therefore, elevated along those margins (Fig. 36) ? Significantly, facing it across the supposed gräben of Baffin Bay and Davis Strait, rise the lofty ranges of Baffin Land as described by Holtedahl, sinking towards the south-west, and also the physiographically similar highlands of north-eastern Labrador with

[1] Gould (33; 35).

rivers escaping eastwards through the up-tilted peneplain-edge, as Odell (33) has made clear. As typical rifted blocks or horsts one can also regard Newfoundland (with tilt from west to east), Cuba, the Iberian Peninsula and also the submerged platform of the Bahamas, while other equally instructive instances could be cited did space permit.

It is essential to notice that none of the masses specified has experienced post-Cretaceous folding, and that the phenomena in question have affected only those sections of the continents which under the hypothesis are viewed as having been torn away from one another. Rifting would preferably, in that case, have followed the grain of the crystalline basement, as in Madagascar, or a zone of previous folding, as on the outer side of the Appalachians. Under any circumstances the forcible tearing apart of the crust, in many instances athwart the grain, and accompanied by definite torsion, must have developed a host of variously oriented fault- and joint-planes arranged in roughly geometrical patterns, which would have lent themselves admirably to enlargement by river, frost or glacier erosion.

Characteristic of these fault-line coasts, therefore, is their rugged character and in high latitudes the marvellously developed fiord systems, which, as Gregory (13), Koch (26), Peacock (35) and others have shown, has primarily been due to faulting, for example in Scandinavia, Scotland, Greenland, British Columbia and Patagonia. Where folded strata have been snapped across, " ria " coasts have resulted, for instance in southern Ireland, Newfoundland, Nova Scotia and the southern Cape.

By actual rupture and not through the mere sinking of down-faulted blocks has originated, so it is maintained, the rugged " Atlantic Type " of coast-line, so called by Suess. On the contrary, where the mass has pressed upon the ocean, or *vice versa*, there is found the " Pacific Type," physiographically youthful, of low altitude, with causally related fold-ranges developing or else developed in the rear, for example in California.

THE HYPSOMETRIC CURVE

Particularly instructive is the study of the curve showing the relative frequency of the various levels of the lands and depths of the oceans. This is a typical double skew-curve giving maxima at +230 m. and -4420 m. (Kossinna), which correspond with the average continental and deep-sea platforms.

Wegener [1] took this to indicate the existence of two undisturbed original levels in the crust and as the proof that in the continents and oceans we have two different layers of the body of the earth, a view strongly criticised by G. V. Douglas, A. V.

[1] Wegener (24), 30-31 ; (29), 36.

Douglas and W. H. Bucher.[1] The curve nevertheless sets out in obvious fashion relationships that have been determined by the relative densities of the land and oceanic areas, the degree of isostatic adjustment, etc., and, if those factors had possessed different values, the face of the Earth and the frequency curve therefor, might well have been unlike those of to-day. The author cannot quite agree with Bucher (his Opinion No. 7) that

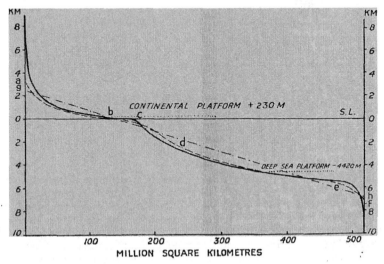

FIG. 37.—Hypsometric Curve of the Earth's Surface (for explanation see Text).

the two frequency maxima are due to two independent sets of forces.

The well-known Hypsometric Curve (Fig. 37) clearly demonstrates the efficacy of isostatic control ; indeed, without such a vital principle the Earth could well have remained covered by a uniform ocean, and there would have been no curve. It furthermore suggests the importance of the paramorphic effect, for without the latter the curve would be less accentuated and would assume a more even position, presumably like that shown by the broken line *a–b–c–d–e–f*. Through it super-elevation *a* of the highest ranges has become possible, over-depression *f* of the abysses, over-loading *c–d* of the continental slopes and flexing *b–c* of the continental margins. The wide departure of the curve from a line sloping evenly, as in *g–h*, supports this idea.

Otherwise we shall have to presume that all the highest chains are today maintaining their altitude by unexhausted tangential compression, and the exceptional sagging of the

[1] Bucher (33), 51-58.

oceanic deeps their depth by persistent tension, which is rather unlikely.

THE " ROOTS " OF MOUNTAINS

The idea that a mountain is maintained by some peculiarity of the crust below has been responsible for several hypotheses, all taking their keynote from the behaviour of the pendulum, although all show greater or less discordances between the observed and the corresponding computed values of Gravity.

In the case of fold-ranges an important clue to their origin is provided by the fact that normally they have not been pushed up by the tangential forces directly, but have been elevated during the latest stages of compression, or, more commonly, *after* that process had ended.

Now a mountain can be viewed as an extra load on a uniform crust that remains undeformed. This, the Bouguer method, gives considerable errors and is little used today.

If applied at the top of such a mountain, the value actually got is about the same as though the mass of the mountain had become zero, its height remaining unchanged. This gives the " Free Air formula," which fails, however, with high elevations.

Under Pratt's conception different crustal columns rest upon and rise from a uniform zone of compensation and owe their varying heights to differing mean densities, that is to say, the higher the elevation the less the density of the crust, and *vice versa.* The continents would indeed project just because they are formed of rocks lighter than those underlying the oceans.

In certain respects Pratt's hypothesis agrees fairly well with geological evidence, only the great diversities of density demanded by it by reason of the geological revolutions are not readily forthcoming under current theories. In spite of this inherent defect it nevertheless forms the basis of several schemes of isostatic adjustment. Hayford by a process of repeated trial found about 122 km. as the best depth for uniform compensation for the United States, which figure Bowie later reduced to 96 km. In 1912 H. L. Crossthwait suggested that *uneven distribution* of density in individual columns might account for the irregularities found in the deflection of the plummet, a view afterwards independently proposed by G. K. Gilbert. Heiskanen obtained better results by regarding the compensation as concentrated at the base of a single layer rather than as distributed uniformly down to a certain depth, as under the aforegoing hypothesis. The single surface which he gets lies at a depth below sea-level of 50 km. in the United States, 77 km. in the Caucasus and 41 km. in the Alps. This fits with Jeffreys' [1] view that compensation should be concentrated either at one or two

[1] Jeffreys (29), 197.

surfaces, and in the latter case at the bottom of the granitic and intermediate layers respectively.

Airy, on the contrary, pictured the columns as of uniform density but varying length standing upon a zone of compensation, the depth of which altered with each column, that is to say the light crust was thin beneath the oceans and thick beneath the continents. Each important elevation would thereupon be off-set by an equivalent downward protuberance pressing into the heavier substratum, wherefore mountains must be possessed of " roots." Admittedly, not every minor inequality could thus be individually compensated—only the larger areas, say 100 km. or more across. In this connection Melton's [1] pointed remarks should be studied. Bearing in mind the relative densities of the Upper and the Intermediate layers, the depth of such roots would be quite considerable.

It has, however, been objected to by many persons, for example Bucher, that such deep roots are not physically possible, since they would experience melting in depth, but this overlooks several important features, as Longwell (35) has made clear in a stimulating paper. First, the roots would not be a mirror-reflection but a subdued inversion of the surface irregularities, that is, the latter would be compensated not locally, but region-ally. Secondly, a spread-out base is demanded on mechanical grounds, as is demonstrated by the constant association of a foredeep or backdeep, whereby a part of the extra load has become transferred through the rigidity of the crust (cantilever action) to the regions adjoining the mobile belt (Fig. 38). Thirdly, fusion of the roots need not necessarily have involved the collapse of the fold-ranges above them, as consideration of the volume-changes and densities of magmas will readily indicate. If melting were to take place, the portion ascending in the cores of anticlines would normally be only the more acid and lighter fraction, the *basic and heavier differentiate remaining below and still providing support.* As Jeffreys has pointed out, regional compensation would tend to raise the summits of mountain ranges because of the greater erosion of their slopes and valleys ; hence the abnormal height of certain chains and peaks. W. T. Lee's [2] arresting account of the super-elevation of the southern Rocky Mountains is worthy of careful study. Airy's concept that mountains possess roots, and its logical consequence that the crust must be correspondingly thickened beneath ranges and plateaux, can therefore be whole-heartedly supported.

Application of the *paramorphic principle* will, however, not only smooth out difficulties inherent in the several hypotheses, but bring about a far better agreement between them. The

[1] Melton (30), 286.   [2] Lee (23), 292-98.

long-desired agency for crustal expansion and contraction is
now available, although admittedly not uniform throughout the
column as required by Pratt, Hayford and Bowie, but restricted
to part of the Intermediate layer.   Heiskanen and Jeffreys' concepts are furthermore met in so far that *two levels of adjustment*
are indeed provided, one—the above-mentioned—within the
paramorphic zone, the other deeper down in the currently recog-

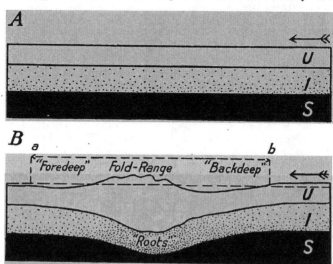

Fig. 38.—Showing diagrammatically the development of " Roots " beneath a
" Fold-Range."   *U*, Upper Layer ; *I*, Intermediate Layer (the closeness of the
stippling corresponding with the degree of paramorphic condensation) ; *S*,
Sima.   Observe how the weight of the elevated mass has been spread out over
the much broader crustal segment, *a, b*.

nised zone of isostatic compensation, that is to say, in the *sima*.
Moreover, because of the paramorphic condensation of matter
beneath them, *the roots of mountains are not so deep, but, on
the other hand, broader than would be demanded under Airy's
conception.*

The new viewpoint indeed postulates *varying density with
varying depth* and is accordingly intermediate between the
hypotheses of Pratt and Airy, while tending furthermore to
harmonise those of Hayford and Heiskanen and also meeting
all Longwell's contentions.   A full mathematical treatment
thereof would hence be welcomed for comparison with the other
formulæ.   The fact that it brings so much nearer together
theses based on such diverse postulates is distinctly in its
favour.

GRAVITY ANOMALIES

Although of high importance in the interpretation of the Depths, the significance of the Gravity Anomalies can only be touched on briefly here. The figures obtained through reduction are naturally based upon various assumptions regarding the depth of the zone of isostatic compensation, the densities of the rock shells composing the crust, etc. Since the attraction of gravity varies inversely as the square of the distance of the centre of mass of the attracting body, the strata directly beneath the station will exercise a far greater effect than the layers situated deeper down It is clear, nevertheless, that some correction would have to be applied to the reduced value if the paramorphic influence be admitted, since the depth of the zone of compensation would not then be the same everywhere. It would be necessary then to correct not only for topographical and isostatic compensation, but for the orogenic history of the area in question. For example, a part of the deviation found could conceivably be due to the presence of unbalanced tangential forces in the crust, since tectonic structures are not necessarily passive. For instance, the simple monocline, which could be produced, either by stretching or squeezing, might still be in a state of tension or compression.

Speaking generally, overloaded areas show positive anomalies, abnormally light areas negative anomalies. Over the lands there would, however, seem to be a tendency towards only moderate anomalies (irrespective of sign), since erosion of a region would not only bring the denser layers nearer the surface, but produce some compensatory expansion in depth ; weighting of the crust or down-folding of lighter strata would in addition produce some compensatory condensation in depth. It is probably for such reasons that gravitational surveys of the continents have not yielded as unequivocal results as other lines of investigation. It might incidentally be observed that, just as the crust can undoubtedly carry large concentrated excess masses, such as volcanoes, without dimpling, so it should be able to endure localised *negative loads*, such as worn-down mountain roots, without bulging.

It is of high interest to observe that from the anomalies themselves E. A. Glennie (36) has derived a correction-factor, called the " warp anomaly," that expresses numerically more or less the above ideas. This is ascribed to deep-seated warping of the crustal layers, up-warping giving positive and down-warping negative values. The application of this factor gives a marked improvement in the final results.

For North America certain broad generalisations have been

brought out by D. White,[1] modified in some respects by Glennie. The pre-Cambrian shield usually gives positive anomalies, though, as Nevin[2] has remarked, much of that may really be due to the low value arbitrarily chosen for the density of the upper crust. . The geanticline of the south-eastern part of the Appalachians gives fairly strong positive anomalies, while the geosyncline of the north-western part—deeply filled with folded palæozoics—gives marked negative values, in contrast, again, with the smaller positive ones of the little-disturbed palæozoics towards the west with higher figures locally. The Coastal Plain of the Atlantic and Gulf regions displays mainly small negative or positive values. Over the Tertiary-disturbed western States the normal warp anomalies are almost entirely negative (Glennie).

For Central Europe Kossmat's[3] map not only shows that the regions of piled-up Alpine chains are strongly negative, whereas the tension-basins of the Adriatic and Mediterranean are markedly positive, but reveals the defect of mass beneath those mountains due to their lighter roots. The wide, elevated East African plateau is almost isostatically compensated (Willis). From his work in Hawaii Goranson (28) concludes that the sub-Pacific crust differs from the continental crust in being uniformly more basic down to a depth of 60 km. The volcanic islands of the Pacific are essentially uncompensated and are hence subsiding loads on the ocean floor.

Over the oceans gravity is seemingly more regular. Along coasts the pendulum generally deflects slightly seawards, which peculiarity can be correlated with the observations by Meinesz that the anomalies increase in the algebraic sense on proceeding from shallow to deeper water, where vast areas show positive values.

The Indian Ocean is weakly negative ; the Eastern Atlantic is moderately positive, while the Mid-Atlantic Rise gives only small deviations so far as is known ; the Mediterranean, Red Sea and Pacific generally are strongly positive, especially the eastern side of the last named. The general supposition that the Atlantic volcanic islands are positive has not yet been confirmed. Meisner (18) has pointed out that the Atlantic coasts appear to be isostatically compensated, but that along the west of Africa the depth of such zone is unusually deep—from 150 to 200 km. This is understandable if the coasts were the fractured margins of crustal blocks. The Pacific coasts are, on the contrary, not isostatically compensated, which must be ascribed to existing tangential forces and general instability.

Attention must be directed to the remarkable persistent belt of strong negative anomalies fronting the Sumatra–Java–Ceram–

[1] White (24), 275.   [2] Nevin (31), 240.   [3] Wegener (24, Fig. 31).

Celebes arc indicated in Fig. 22 lying between the blocks of
Asia and Australia. The axis thereof does not coincide with
the tectonic features, since along particular sections it follows
oceanic trough, submarine rise, continental shelf or island
margin. Meinesz (31) believes this to be due to the pressing
down into the denser basement of the lighter crust, which latter
has also been thickened by upward-folding in its top portion,
with probably some overthrusting as well. This line is indeed
characterised by numerous earthquake epicentres, which indicate that displacement of the ridges and furrows is actively
in progress, corroborated by the accompanying strong volcanicity. It is significant that the belt is quite narrow—100 km.
generally—and hence only a few times the inferred thickness
of the entire crust. Although the latter is mostly covered by
ocean and is furthermore attached to the underlying sub-crust,
it can be viewed in the mechanical sense—in the vertical plane,
that is to say—as a short strut fixed at either end and bent by
horizontal pressure into a double curve, namely an arch joined
to a trough. These two elements cannot be dissociated from
one another, since they really form complementary parts of a
single structural unit. The precise shape taken up, as well as
the relative position of the point of inflection, could, therefore,
change along the strike without affecting such structural unity.
The tract of negative anomalies can hence be visualised as marking
the general deficiency of mass within the confines of this unit,
and the axis thereof could, therefore, fail to coincide with either
ridge or adjacent furrow. Of course, such could only hold good
where the geosyncline was quite *narrow*.

An alternative explanation is that underneath this belt—
between the Malay States and Halmaheira—lies the Tertiary
Gondwana Front, weakly developed and so overwhelmed by
that of Asia, such corresponding, in fact, to the double structure
underlying Tibet. Both explanations might quite possibly apply
here.

THE MOBILE ZONES

Crustal deformation must be a very complex process.
Originating primarily through alternate crustal tension and
compression, the evolution of such a zone would be governed
by ever-varying factors of tangential and radial pressure,
erosion, sedimentation, radioactive heating, magmatic activity,
etc., with a striving towards isostatic adjustment at every stage.
While geosynclines are of several distinct kinds, as stressed by
Schuchert and Stille and as mentioned in Chapter III, their
individual histories fortunately follow rather similar lines.

During the diastrophic cycle the Paramorphic influence
would assist in maintaining the crest of the deforming belt,

in promoting orogenic sedimentation on either one or both sides thereof and in assisting subsidence under the changing load. The down-folding process would be facilitated by condensation in the Intermediate layer beneath the fold-belt and to some distance on either side, with related movements in the Sima as well. The contrast between height and hollow would accordingly become accentuated.   Under a one-sided pressure plastic matter would become transferred back from the forward to the rearward side of the " roots," where the compression was less, thus enabling the structure as a whole to *migrate*. In such sheltered position—or else along shear-planes—melting would be promoted with eruption, leading up to the surface some short way behind the active front. That in such an orogenic scheme radioactive heating must be a vital factor cannot be doubted, as has been clearly set forth by Joly and Holmes.

With cessation of pressure vertical movements would generally manifest themselves, leading normally to upheaval, more particularly of the fold-region, and to tension-faulting. Much thereof might furthermore be due to re-solidification of the supporting sub-crust. The actual picture would, of course, be far more elaborate than the simple outline here presented. For example, renewal of pressure or alternating pressure and tension would introduce appreciable complications.

Sufficient will, however, have been said to indicate the importance of the Intermediate layer in the Orogenic Cycle. The postulated paramorphic transformations would naturally be regulated not only by dynamical but by thermodynamical principles.

GEOTHERMAL GRADIENT

Daly[1] has done good service by drawing attention to the scant regard given to the Geothermal Gradient as a measure of the Earth's internal heat, and by discussing the factors that might serve to explain the considerable variations thereof that have been measured within the lands, the value beneath the oceans remaining of course unknown.

The most important of these—and the one which is so frequently ignored—is the presence of radioactive substances in the rocks, and, since these cannot be uniformly distributed either vertically or horizontally within the crust, such could well, as De Lury has shown, account for much of the diversity revealed by the measured geothermal gradients.

It is not insignificant that the lowest values—1° C. per 45-109 m.—are from the shield areas, e.g. Iowa, Michigan, Brazil and South Africa, while nearly all the high ones—1° C. per 20-40 m.—are, on the contrary, from regions of marked mesozoic

[1] Daly (33), chap. x.

and tertiary sedimentation, deformation or magmatic activity. Another manifestly important factor would be the horizontal or vertical transfer of heat in the sub-crust by convection currents.

It should not be overlooked, however, that even simple epeirogenic movement could constitute another, and possibly a not negligible, factor through demanding the production or the absorption of heat within the Intermediate layer.

CHAPTER XIII

PAST CLIMATES AND THE POLES

Introduction. Climatic Girdles. Criteria for Climatic Zoning. Polar Positions. Pole-wandering in Laurasia. Pole-wandering in Gondwana. Reality of Drift.

INTRODUCTION

THE existing climate of the Earth, its distribution and particularly its variations with latitude and altitude, naturally constitute our available standard for comparison in this knotty problem. Unfortunately, the present rather tends to colour our outlook upon the past despite the fact that the stratified rocks provide testimony of profound changes of climate, at times in open conflict with preconceived ideas on the subject.

Huge areas in the Southern Hemisphere far removed from the present South Pole were, for instance, heavily ice-capped in the Permo-Carboniferous, although equivalent refrigeration is almost unrepresented in the Northern Hemisphere; well-developed floras indicative of milder environments are known from high latitudes, during the Permian from close to the South Pole, during the Jurassic from Graham Land and during the Cretaceous from Greenland; in the Triassic aridity affected extensive regions in both hemispheres, and so on.

Such climatic incongruities form one of the major problems of geology, since they are difficult or impossible to account for satisfactorily under current theories with the lands and poles fixed as at present, and that, too, despite the invoking of pretentious land-bridges, ocean currents and winds. Orthodoxy is indeed endeavouring to defend a wholly untenable position. As Wegener has caustically remarked, the general failure to admit the hypothesis of drift has absolutely crippled the science of Palæoclimatology.

The many-sided problem of past climates is too vast to be dealt with here, and can only be outlined, although of the greatest moment in our polar reconstructions. The illuminating monograph by Köppen and Wegener should be carefully studied by the reader, as well as the writings of Dacqué, Neumayr and

Arldt, Kreichgauer's suggestive work, Brooks' valuable *Climate through the Ages* and the instructive papers by Simpson and Davidson Black.

CLIMATIC GIRDLES

The Earth's surface is today divided into the seven climatic girdles : the low-pressure moist equatorial zone, the two bordering high-pressure zones with much lower rainfall—in which are located almost all the desert regions—the two temperate zones and the two polar circles. Had exactly the same arrangement persisted throughout the past, then even under orthodox theories glacial phases should have accompanied most of the formations of high latitudes. On the contrary, such types are conspicuously lacking there, a notable exception being the Eo-Cambrian tillites of Spitzbergen and eastern Greenland, whereas glacials are strikingly developed in moderate and even in low latitudes in various epochs.

The same general order, with its decrease of mean temperature from equator to poles, must, however, have held good, though great climatic changes would have been attended by appreciable expansion or contraction of certain of these zones —though not necessarily in the same ratio or even in the same sense—with widening or shrinking of the polar caps and with correlated variations in the other zones, but not necessarily to the same extent, in both hemispheres. In extreme cases the poles may perhaps have been practically free from ice.

While there is a close correspondence between the two hemispheres today, the heat equator does not coincide everywhere with the geographical equator, and geological considerations strongly suggest that at times in the past such deviation may have been far greater, thus leading to more marked contrasts between north and south. That is to say, the intensities of glaciation, aridity or humidity respectively could have been very different in the northern and southern hemispheres at any particular instant.

One could furthermore anticipate differentiation into climatic " regions " according not only to latitude but position within the continent itself, thus giving on the western side of the latter a regional sequence such as ice, tundra, wet-temperate, warm dry summer, steppe, desert, periodical dry savanna and hot forest, and on the eastern side one such as ice, tundra, dry cold winter, wet cold winter, wet temperate and damp hot forest. By its radically different arrangement of land and sea Continental Drift will necessarily involve atmospheric and oceanic circulations—and, therefore, climates—quite different to those deduced under current concepts.

There is a tendency to overlook this basic zonal distribution of climate and to picture much of the earth as enjoying at any one period the same general sort of climate, which was supposedly warm and moist in the Carboniferous, arid in the Triassic, humid in the Jurassic, warm in the Miocene and frigid in the Pleistocene. In mitigation thereof it must be admitted that the supposed drifting asunder of the Lands, by moving them into other climatic settings, has done much to obscure their true relations in regard to the meridians and parallels of the past. This is not the place to discuss the causes of such climatic " revolutions " about which so much has been written ; our essential object is to trace out beneath these almost-world-wide and obscuring undulations of climate the earth's underlying zonal pattern and to deduce therefrom the approximate course of the equator of the time, and in that way the positions of the corresponding poles. Throughout this discussion the term "*pole-wandering* " or " *polar shift* " will explicitly refer to the creeping of the thin and distorting crust over its rigid core that has been rotating upon a *fixed axis* and not to a change in the direction of that axis in stellar space.

CRITERIA FOR CLIMATIC ZONING

The chief of these are (*a*) glacial deposits, (*b*) coals, (*c*) fossil soils, (*d*) desert sandstones, salt and gypsum, (*e*) coral limestones and (*f*) plant and animal life. The significance of most of these had been pointed out so far back as 1902 by Kreichgauer, while the subject has been ably dealt with in detail by Köppen and Wegener (24).

(a) *Glacials*.—Keyes has pointed out that ice-capping must have been far commoner than the imperfect geological record would suggest, that future investigation must extend the number and range of such climatic episodes and that each important glacial period must include a series of shorter rhythms. Whether the Earth was ever quite free from such frozen coverings is doubtful, but glaciation seems to have been reduced to a minimum during the Ordovician, Triassic, Jurassic and Miocene.

Such, of course, does not refer to glaciers of alpine type which could have formed in most latitudes and at many times, but to continental ice-caps that spread out over moderately high or even low-lying ground, in some cases covering archipelagos, and discharging into seas or lakes and producing extensive sheets of ground-moraine or till, that may alternate with inter-glacial deposits. The common view that such caps must have been developed symmetrically about the poles is borne out neither by the present Arctic nor by the Pleistocene glaciation, so that at various times in the past their positions might have been appreciably *eccentric*, though we may provisionally doubt whether the principal ice-margin would at any time have exceeded the distance of some 50 degrees from its pole. The entire region thus glaciated could moreover have embraced a series of over-lapping areas that were successively occupied and then abandoned by one or more migrating ice-caps. On meteorological grounds such migration should have been from west to east, as was indeed the case in North America in the Pleistocene and in South Africa in the Carboniferous. Under such circumstances the successive polar " positions " would have tended to follow an arc.

That the Alpine type was also represented during the past is indicated by certain occurrences of peculiar character, limited extent and proximity to growing mountain systems from which they derived their erratics, for example the sporadic low-level glacials of the Devonian and the Carboniferous of the eastern United States, both connected with the Appalachian chains and both situated paradoxically in what are believed to have been warm environments. Given sufficient elevation, however, such glacials could have been formed even in the equatorial zone, just as with Ruwenzori today.

(b) *Coals.*—Modern researches have shown that coals must have been accumulated under varied conditions, some *in situ* like modern peats, others in estuaries and lakes, and, further-more, from widely different kinds of vegetation. The climates in question were apparently as variable themselves though the botanical evidence thereon is surprisingly inconclusive, the least indefinite being the presence or absence of annual rings in associated fossil stems. " Luxuriant growth is not a feature exclusively tropical, it is more a question of water supply at the proper time " (W. T. Gordon). *Coals*, it would seem, *have been formed under both tropical and temperate climates*, and, curiously, in numbers of instances in proximity to areas showing low rain-falls. When dealing with our problem, therefore, each particular coalfield requires a separate study to decide its probable mode of origin.

In a rough fashion we can nevertheless regard coalfields as marking out *belts bordering closely either one or else both sides* of the drier sub-tropical zones.

(c) *Fossil Soils.*—Ancient soil profiles may prove good in-dicators as exemplified by the laterites, bauxites and kaolinites occurring in some of the coalfields of Laurasia that, according to Harrasowitz, originated in the equatorial zone. Certain pale sandstone formations, such as the Table Mountain sandstones of the Cape, recall the bleached podsols of high latitudes.

(d) *Desert Indicators.*—Of high significance are deposits of salt and gypsum from the evaporation of shallow seas (" evaporites ")—mainly within periods of regression—dis-seminated salts in the terrestrial accumulations of interior basins,

T

æolian and dune sandstones showing strong false-bedding and well-rounded grains, wind-facetted pebbles (dreikanter), breccias, strongly-coloured and ferruginous strata, loess deposits, superficially silicified rocks (silcretes), etc., indicative of an arid environment. Some discretion is, however, required in our interpretation, since red clays can be the product of tropical weathering as well.

The steppe and desert regions of today reach the oceans only along the western sides of the continents and, while they generally follow the high-pressure belts or dry zones—that is to say, latitudes 20°-30°—their extreme range falls between latitudes 12° and 50°, the second figure being only attained on the lee-sides of mountain chains. They can be divided into two kinds, hot and cold, and, furthermore, occupy quite a considerable proportion of the land surface. That ratio may, however, have been respectively greater or less during past epochs.

Brooks (26) has aptly commented on the former existence of "biological deserts," in which the vegetation was as yet unfitted to protect the surface, e.g. in the epoch of the Old Red Sandstone, and, doubtless, in other periods as well.

(e) *Coral Limestones.*—The living reef-building corals are restricted to the warmer waters, having a temperature of not less than 18° C., and occupy an equatorial zone that narrows on approaching the western sides of the continents, one that, in the case of the Pacific, is fully 45 degrees in width.

Various geological series show huge thicknesses of limestones that have been presumably built up in the same way. In the Cretaceous a similar part was played by certain mollusca, the Rudistids, and in the Tertiary by certain foraminifera that throve in the warm seas.

(f) *Animal and Plant Life.*—An abundance of life nevertheless characterises the cold waters of the ocean, and presumably the same was true during the past. Certain mollusca constitute indicators of cold conditions, such as the Pleistocene *Tellina* and the Cretaceous *Aucella*. Dwarfed forms commonly denote highly saline seas with the suspicion of an arid setting.

Unfortunately, not much weight can be placed upon the vertebrates, as a little consideration will show. Curiously the Permo-Triassic amphibia mark out strata of the semi-arid "red beds" facies.

Although it has been said that "plants are less trustworthy as instruments for measuring changes in climate than is generally supposed, even though the extinct species may be closely related to living plants," associations of fossil plants are distinctly valuable as geological "thermometers." Seward's [1] reviews of the Arctic floras are eloquent of the climatic changes that have

[1] Seward (30), 212; (31), 531.

taken place in that great region in various epochs, with a final chilling that pushed the poleward limit of trees far to the south of its former position. Irmscher indeed regards the trees along that floral limit as descendants of tropical ancestors. A useful criterion is the fact that most trees in the equatorial zone do not show marked annual rings.

POLAR POSITIONS

Our attempts to fix the poles of the time will be governed by several important considerations : (*a*) Laurasia and Gondwana, together with the ever-varying width of the Tethys between them, must be viewed as *separate and in certain respects as independent masses* belonging to the Northern and Southern Hemispheres, and their displacements, with reference to their corresponding poles, as having been not necessarily equal nor simultaneous. That is to say, an apparent movement of the North Pole along an arc would *not* imply an equivalent apparent shift of the South Pole in the opposite direction—as would be compulsory under any theory postulating fixed land masses, such as that of Kreichgauer, or, indeed, of current geology. This removes a serious stumbling-block met with in all other interpretations. (*b*) It is further necessary to discriminate between changes of climate brought about in a continent through (1) drift *en masse* and (2) drift of a portion thereof during dispersal. (*c*) It has not been established that one or both of the polar regions were ice-capped throughout geological history. (*d*) Such cap or caps may, as noted above, have been markedly eccentric. (*e*) Glacial periods have not been simple, but compound, with progressively shifting centres, such movement being in part connected with the general scheme of crustal creep. Sporadic glacials in low latitudes could probably occur only under abnormal conditions. (*f*) Coals, having been formed not only under a tropical environment, but under a temperate one as well, and under widely differing conditions in various places and times, may mark out either the tropical or the temperate zones. (*g*) The sub-tropical low-rainfall zones may have varied in width and position, but their latitudinal development would have been aided by the elongated shapes of both Laurasia and Gondwana.

Until the positions of the lands, when restored, can be fixed with more assurance than as yet, the various glacials correlated more closely, their former limits better determined and the importance of the climatic girdles more satisfactorily evaluated for each epoch, the successive locations of the poles and their apparent shift can only be deduced very approximately. It would indeed appear premature to compute, as Kerner, Köppen and Wegener have done, the geographical co-ordinates of places to within a few degrees for the pre-Tertiary epochs, while the same criticism would apply, though with less force, to Staub's latitude calculations for Europe. It cannot be too firmly impressed that this Climatic Method is not sufficiently sensitive to do more than establish the *major polar shifts* ; the minor movements can only be deduced from orogenic considerations.

While our polar restorations follow in a general way those of Köppen and Wegener, they depart materially therefrom, partly through the somewhat different fitting together of the lands and the different manner of their partition, but, above all, through the independency of horizontal movement postulated for the two great continents.

POLE-WANDERING IN LAURASIA

Whereas the climatic zoning can readily be made out, the fixing of the polar positions is fraught with difficulties, partly because of the intense repeated compression from the palæozoic onwards, partly because of the serious distortion thereby introduced, particularly in Central Europe and Central Asia. The originally wide belt of Carboniferous sediments extending from North America to China was, for example, forced under the Hercynian compression to occupy a space much narrower in the N.–S. direction.

The most ancient glacials are suggestively confined to North America—to Ontario, Quebec and Lake Huron,—but tillites are locally present at the base of, or just within, the Lower Cambrian at widely separated points—Utah, eastern Greenland and Spitzbergen, Finmark, Siberia (Yennesei River—Lat. 61° N.) and China (Yangtze River—Lat. 31° N.),—which are puzzling to explain under any scheme of climatic distribution whatsoever.

Zoning first shows up clearly during the late Silurian, when the continent was emergent, the Caledonian chains proceeded to rise and the climate became drier. Ohio, New York and Ontario fell within the dry zone, shown by the Salina formation with its red beds, salt, gypsum and æolian sandstones—also the Baltic with similar strata and probably the Lake Baikal region in Siberia. Warm seas lay to the south and colder ones to the north. The Silurian North Pole must therefore have lain far out in the North Pacific, a conclusion supported by the evidence that land lay in that direction and by the presence of an actual tillite in south-eastern Alaska (Kirk).

During the Devonian the dry zone expanded—its southern side well defined by the Downtonian (Chapter VII)—as exemplified in the Lower Old Red Sandstone with its breccias, conglomerates, oxidised sands, deltaic and lacustrine accumulations in interior basins bounded by high ranges—on north the Taconian, diagonally through the centre the Caledonian, on south the Appalachians, Ardennes and Suedetes, and in the far east those chains curving around the Angara platform. A warm, uniform climate is generally conceded for this period, though not necessarily one with a very low rainfall, since the vegetation was certainly not yet efficient for protecting the rocks from erosion. The absence, however, of annual rings in those trees shows an even climate. The later Devonian saw profound extension of the " Red Northland " till it embraced a long oval stretching from New York to Angaraland and from the Baltic across eastern Greenland to Ellesmere Land, a tract which was moreover crossed by the Ural and Mongolian seas (where the strata deposited carry marine intercalations). Thin coals were formed at various points, and also warm-water coral phases, along its southern side at intervals between France and China. From a study of the change of growth rate in the Devonian tetracorals Ting Ying (36) has concluded that the equator then passed from Scandinavia to Australia.

The Devonian North Pole must have been situated some 60 to 70 degrees north of Scandinavia, not far from that of the Silurian. In seeming conflict with such a conclusion are certain authentic glacial phenomena at sea-level from Pennsylvania (Huntingdon County), New York State and just possibly Quebec.[1] It can, however, be suggested that the neighbouring and lofty Appalachians could well have borne small glaciers even in the low latitude deduced, just like the Andes today. Furthermore the eastern part of North America might have been appreciably cooler than the region to the east.

The southerly drift with anti-clockwise rotation of Laurasia, thus deduced, was continued into the Upper Carboniferous before becoming reversed. This is indicated by the submergence of the Devonian wastes beneath the warm waters of the Lower Carboniferous seas, that, following closely the present 50th parallel, stretched continuously through Ireland, south Russia and Mongolia to the Sea of Japan, " red beds " conditions persisting for a while in Nova Scotia and Scotland.

The high-rainfall zone is furthermore defined, as emphasised by Wegener, by the broad belt of terrestrial deposits with abundant coals occupying a similar arc that extended from Alabama and Missouri through Pennsylvania, New England, Nova Scotia, Britain, France, Germany and south Russia to the Ural Sea, and beyond that, again, between Kansu and Shantung in north China, the latter along almost the present 40th parallel. Indo-China and Malaysia lay on the *southern side of this zone*. The multiplicity of coal seams, each usually reposing upon its

[1] Willard (35).

fireclay floor of lateritic composition, the luxuriance of the component vegetation, the absence of annual rings in the associated fossil stems and other botanical features which have been pointed

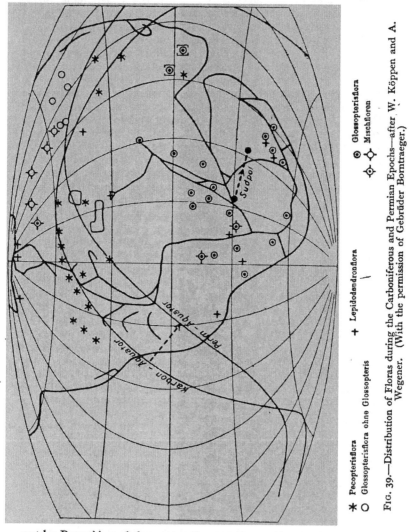

Pecopterisflora
Glossopterisflora ohne Glossopteris

+ Lepidodendronflora

Glossopterisflora
Mischfloren

Fig. 39.—Distribution of Floras during the Carboniferous and Permian Epochs—after W. Köppen and A. Wegener. (With the permission of Gebrüder Borntraeger.)

out by Potonié, and the mass of supporting evidence unquestionably favour a uniformly warm and moist climate and thereby allot these coalfields to the *equatorial rainfall girdle*. Such is well brought out on Köppen and Wegener's map, reproduced here (Fig. 39). It will be observed, too, from our somewhat

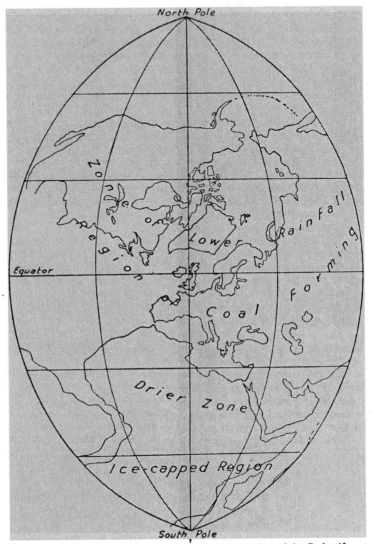

FIG. 40.—Approximate setting of the Continents at the close of the Carboniferous Epoch. No allowance has, however, been made for the several palæozoic compressions, which would appreciably alter the position of Laurasia.

different reconstruction (Fig. 40) that more of the coalfields lay to the north than to the south of the geographical equator of the time. Such is not peculiar, since the "heat equator" would have experienced some northerly displacement through the presence of the contemporaneous ice-cap of Gondwana. It must be noticed that this lengthy belt of coalfields does not coincide with Kreichgauer's "Carboniferous Equator."

It is most impressive to realise that, where the present 45th parallel now runs—from North America eastwards to Central Asia—there ranged belts that from the Silurian to the Carboniferous probably never oscillated in their latitude outside of 0° and 25°.

The progressive migration towards the south of coal-forming, down to the early Permian, denotes the persistence of tropical conditions within this girdle. In eastern North America as in the Devonian, local glaciation occurred, represented by the celebrated Squantum Tillite and Roxbury Conglomerate of Boston (Sayles), but it is doubtful whether the "exotic blocks" of the Caney Shales of Oklahoma are of glacial origin. Suggestively Permo-Carboniferous tillites have been reported from Alaska and the Alaska-Yukon boundary.[1]

The closing Carboniferous and the Permian show the gradual return of "Old Red Sandstone" conditions, namely the cessation of coal-forming in the western regions, the universal spreading of the "red beds" facies, the formation of curious breccias and the presence of salt and gypsum—best exemplified by the salt deposits of Germany. This dry zone ran from Texas and Kansas through New Brunswick, Britain, Spain, Germany and the Baltic eastwards to the Ural Sea and northwards to eastern Greenland, the equatorial zone being signalled by limestones of the Tethys in the Mediterranean region and eastwards.

On the contrary, coal-forming went on through the Lower Permian in north China and Manchuria along a belt trending a little north of east through Shansi, Shantung and Mukden, the seams being associated with fossil laterites that were formed under a hot and humid climate. In contrast, the Upper Permian includes red beds, lagoonal deposits, æolian sandstones, salt and gypsum, not only in north but in central China (Szechuan and Hunan) as well. The widespread Kusnezk coals of Angaraland, occupying a region stretching from the Kirghiz steppes to the Arctic shore, must, like those of Gondwana at this very period, have originated in the *wet temperate zone*.

All this indicates a *reversal* in the drift of Laurasia and a movement to the north, somewhat greater, too, in Asia than in America, and hence some continued although slight anti-clockwise rotation.

The Triassic saw the repetition of "Old Red Sandstone" conditions to a wonderful degree with its climax in the Keuper,

---

[1] Coleman (26), 180.

only the dry region lay farther to the south. It then stretched from western North America through Nova Scotia, Great Britain and Central Europe to the Ural Sea, failing to reach Spitzbergen, but, on the contrary, extended across Spain and Portugal and the Atlas region and thus enveloped the northern edge of Gondwana. Typical are its torrentially-formed deposits, cross-bedded and oxidised sandstones, red marls and clays, salt, anhydrite, and gypsum layers and wind-faceted pebbles, and, more rarely, the wind-etched landscapes of the time (Charn-wood Forest, Leicestershire). Similar strata with salt occur in Turkestan and China (western Shansi). Over large areas an intense aridity prevailed, fostered by the immense area of con-tinuous land, the Appalachian-Hercynian barrier ranges and temporary obstructions across the Tethys channel that hampered oceanic circulation. In impermanent lakes, chiefly along the southern margin, coals were formed, that attained a greater importance in the late Triassic in the far east—central and southern China and Tonking, where they are nevertheless frequently associated with red beds. Luxuriant Rhætic floras, however, characterised eastern Greenland and Sweden.

The equatorial zone is signalled by coral and algal limestones in the Tethys geosyncline between the West and East Indies with sharply defined northern limit in the Alpine region, This is brought out on Köppen and Wegener's map, reproduced in Fig. 41. The northward drift of Laurasia must have brought the pole not far from Alaska, though no corresponding glacials have yet been recorded from that quarter

The universally greater humidity in the Jurassic weakened, where it did not destroy " red beds " conditions, but the latter persisted over a broad zone extending from New Mexico to Nevada, wherein thick desert sandstones were deposited, and in a milder form over wide areas in Central Asia. Coals were abundantly developed in continental or paralic formations in the Tianshan, north and south central China, Manchuria, Korea and Ussuriland, commonly associated with some red beds.

The floras, with their abundant cycadophytes, are distinctly uniform throughout the Jurassic and into the early Cretaceous, this pointing to mild temperatures even in such high latitudes as eastern Greenland, Spitzbergen, Franz Josef Land and New Siberia, thus recalling the conditions during that epoch in the Southern Hemisphere.

During this period the marine faunas became, on the contrary, more diversified and reflect in their distribution the influence of climatic zoning, as long ago made clear by Neumayr. Thus the Boreal realm reached southwards from the Arctic to California, Volga region and Japan, while adjacent thereto extended the

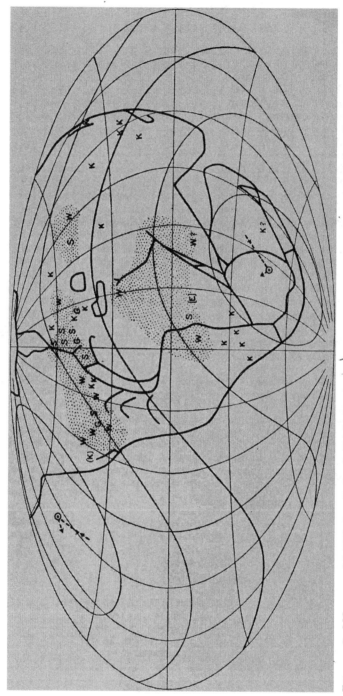

FIG. 41.—Coal Measure and Arid Regions during the Triassic Epoch—after W. Köppen and A. Wegener. *E*, Glacials; *K*, Coal; *S*, Salt; *G*, Gypsum; *W*, Desert Sandstone; stippled, Arid Regions. (With the permission of Gebrüder Borntraeger.)

narrow Caucasian and to the south thereof the warm-water Mediterranean realm. The Jurassic equator must therefore have run through Central America, Sahara, Mediterranean, Persia, Tibet and Indo-China, which would imply a continuance of the northerly drift of Laurasia.

Rather similar conditions persisted into the Cretaceous. The full width of the equatorial zone is defined by the distribution of the Rudistid limestones that stretch in a belt some 30 degrees in width from Mexico [1] through North Africa, Southern Europe, Persia and Himalaya, the small apparent breadth being due to narrowing through the Alpine "packing." The supposition of a general ample and uniform rainfall finds support in the restricted development of steppe deposits, such characterising particularly Shantung and Mongolia (between Lat. 35° and 45° N.). The luxuriance of vegetation resulted in widespread coals, formed mainly within the wet temperate zone—from the Rocky Mountain region through British Columbia, Alaska and Eastern Asia into Mongolia. Confirmation of this deduction is secured from the associated marine faunas of boreal affinities. Reference can be made to Köppen and Wegener's map depicted in Fig. 42.

During the Cretaceous, North America and Eurasia proceeded to draw away from one another, the poleward motion of the whole being converted into movements of the two portions towards the west and east respectively, otherwise "Laurasia" appears to have remained in more or less the same latitude, though for a time it may even have acquired a slight southerly drift.

From the Cretaceous onwards we can accept a series of polar "shifts" like those worked out so thoroughly by Kreichgauer and by Köppen and Wegener—namely across Behring Strait, reaching Baffin Land by the beginning of the Pleistocene, and thence across northern Greenland to its present position. Such a course would agree well with Davidson Black's enlightening researches for China (31), which denote a practically similar shift. A general movement can therefore be deduced for North America and Eurasia at first north, then north-east, thereafter north again and finally east, modified to some extent by the continued *divergence of the two continents.*

POLE-WANDERING IN GONDWANA

The rigid and undistorted nature of nearly all the interior of Gondwana from the mid-palæozoic onwards, on the contrary, enables its climatic zoning to be done surprisingly well.

[1] Bordered along its northern edge in Mexico by important gypsum and other deposits suggestive of the high-pressure zone (Imlay).

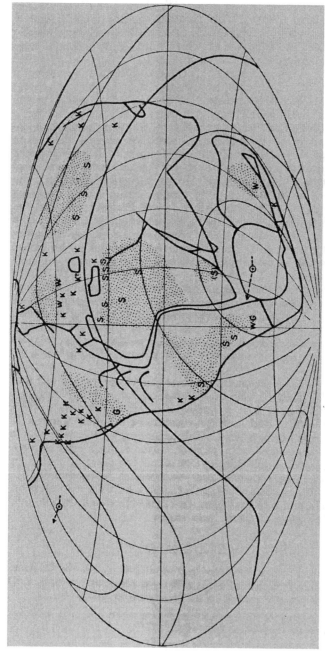

FIG. 42.—Coal Measure and Arid Regions during the Cretaceous Epoch—after W. Köppen and A. Wegener. *K*, Coal; *S*, Salt; *G*, Gypsum; *W*, Desert Sandstone; stippled, Arid Regions. (With the permission of Gebrüder Borntraeger.)

The evidence suggests that throughout much of geological time the South Pole wandered within or close to the confines of that continent. Astonishing is it to find in the southern half of Africa so many occurrences of tillite—some having a great extension—of ages ranging from the Archæan to perhaps early Cambrian : for example, the extremely ancient ones of South-west Africa (Chuos), that of the Witwatersrand, the late pre-Cambrian ones of Namaqualand (Numees), southern Angola, Lower Congo, Katanga and Uganda and the possibly Cambrian ones of Griqualand West, Transvaal and Southern Rhodesia, while outside of Africa exist the probably Eo-Cambrian tillites of South Australia (Sturtian) and certain possible fluvio-glacials in Minas Geraes, Brazil (Fig. 9). Now these regions are precisely those known or believed to have again been ice-covered during the Carboniferous, and, as our basic assumption must be the minimum of pole-wandering, we must presume that from that time backwards into the pre-Cambrian the pole was probably situated somewhere in or near to South Africa. This view finds support in the early Devonian glacials of the western Cape and the cold climate aspect of the associated sediments and faunas around western Gondwana generally.

A regular arc can be drawn linking together the areas with warm-water marine faunas in western Venezuela, Atlas region, Asia Minor, Himalaya, northern Burma, North-western Australia, Queensland and New Guinea, that manifestly signals the southern edge of the Devonian equatorial zone.

From the fact that glacial conditions spread so suddenly and widely in the Carboniferous, it would not be unreasonable to conclude that just prior thereto, that is to say in the late Devonian, the drift of the continent must have been generally polewards, i.e. *southwards*.

For this great palæozoic ice-age the position of the South Pole can be fixed tolerably well, and that, too, irrespective of whether the bulk of the glacials are viewed as of Upper Carboniferous or, as maintained by Schuchert, of Middle Permian age. The areas glaciated, the directions of ice-flow and the overlapping of the various and probably migrating centres collectively point (Fig. 9) to some spot just off the coast of Natal or southern Moçambique as the most likely mean polar position during the closing Carboniferous. New discoveries may help to secure a more precise location, though the possibility of an eccentric relation of the pole to its ice-margin should not be overlooked. The later Permian glacials of New South Wales, and the ice-borne erratics in the Permian of North-western Australia (Condit) and Queensland, suggest that any polar shift was from Africa towards Australia and hence in a south-easterly direction. From the fact that the Lower Permian faunas of the

Malay archipelago were warm-water ones, it has been argued
that such is incompatible with the climatic conditions deduced
for Australia.   This is not really the case, for in that epoch the
Malay region lay under our view much farther to the north, its
proximity to Australia today having been determined by their
mutual approach during the Tertiary.   Furthermore, the characteristic fusulinids of the Malay region are not represented in
Australia (Chapman) though the general profusion of life in the
Lower and Upper Marine Series would suggest that the ice may
have discharged into warmish water.

The Permo-Carboniferous low rainfall zone is traceable for
only limited stretches in Bolivia, northern Argentina, northern
Brazil and western Sudan—the latter area being marked by
continental sandstones—the rest of the northern margin of
Gondwana being lapped by ocean with warm-water faunas,
while Morocco formed an island or more probably belonged to
the southern edge of Laurasia.

The Permian saw the development over the very region previously ice-capped—including in this the eastern side at least of
Antarctica—of vast lakes, swamps and forests, and of coal-
forming from the Glossopteris vegetation, more particularly at
two periods.   That the climate was not only wet but cold, at
least in many places, can be deduced from the inferred proximity
to the pole, the presence of local ice in Australia, the limited
development of fire-clays beneath the coal-seams, the presence of
fossil wood showing annual rings and the abundance of unde-
composed felspars in the associated sandstones, certain of which,
as in Natal and Antarctica, are coarse arkoses.   A wide extension of the South Temperate zone must accordingly be presumed.

Some southerly or south-easterly retreat of the pole is suggested by the appearance of red beds—at first scantily—in
central Argentina, east-central Brazil and South-west Africa,
with considerable spreading during the Lower Triassic to the
Karroo and thence through East Africa to India.   Such was
aided by the throwing-up of the Gondwanide chains that cut
off moisture from the interior region.   A further shift of the
pole to the south-south-east or south-east, thereby throwing it
into the Pacific, is signalled during the late Triassic by the extension of low-rainfall conditions over a zone extending north-
east to east-north-east from northern Argentina across to central
India and reaching the ocean at either end.   Gypsum and salt
characterise its equatorial side in Eritrea, Arabia and Persia.
The red beds of Central Brazil are in places gypseous.   From
its median line there can be made out a general gradation
both to north-west and south-east from desert to temperate
types of sediments with sporadic coals, the last being developed
mostly along its southern side.   The vertebrates, and even the

amphibians, followed this low-rainfall zone and must have been specially adapted for a steppe-like environment. A belt some 40 degrees wide separated this zone from the edge of the corresponding dry belt of the Northern Hemisphere, which appeared in the Atlas region with its " germanic " facies of the Triassic. The Triassic equator must therefore have run in a direction a little north of east through the southern Sahara and Arabia, as the warm-water faunas of northern Peru, Persia, Himalayas, Yunnan and Malay Archipelago will confirm. The insect fauna of the Ipswich Series in Queensland betokens a warm climate. Reference can particularly be made to Figs. 11 and 41.

In East Africa and Madagascar and perhaps in southern Tunis steppe conditions lingered on into early Jurassic times, but a more humid climate became almost universal in that epoch, as shown by extensive lacustrine and deltaic deposits in the Congo basin, West and East Australia and Indo-China. Medium to warm temperatures are indicated by the absence of glacials and by the striking uniformity of the floras with their ferns and cycads even from places so far removed as Graham Land, Patagonia, India and New Zealand. Coals occur in Victoria and Queensland. The zonal distribution of the marine faunas was long ago pointed out by Neumayr, significant being the cold-water forms of the long even shore-line that stretched eastwards from Patagonia to New Zealand. In the Jurassic the South Pole must have lain well out in the southern Pacific.

The above conclusions accordingly disclose a progressive drift of Gondwana from the Carboniferous at first north-westwards and thereafter northwards and, furthermore, with anti-clockwise rotation until the close of the Jurassic, when the motion was *reversed*, doubtless because (*a*) a considerable proportion of the continent had by then moved beyond the equator of the time and (*b*) the fossa extending between Chile and New Zealand had become unstable, both of which influences would now be producing horizontal forces directed southwards, i.e. polewards. In the early Cretaceous Africa tore itself free, renewed its northerly drift into warmer climes, came again into conflict with Eurasia and proceeded to push up the Atlas-Persian ranges.

Attention need only be directed to the widespread " Nubian Sandstone " facies of the Cretaceous arid zone in Nigeria, Sudan, Egypt, Arabia and Somaliland with the equatorial belt marked out to the north of it by the E.–W. trending tract of Rudistid limestones in Morocco, Tunis, northern Sahara, Egypt, Syria and Persia (Fig. 42). The movement of Africa was thereafter intermittently northwards, with the consequent squeezing and overriding of Eurasia, and need concern us no further.

The irregularly-shaped remainder of Gondwana, bounded on the north by the U-shaped fracture corresponding to the periphery of South Africa and on the south by its marginal migrating fossa, crept southwards, i.e. polewards, the consequences of which are to be seen in the cold-water Cretaceous faunas with *Aucella* of its southern side, the ice-transported erratics derived from a southerly source in the mid-Cretaceous of South and Central Australia with associated fossil wood showing narrow annual rings.

Upon the breaking away of the " wings " made by South America, India and Australasia, the Antarctic nucleus continued on its southerly drift through the Tertiary and thus came to acquire its present co-polar position. The radial dispersal of the three fragments, on the contrary, took them back into progressively warmer latitudes, their climatic histories thereafter proving consistent with the main idea expounded above. For instance, significant for the Cretaceous are the widespread continental deposits of central South America, the dry phases of the north Andean region, the steppe-like land surface beneath the Lameta Beds of the Deccan, the coals of Queensland and New Caledonia, the Rudistid limestones fringing the north of South America and those of the Himalayas and Borneo (Fig. 42). The biological evidence relating to the Patagonia–New Zealand belt is in conformity with the above scheme.

THE REALITY OF DRIFT

While the data used are admittedly much generalised and often fragmentary, they are drawn from stratified formations *deposited mostly close to sea-level*, so that the altitude factor—so disturbing in present-day zoning—becomes of minor importance. Nevertheless no pretension towards great accuracy is, nor can be, claimed.

An excellent test of the above reasoning is nevertheless provided by the vital fact that the polar positions of Laurasia and Gondwana have been arrrived at *separately and quite independently for each mass*, yet prove on comparison to be mutually consistent. W. R. Eckhardt [1] has incidentally made the strong point that during late geological times the evidence would suggest very definitely an apparent polar shift to the north-east for the New World but to the north-west for the Old one, a contradiction only explicable if the North Atlantic had simultaneously been widening through Drift.

Even making allowances for the many uncertainties in this

[1] Eckhardt (21), 262.

problem, it will still have to be admitted that the positions of the present continents must have experienced radical changes in their distances from the poles through geological time. Since polar movement, in the sense of a changed axis of rotation, is generally denied by geophysicists, it assuredly follows that those lands must have suffered displacement across the surface of the globe with reference to that fixed axis. Indeed, from the mid-palæozoic onwards the lands must have crept *northwards* for thousands of kilometres to account for their deduced climatic vicissitudes. *Such, indeed, constitutes the most telling demonstration of the reality of Continental Drift.*

U

CHAPTER XIV

BIOLOGICAL RELATIONSHIPS

Similarities. Evidence of Marine Faunas. Palæontological. Land-bridges. Life in relation to Past Land Connections.

SIMILARITIES

THE strong similarities displayed by the terrestrial life of the continents, unquestionably more marked, however, during the past, forms one of the great facts and marvels of Nature. Such, indeed, constitutes the basis of our conception of former Land Connections.

Many seeds and certain other organisms could have survived transport by ocean currents for perhaps considerable distances, while a few invertebrates might have been carried by floating timber, but the " rafting " of larger vertebrates, advanced in all seriousness by a few scientists, overtaxes one's credulity, wherefore revolutionary changes of land and sea have had to be invoked to explain the observed biological affinities.

The vast body of data to hand brings out in striking fashion how *out of accord existing terrestrial life is with present land-relationships*, more particularly in the case of those groups or families that appeared some distance back in time. Such incongruity is brought out by (*a*) numerous and surprising resemblances in the life of most of the now widely parted lands, e.g. South America and Africa, (*b*) the lack of agreement at certain periods between areas quite close together, e.g. Madagascar and Africa, (*c*) the " concentric " distribution of long-lived families in which differentiation is marked, e.g. the snails and scorpions, and (*d*) the migration of assemblages by routes wherein the climatic conditions are today adverse or prohibitive e.g. Patagonia and Australia. Whereas the relationship (*a*), which is the most conspicuous one, is readily explicable by land-bridges in the current sense, (*c*) and (*d*) favour, on the contrary, some different and closer geographical fitting of the lands. The hypothesis of Drift can successfully interpret, furthermore, many of the difficulties met with in the division of the Earth into the various " realms " of Zoogeography

EVIDENCE OF THE MARINE FAUNAS

It is probably overlooked by many persons that such faunas are scarcely less important than the terrestrial ones for indicating former continental unions. Excluding pelagic forms of life—such as the radiolarians and certain cephalopods—the echinoderms, bryozoans and most of the mollusca demand shallow or at the most moderately deep sea. The spread of these latter types of organism would take place either along the shore upon the continental shelf or else over the bottom of a shallow sea.[1]

Throughout geological history we find innumerable instances of specific or allied fossil forms in the equivalent marine formations of countries now widely parted by deep ocean; indeed such a fact constitutes the basis of palæontological correlation. Furthermore, where the two areas are situated in corresponding latitudes, an assumed migration by round-about courses to the north or south might be negatived by the differences in the characters of the faunas *en route*. Unless the unwarranted assumption be made that most of the ocean deeps came into existence during the late Tertiary, we must postulate either shallow seas or continuous and relatively direct shore-lines between the continents in question to explain such related fossil faunas. Identity of species is not demanded, since differentiation would have occurred wherever the (sub-aqueous) distances were considerable. A good example is made by the Mediterranean affinities of some of the Tertiary faunas of the West Indies, pointing at the least to a submarine ridge right across, covered at times by only shallow water. Migration, being influenced by currents, could have been more effective in the one direction. Of high significance is the recognition of faunal facies and provinces. Faunas can be used also to prove the separation of land-masses. Thus the crayfishes of the family Parastacidæ characterise the Southern Hemisphere but not Africa (Ortmann), while a similar relationship holds for the corals and primnoids (Versluys), which is in full accord with our hypothesis of the earlier breaking away of Africa from the rest of Gondwana.

The laws governing the distribution of the fishes are doubtless peculiar, for they do not appear to spread freely although the sea is open to them.

PALÆONTOLOGICAL

When comparing the former terrestrial life of the now far-spaced lands, too much weight should not be placed upon the

[1] See Schuchert (32), part iii.

apparent absence of particular forms because of the limitations in fossil collecting and because of the obscurity of the factors determining the spreading of organisms.

Again, forms that could at the most be regarded as "varieties" would too frequently, under current views, be described as distinct "species," *mainly because of the distance separating the lands in question.* Furthermore, a specific identity would not be demanded under our Hypothesis, for the nearest exposures in the opposed lands must, in any case, have been originally some distance apart (p. 55).

In the case of Laurasia, only broken across by the North Atlantic–Arctic Ocean, periodic connection is demonstrated by the Old Red Sandstone fishes and primitive plants, by the Carboniferous and Permian amphibians, fresh-water mollusca and flora of lycopods and pteridosperms, by the Triassic reptiles, the Jurassic and Cretaceous dinosaurs, ferns and cycads and by the Tertiary mammals and angiospermous vegetation, examples having been cited in Chapters VI and VII.

Although Gondwana was more widely dispersed, the resemblances in its case are not less striking as displayed by the uniform Permo-Carboniferous Glossopteris flora, the Permo-Triassic "Karroo" reptiles, and the Jurassic and Cretaceous dinosaurs, ferns and cycads. Separation thereafter led to general isolation and to evolution upon different lines in the several portions during the Tertiary.

In both hemispheres the resemblances are often not only generic but specific, and such agreements have, on the whole, been strengthened by each new discovery. Arldt (17) has dealt very fully with this fascinating subject, more particularly in reference to the probable or possible positions and times of the presumed land-bridges. An outstanding inductive study is the *Palæogeographie* of F. Kerner-Marilaun.

LAND-BRIDGES

Palæontologists, zoologists and botanists have one and all been forced to accept the hypothesis of more or less temporary connections between the lands—ranging in width, according to the ideas of each person, from bodies occupying the full length and breadth of the intervening ocean, through isthmuses, to mere chains of islands—which ultimately sank beneath the waves. Too often the points and times of such linkages are left vague.

In the majority of instances the grounds advanced are weighty or indisputable. One has only to study the writings of Neumayr, Blanford, Sclater, Suess, Arldt, Dacqué, Gregory, von Ihering and Schuchert to appreciate this.

In opposition thereto the majority of geophysicists have maintained that the principle of Isostasy negatives any deep or permanent sinking of areas of continental, and perhaps even of isthmian, dimensions.

A few, among whom stands Gregory, hold that such an attitude is not justified by the biological facts. The writer regards the assertion that " land-bridges cannot be sunk " as altogether too sweeping. If the area in question fall within a region of epeirogenic movements, sinking of the block to profound depths would appear to be physically impossible. When, however, the bridge is made by a geanticlinal structure or welt, its crest could become depressed beneath the ocean by subsequent *crustal tension* not only across but along its length ; in the latter case it could even become broken into segments to form an island-chain before vanishing. In this manner, it is maintained, links were intermittently built up and destroyed during the Cretaceo-Tertiary through stretching in the direction of their lengths while they were still being compressed by forces at right angles thereto. " Advance folds " could also have played a similar part (p. 43). As examples are the island-arcs of the Pacific and the connections between the West Indies and the Old World. These methods of linking and unlinking between the lands are indeed regarded as having been operative throughout geological time.

Those who have maintained that vast blocks—" inter-continents " they may be called—have subsided to form the oceans should note Nissen's penetrating observation that in such a case any close correspondence between the outlines—and, we may specially add, the geological structures—of the opposed shores would be improbable. Let us imagine North America divided into three strips and the central one, say of ten degrees in width, submerging. The amount of structural agreement between the opposed edges of the two outer strips would be good only if the new ocean ran E.–W., being fair if the latter trended N.E. or N.W. but trivial if it extended N.–S.

Too many of the postulated bridges possess the inherent weakness of a vagueness concerning their precise route or a lack of supporting geological evidence at either end. Bailey Willis merely makes his " isthmian links " follow the shallows of the existing oceans. On the contrary, under our conception of unstable geanticlinals functioning as bridges the positions of such links become positively fixed at either end, for example, the connection visualised as joining Venezuela to the Atlas in the Tertiary.

Biologists have, moreover, been demanding an inordinate number of such links ; indeed, were all the deduced connections to be plotted on a globe, they would recall the criss-crossing system of deep-sea cables and bear as little relation to one another in time and space ; few parts of the oceans would not at some period have been thus spanned. Such a haphazard network, being palpably absurd, indicates that current ideas of continental connection are *fundamentally unsound* and so induces us to accept the principle of Drift. This in turn wipes out straightaway all those mythical lands, such as Atlantis, Archhelenis, Archipacificis, Archinotis, Lemuria, Tasmantis, Flabellites Land, etc., that have been conjured up to explain the life-distribution of the earth.

As Baker has remarked, land connections fall into two categories ; those that existed *before* and those developed *after* fragmentation had begun, and both of these can be accounted for under our hypothesis. The scheme adopted by certain persons, for instance Bailey Willis, of aligning such bridges along the higher elevations of the present ocean floors is not beyond criticism, since the oceanic deeps are relatively youthful, while profound changes in elevation have in many places been produced by forces acting so late as the Pliocene.

Two biological aspects of narrow land-bridges need stressing : that (1) migration along a link need not be equally effective in both directions, whereas our palæontological studies commonly favour a two-way interchange ; and (2) faunal differences at the ends of a link, as von Übisch (quoted by Wegener [1]) has pointed out, ought to be appreciable owing to the differentiation that must have occurred along the bridge itself, whereas a close specific relationship or even identity of forms may indeed characterise the life of the now-opposed lands.

LIFE IN RELATION TO PAST LAND CONNECTIONS

The undoubted likenesses between the fossil or living forms of certain ocean-parted lands has, and not unreasonably, been ascribed by some workers to *parallel development* or *convergent evolution,* but, the moment that other lines of evidence point definitely towards Drift, such objections fall away and the observed biological resemblances constitute instead an important argument in support of that hypothesis. Particularly would that apply in the case of identical successions of whole faunas and floras and of unrelated species belonging to different groups— " *multiple relationships* " Baker [2] has aptly termed them.

Of even greater weight would be the presence in the two lands of organisms *parasitic* upon such comparable forms, an association emphasised by von Ihering, Metcalf and Harrison (as cited by Gregory),[3] well instanced by the southern Frogs.

[1] Wegener (24), 20 ; (29), 105.   [2] Baker (32), 186.
[3] Gregory (30), cxiv.

Whereas the resemblances between the hosts could conceivably be regarded as a possible example of parallel development, it would be in the highest degree improbable that the same could be true for their parasites as well.

Baker has pointedly remarked that the continental fragments cannot always be viewed as " biological units," since each may have been separated at times into two or more parts which became subsequently welded together, wherefore all the continental areas are today *faunal and floral composites*. Again, on comparing continents one is faced with great extensions in

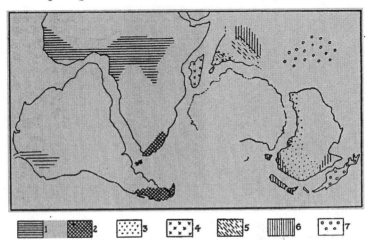

FIG. 43.—Showing the distribution of the living Rain-worms of the Family Megascolecina. 1. Dichogaster; 2. Chilota; 3. Megascolex; 4. Howascolex; 5. Octochætus; 6. Perionyx; 7. Pheretima. (Adapted from Michaelsen.)

latitude and hence large differences in their environment. Baker [1] has furthermore stressed the differences that would result from either direct, oblique or indirect linking, and has quoted numerous important examples thereof.

Of outstanding significance in our problem are certain maps that have been drawn to show the distribution of particular organisms, such as those for the snails by J. W. Taylor, for the scorpions, snakes and lizards by J. Hewitt (23), and especially for the earth-worms by Michaelsen (22), one of which is reproduced (Fig. 43). These depict geographical groupings for the families that are not only in strict conformity with the Drift Hypothesis but not readily explicable on any other basis. In his analysis Michaelsen shows himself a definite supporter of Wegener. Remarkable in the case of the snails is the concentra-

[1] Baker (32), 197.

tion of the primitive groups into the extremities of South America, Africa, Madagascar and Australia with the more specialised ones occupying successive zones to the north; the scorpion distribution as set forth by Hewitt (25) follows similar lines (Fig. 44). Such would strongly suggest an evolutionary centre situated in the south with successive waves of dispersal towards the north, but, even if we accept the contrary views of Taylor and Hewitt that the primitive groups have been forced southwards, the *arrangement is nevertheless still in accord with our geographical restorations.*

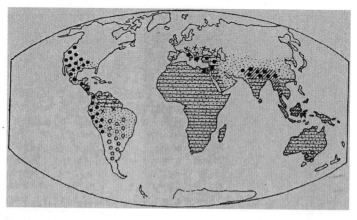

Fig. 44.—Distribution of the Scorpions throughout the World: 1. Buthidae; 2. Chærilidæ; 3. Vejovidæ; 4. Charactidæ; 5. Scorpionidæ; 6. Both-riuridæ. After Hewitt.

Biological evidence is valuable also in the case of "border-lands" where two impinging continental margins have extended themselves alternately over an intervening and fluctuating sea, such as the Indo-Chinese region during the interplay of Laurasia and Gondwana in the mesozoic era.

It must be emphasised that the outlook of Zoogeography, that is to say, the division of the earth into "realms" and the origin of the life therein, would be profoundly altered by the recognition of Drift. The mass of information available is already enormous, and only the briefest of references thereto can be made, the reader being directed particularly to the reviews by Irmscher, Wegener (fourth edition), H. B. Baker and J. W. Gregory.

We can deduce therefrom not only the former unity of

Gondwana but the order of its partition. Thus the blind snakes (Typhlopidæ) are confined to Gondwana, Cichlid fishes to its northern margin: the soft-bodied Peripatus, Phreodrilus and Phreatoicus to the southern side ; the widespread Acavid snails [1] are represented in Laurasia only in the Oriental region : certain primitive frogs occur only in South America and Africa and the fresh-water turtles in them and Madagascar.

The drawing-away of Africa is reflected by the presence only in South America and Australia of the cystignathid frogs and the diprotodont marsupials, of certain kinds of snakes and iguanas in South America and Madagascar but not in Africa, and by many genera and species of plants with similar peculiar distribution (Andrews ; Oliver). The objection advanced by Andrews (16), that plant migration could not have been possible *via* Antarctica, loses much of its force when it is recollected that under our hypothesis Antarctica lay well to the north of the pole during much of the Tertiary.

The separating of Madagascar is shown by the fact that, despite its nearness to and former association with Africa, its living mammal and bird faunas are more with the Oriental than the Ethiopian realm (Lydekker).[2]

Through its drift and isolation Australia became the home of the monotremes, and was moreover prevented from receiving the higher eutherian mammals from Asia. The zoological boundary between the Oriental (Malay) and the Notogæic (Australian) region, known as " Wallace's Line," is fairly sharp and, with slight modifications, follows closely the tectonic Banda arc.

While the floral resemblances betoken the former union of Australia and New Zealand, their separation is evinced by the fact that certain genera of microlepidoptera common to both Australia and South America are absent from New Zealand (Meyrick (25)).

Connection between Brazil and North Africa—by way of the Tertiary ranges between Venezuela and Morocco according to our view—is shown by that mammal the manatee (now living in estuaries), the fishes, spiders, microlepidoptera, fresh-water bivalves, land mollusca (Pilsbry ; Germain), earthworms (Stephenson) and plants (Engler). A similar and parallel linking between the West Indies and Spain (giving perhaps an occasional mixing with the first route) is indicated by the interchanging of forms such as the fossil *Hipparion* and certain antelopes, a view supported by the shallow-water molluscan faunas. Gregory's [3] presentation is well worth studying.

---

[1] The South African Acavidæ are more closely related anatomically to the South American than to the Australian forms (Watson).

[2] Lydekker (11), 1009.   [3] Gregory (29), xcix.

Outstanding are the relationships in the case of Laurasia, that is to say, those between North America, Greenland and Europe on the one hand and Eastern Asia and Alaska on the other, which have indeed caused zoologists to place North America and North Eurasia in a single realm—the Holarctic.

Connection *via* Behring Straits enabled elephants and rhinoceroses to enter North America from Asia and horses to pass in the reverse direction during the Middle Miocene, which link became broken in the late Miocene, though its temporary restoration during the Pliocene permitted the camel to enter Asia.

Similarly there was direct exchange of mammalian life between North America and Europe in the Oligocene, a migration terminated by the intense rifting that took place in the North Atlantic–Arctic region, though Gregory believes that the last link—*via* Iceland—was not snapped until perhaps early Palæolithic times. Supporting evidence can be got from certain fresh-water fishes, the earth-worms (Michaelsen ; Stephenson), insects (Handlirsch) and plants (Fernald). H. E. Forrest, citing long lists of such common or allied plants, points out that the Arctic flora is American rather than Siberian, and that the presence of various temperate forms would suggest a southern as well as a northern route between North America and Eurasia. Incidentally, the collective evidence presented by Forrest[1] is strongly in favour of his stimulating hypothesis of the existence of a central " Atlantean Continent " down to the Pleistocene, which is worthy of a close study

[1] Forrest (35), 243, 251.

CHAPTER XV

GEODETIC EVIDENCE

WEGENER claimed that his hypothesis had the merit that it could be verified by astronomical observations, and the figures submitted by him undoubtedly made a deep impression upon the public mind. Newer and more compelling information was given in his fourth edition (chap. 3). So far back as 1886 " Oldham drew attention to the recorded small changes of latitude at certain observatories and to the probable changes in azimuth in the Pyramids of Egypt." [1]

Wegener quoted J. P. Koch as the first observer to discover evidence for a geographical change of position emerging from the repeated longitude determinations of Sabine Island in eastern Greenland made by Sabine (1823), Börgen and Copeland (1870) and Koch (1907), which appeared to show a progressive movement to the west amounting to some 1610 m. in all. It was recognised that the same spot in Sabine Island may not have been occupied in 1870 as in 1823, while Koch's measurements were made to the north at Danmarkshafen, which was ultimately connected with Sabine Island by a triangulation.

Critics at once seized upon the fact that the earlier determinations were based upon observations of the Moon—the accuracy of which would be far less than those of modern times using the stars in combination with radio time-signals—while the reliability of the lengthy triangulation network was also questioned —especially by Penck—despite Koch's contention that the probable errors introduced thereby were less than those involved in the longitude observations themselves. In 1932 Jelstrup accurately re-determined the longitude of Sabine Island and, rejecting the figures of 1823, found for the period 1870–1932 an apparent westerly shift of some 615 m., or about 10 m. per year, which, after applying other probable corrections, became reduced to some 250 m., or 4 m. per year.[2] Similarly the observations made at Kornok (Orornoq) in the Godthaabfiord in western Greenland by Falbe and Bluhme (1863), von Ryder

---

[1] Holland (15), 350.
[2] The writer desires to thank Prof. F. Debenham for kindly supplying extracts from the orginal papers which were not otherwise obtainable.

(1882–83) and Jensen (1922)—the last-named employing stars and radio-signals—showed an apparent westerly drift of some 980 m. It is important to note that, even if all the measurements made prior to those of Jensen be rejected, the reoccupation of the station by Sabel-Jörgensen in 1927, only 5 years later, showed a difference in longitude far in excess of the probable errors, one corresponding to a westerly movement of some 36 m. per year.

Furthermore the repetition by Tollner-Kopf [1] in 1932–33 of the measurement made by von Basso in 1882–83 of the longitude of Jan Mayen Island gave a difference corresponding to an apparent drift, also to the west, of about 25 m. per year, which is more than ten times the probable errors of both sets of observations. In each of these cases no appreciable changes in latitude were detected.

As Wegener has emphasised, it would indeed be extraordinary if the systematic errors of the various observations invariably fell in the one direction, and it must therefore be concluded that a positive shift of crustal matter has been instrumentally demonstrated. Momentous is the fact that such shift is precisely in the direction indicated by the geological evidence. The reluctance of so many persons to concede the probability of such movement is hard to understand in view of their readiness to admit tectonic movement generally, and the more-so since the region is one expressly marked out by the sinking of crustal blocks during the later Tertiary.

Wegener furthermore mentioned Littell and Hammond's review of the longitude-differences between Paris and Washington, which would point to an increase of 0·32 m. per year over a period of 13–14 years, and cited figures that would suggest an appreciable easterly movement of Madagascar.

Illuminating are the successive longitude determinations of Sydney Observatory, New South Wales, detailed by Baracchi.[2] Excluding all the older and less reliable ones, there remain those of 1883 (by telegraph) giving 10 hr. 4 min. 49·54 sec. east of Greenwich and of 1903–4 (by cable, both east and west) 10 hr. 4 min. 49·32 sec. With these can be compared the measurements of 1926 (with radio signals) giving 10 hr. 4 min. 49.195 sec. We thus find progressive decreases of 0·22 sec. over 20 years and of 0·13 sec. over 23—a total of 0·33 sec. for 43 years. This corresponds with an apparent *westerly movement of 161 m., or an average of 3 m. a year between 1883 and 1926.*

Such a westerly drift would not be incompatible with our interpretation, which pictures Australia as first attached to Antarctica and moving polewards, then breaking away from the latter and finally losing its eastern fold-margin which continued

---

[1] Tollner (34).   [2] Baracchi (14), 366-73.

eastwards and northwards, whereas the mainland, released from such a drag, retreated westwards, any northerly tendency being checked by the Malay arc. No appreciable change in the latitude of Sydney has been noted.

In the case of Hawaii no certain difference in longitude has been obtained over the period 1926–33 (Lushene), but such drift would scarcely be looked for right out in the Pacific.

As regards latitude, measurable differences have been found from the older determinations in a number of cases, some of which may be real and not due to the increased accuracy in measurement alone.

An international scientific body has been recently established to investigate this vital problem, and more positive evidence ought before long to be forthcoming on the subject.

Based on the fact that the Great Pyramid of Cheops in Egypt lies 1′ 5″ to the south of the 30th parallel with its axis oriented 3′ 42″ to the west of north, C. L. Sagui (32) has developed the novel hypothesis that such deviations have been produced by a southerly movement coupled with slight (anti-clockwise) rotation of North Africa since the pyramid was built, the difference in horizontal distance amounting to about 4000 m. He points out that the Straits of Gibraltar are stated by Pliny to have had a width at their narrowest which would correspond to 7408 m. as compared with about 13,000 m. today, and the Straits of Messina 2229 m. in comparison with about 3100 m., which, after making allowance for some erosion as well as the aforegoing rotation, would agree very well with the movement within the historic period deduced by measurement from the pyramid.

In order to demonstrate the low rate of the presumed drift Wegener [1] has compiled a table showing the distances between continental coast-lines that have drawn apart and the assumed time for such process, from which he has obtained figures ranging from 0·2 up to 36 m. per year. Even with rather different values for the distances and times, such as could in certain cases be argued, the calculated rates are significant in not exceeding the figures thus far obtained from astronomical measurements.

Assuming, for instance, that Argentina started to move off from the Cape at the beginning of the Cretaceous—i.e. at least 50 million years ago—it would in travelling the 6000 km. have attained an average rate of drift of only 0·12 m. per year. Such movement would, however, have embraced short pulses of rapid advance and longer intervals of little or no progress. Even if we assume that the active periods represented, say, only one twenty-fifth of the whole, the mean rate of the actual drifting need only have been 0·12 × 25, or 3 m. per year.

[1] Wegener (24), 114; (29), 25

CHAPTER XVI

THE PATTERN OF THE EARTH'S OROGENIES

Introduction. Inter-continental Spacing. Basic Plan. Symmetry in Evolution. Earlier Orogenies. Hercynian Orogeny. Post-Hercynian but Pre-Alpine Orogenies. Alpine Orogeny. Conclusions.

INTRODUCTION

WHEN one considers the difficulties introduced by the over-lapping of one orogenic belt upon another, the interference produced by such later movement, the emplacement of great masses of intrusive rocks, the erosion of folds sometimes down to their very roots, the concealment of such truncated formations by younger sediments or by the ocean and so on, it is a matter for congratulation that so much has been achieved towards disentangling the tectonic complexities of the Earth.

Structural maps have indeed been prepared for particular countries and even for whole continents, such as for Europe (by many persons), North America (Keith), United States (King), Asia (Argand; Lee) and Australia (David), but for the entire Earth there has appeared since the general review by Suess only Staub's masterly synthesis.

Unfortunately, along with greater knowledge has come increasing divergence in opinion regarding the *interpretation* of such crustal movements. On one side stands " orthodoxy " as represented by Dana, Haug, Suess, de Lapparent, Heim and Bertrand, modified in various particulars by Willis, Barrell, Hobbs, Keith, Stille, Born, Bubnoff, Bucher and Cloos, to mention only a few names.

Under that attitude the continents remained, on the whole, stationary without appreciable internal distortion, while geosynclines developed across them, or upon or next their margins, that were subsequently crumpled up, such action being repeated from time to time. A certain, though usually a distinctly limited, amount of horizontal movement of the mass or masses is at the most conceded. The precise mechanism involved, and the causes of such periodical deformation as currently presented, are neither particularly clear nor convincing, though some

302

contraction of the Earth is usually favoured or postulated, but, when the Earth is treated as a whole the movements deduced become at times arbitrary, conflicting or inexplicable.

On the other side is ranged " heresy " in the ideas of Taylor, Wegener, Argand, Molengraaff, van der Gracht, Holmes and the writer, who regard the continents as subject to great horizontal movement and in a limited degree to plastic deformation, and who consider the development and crumbling-up of the crossing or bounding geosynclines as primarily due to the drifting of the masses themselves. By the author all major plications are viewed as due to " drifting " in the wider sense of the term, that is to say, the moving forward of blocks of sial over a stationary sima, or their transport by currents in the sima itself, or else a combination of both these processes.

Not only is the mechanism involved simple and straightforward, but, when applied to the globe, it reveals a wonderful and orderly design in the architecture of the Earth, all springing from a *single controlling principle*. One of the merits of that hypothesis, for example, is the neat way in which various ill-spaced and oriented sections of major orogenic belts can be got to link-up into shorter and more regular arcs that approximate to great or small circles (associated sometimes with conjugate sets), for instance the otherwise heterogeneous " Samfrau " structure.

Lastly, there are several hypotheses differing from the above in some vital aspect, for instance those of Joly, Kober, Haarmann, Baker, van Bemmelen and Gutenberg, to none of which the author is able to subscribe save in limited degree.

There are, however, a number of persons who are partial supporters of the Displacement Hypothesis, as well as others who have not yet become convinced of its probability although recognising its merits and particularly its ability to explain the distribution of the Earth's past climates and life. It is for such that these pages have been prepared.

It is no easy task to synthesise the overwhelming masses of information on record about the Earth's crust and weld it into a simple and consistent whole. By applying the principle of Continental Sliding certain broad generalisations nevertheless emerge, which are put forward tentatively in the hope that they may prove worthy of further development in the future. Applying them, we are now able to discern some kind of order or regularity in the wrinkling on the face of the Earth.

Fold-systems are not composed entirely of parallel or sub-parallel anticlines and synclines, for other types of plication are commonly present in greater or less degree. In grouping together related foldings it is of vital importance to recognise, as has so lucidly been expounded by J. S. Lee (29), that widely divergent fold-bundles, échelons, spirals, " epsilon," " eta " and other structures, sometimes having widely different trends and strongly oblique to the dominant crumpling, may all be possible manifestations of a *single compressive deformation*. For such apparently diverse and unrelated structures, which so noticeably mark out the major orogenies, a combination of rectilinear and a greater or less amount of torsional shear against some resistance—not only frontal but lateral as well—must have been responsible. Since their production would be the consequence of strong differential horizontal crustal movement with distortion, their widespread character can be regarded as a powerful argument for Continental Drift

INTER-CONTINENTAL SPACING

Having in Chapter XIII fixed for successive epochs—and that independently,—the absolute movements of Laurasia on the one hand and Gondwana on the other, each with reference to its own particular equator and poles, we are at last in a position to determine their true relative motions, that is to say, their mutual advances and retreats under which the intercontinental space formed by the " Tethys " became respectively narrowed or widened.

From the close of the Devonian onwards their absolute movements agree as a whole, namely a general southerly drift until the later Carboniferous, a northerly one until the Jurassic, a pause with some reversal in the Cretaceous and a renewal of northerly creep until the Pleistocene. Both masses suffered *anticlockwise rotation* as well—though not necessarily to the same degree—down to the beginning of the Mesozoic, after which such twisting can no longer be clearly established. While the dominant motions accord well, closer study reveals certain disagreements at particular periods, indicating that the one mass apparently moved faster or farther than the other, or reversed sooner, which fits in with the conclusion reached on stratigraphical grounds, that in its evolution Gondwana took the lead up to the late Devonian, but thereafter lagged behind Laurasia, down to the Rhætic at least.

This deduced lack of phase or inequality in drift will form the measure of tension or compression developed within the double mass, which in practice would have become concentrated within the weaker zones, that is to say, the intra- or inter-geosynclinal depressions. As calculated by Kreichgauer, such zones had a strong tendency to parallel the equator of the time and hence to become subject to latitudinal folding.

Our picture embodies not only discontinuous movement, but numerous minor rhythms, the continents being indeed regarded, to use a crude expression, as having staggered with many tremblings along their respective and interfering paths, their erratic progress being punctuated by volcanic eruptions and earthquake shocks. Sheets of sial of colossal extent, mass and momentum were involved and the mere slowing-up of any large portion of one might, as in an ice-field, have been sufficient to develop pressure-ridges, and the reverse to have produced depressions or even rifts. We might with advantage keep in mind the two analogies, namely that of ice drifting across quiet water, or else being borne along by currents.

As already stressed, asymmetrical structures and underfolding and underthrusting (in the current sense) are here viewed as usually characterising the normal compressive cycle.

BASIC PLAN

Three fundamentals follow from Chapter XIII : (a) That the inter-continental trough parting Laurasia from Gondwana functioned throughout as an integral portion of those rudely oval masses and shared in their creep, as could well be anticipated upon the natural assumption that it was underlain essentially by sial. We are thus brought back some way towards the Wegenerian concept of a *single, original continental mass or "Pangæa."* (b) That from the later palæozoic this composite mass strove to escape from a position mainly within the Southern Hemisphere—where part of it covered the South Pole—and was so far successful that, following its dispersal, the bulk of the consequent lands are today situated in the Northern Hemisphere —surrounding the North Pole. (c) That in their later history such behaviour of the blocks can furthermore be regarded in a totally different light, namely as an urge towards the antipodal side of the globe, that is to say, towards the great basin of the Pacific or " Panthalassa." (d) That Laurasia in moving from its more or less equatorial position towards the North Pole— where the meridians were *convergent*—became subjected to an *E.-W. compression*, revealed in its orogenies from the Carboniferous onwards and especially in that of the Urals, whereas Gondwana moving towards the equator—where the meridians were *divergent*—became put under *tension in an east-west direction*, which ultimately lead to its fracturing and wide dispersal. Such would explain the vital difference, why, in contrast to their similar " equatorial " and peripheral crumplings, *Laurasia should display dominantly meridional folding but Gondwana meridional rifting.* (e) That their plan would correspond diagrammatically to a figure 8, or, in a way, perhaps, to two epicyclic gear-wheels, since both masses experienced appreciable anti-clockwise rotation about their locus of mutual

X

though intermittent contact. Such impact became extended to the east and west only at a relatively late date. The transmission of such crustal pressures could have taken place through the medium of the sial filling the " inter-continental space " as well as by actual contact between the two continental margins, and the term " impact " used here should be interpreted in this

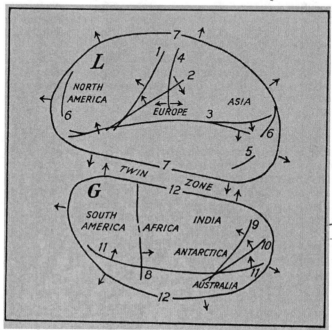

FIG. 45.—Schematic plan of the major Fold-Systems of Laurasia and Gondwana : 1, Taconian ; 2, Caledonian ; 3, Hercynian ; 4, Ural ; 5, Triassic ; 6, Jurassic ; 7, Alpine ; 8, Post-Nama ; 9, Early Palæozoic ; 10, Carboniferous ; 11, Gondwanide ; 12, Alpine. The arrows show the directions of dominant pressure.

wider sense. Spain seems at more than one time to have played the part of pivot or roller. And (*f*), that the determining or controlling forces were the same throughout the existence of this pangæa, a matter that will be dealt with in Chapter XVII.

SYMMETRY IN EVOLUTION

The outstanding feature of our hypothetical Pangæa from the Devonian onwards is the wonderful symmetry of the double-mass, indicated schematically in Fig. 45. Thus the southern half shows the central rigid shield, which in the south-east radiates folds, the E.–W. trending Samfrau girdle deflected northwards at either end and the encircling Cretaceo-Tertiary ring-folds of the marginal fossa. The northern half displays a composite shield traversed by diagonal tectonic zones passing through the Arctic region, the central N.–S. Ural structure, and the E.–W. Appalachian-Altaide girdle deflected northwards at either end, where it has become involved in Jurassic and Cretaceous folds, the whole enclosed by the ring of younger crumplings of the fossa.

Conspicuous in both halves is an outward migration on the whole of the successive crustal waves with some stretching of the masses in the east–west direction. Striking, furthermore, is the regular change in the orientation of the major girdles with time, and that in the *clockwise direction*. Thus we observe across the great interior of Laurasia a N.–N.E. trend for the Taconian orogeny, a N.–E. one for the Caledonian, a nearly W.–E. one for the Hercynian, and an E. by S. one for the Alpine. In this scheme the Appalachians attain an extreme importance through becoming the meeting point or virgation of these divergent systems. Since they experienced compression fully half a dozen times during the palæozoic, such is quite understandable. Again in the Australian block we note a similar clockwise swing of the foldings from the pre-Cambrian to the Triassic, all related to an hypothetical virgation situated in Antarctica; the later ones form part of the outer ring-system and are deviated northwards. A similar plan can also be made out for eastern Brazil.

Such progressive change in direction can be ascribed to the varying incidence of the more or less equatorial squeeze between the two intermittently drifting and slowly rotating land masses. In western Europe the movements were oldest in the north and youngest in the south, the same being roughly the case in Central Asia. The permanence of the Ural system through so long a period is noteworthy.

EARLIER·OROGENIES

Reference can first be made to the ancient Brasilides that run N.–S. from the Sahara down both sides of the southern Atlantic with a conjugate branch in north-eastern Brazil. Thereafter come the Taconian structures stretching north-north-east from Virginia to eastern Greenland and probably far beyond. The Caledonian, in turn, strike north-east from eastern North America to Finmark and thence north to Spitzbergen (with branches into Timan and northern Greenland), the corresponding chains that curve in a wide " V " around the Angara platform being either an extension or, more probably, a separate system (Fig. 18). Of significance is the divergence of the zone of the Ardennes from that of the Caledonian, an anticipation, it can be suggested, of the Hercynian movements.

The Acadian disturbance was localised, and followed closely the previous zones of weakness.

HERCYNIAN OROGENY

Though in many ways less obvious, the mighty Carboniferous and Permian orogeny, partly detailed in Chapter VIII, may not have fallen far short of the Alpine one in magnitude since it stretched in a general W.-E. direction from Mexico to eastern China, a distance under our reconstruction of over 17,000 km. and covered in its several phases a belt of amazing width. As pointed out in Chapter XIII, such coincided approximately with the *equatorial* or, at least, with the *north tropical region of the time*. While the width is relatively small in Europe—in places only 100 km. (as measured, however, in its subsequently compressed state)—it expands enormously to both west and east, attaining an extreme breadth in western North America of perhaps 2500 km.—from New Mexico to Honduras and Cuba. In Eastern Asia the width is at least 1800 km.—and that, too, after the Cretaceous compression—while certain structural lines still farther to the north and the south (Annam) also belong thereto. These eastern elements, which have been termed " the main girders in the edifice of Asia," form three sub-parallel belts : the Tannu-ola in the north (Lat. 50°) (Backlund), the Tianshan–Alai–Kuenlun, and the Hindukoosh–Pamir–Karakorum–Himalaya systems, of which the two northern ones lie outside and the southern one inside the palæozoic Tethyan geosyncline. Far to the south are moreover certain contemporaneous disturbances in the East Australian sector running in a more or less parallel direction.

Conspicuous is the weak development in Eastern Europe, just where the conjugate branch of the Urals takes off, at first to the north-west and, finally, to the north-east through Novaya Zemlya and is lost in the ocean, though possibly linking, as Bubnoff (26) has suggested, with the Verkhoiansk geosyncline, the Timan structure being an offshoot to the north-north-west. A second branch takes off from the southern side of the main belt in the Moroccan Meseta and strikes south-south-west to south for some distance into the north-western Sahara, so far south at least as the Ahoggar region (Kilian), which can be regarded as homologous with the Urals.

The Hercynian involved a series of phases spread through the entire Carboniferous and Permian, which build up and thereafter decrease in intensity, and with which those of the

Urals synchronised. During the slacking-off of pressure the trough or troughs on one or both sides of the newly-made ranges were enabled to re-form, and in a rough way there was a progressive shifting of activity from north to south with increase in the amount of metamorphism and granitic intrusion in that direction.

The earliest or Bretonic phase affected a limited region mainly in the north—New England to Newfoundland, Brittany, Suedetes, Urals and Tianshan. The powerful Suedetic and Asturian phases (*sensu lato*) disturbed the entire region from west to east, while the Saalian was weak in Europe and the Urals, but strong in both west and east—in the Caribbean and Appalachians and along the southern margin of Asia. The latest or Pfalzian seems to have been practically restricted to France and the Saar.

Now it has been deduced that America–Europe drifted southwards until the closing Carboniferous, while Asia continued that movement until the middle of the Permian, whereas Gondwana moved north-north-westwards from at least the Upper Carboniferous onwards. The Hercynian orogeny could clearly then have been due to the mutual approaching of the two masses, and those far-flung waves of deformation to their repeated impact until Laurasia had been thrust back by its more rigid opponent, for which the tectonic evidence speaks eloquently.

We have indeed here a magnificent example on a colossal scale of an "*Epsilon*" (ϵ) *type of structure* with bilateral symmetry, so well described by J. S. Lee (29) or, as the writer would prefer to call it, a "*Bracket*" ⌐—⌐ *structure*—with its curving American and Asiatic wings and meridionally-set axis of the Urals. Such had already been suggested by Lee for Eurasia, but the author will have to extend it to include North America as well. By experiment that type of deformation results from direct pressure concentrated over a limited front, which, in this case, was provided by the Afro-Arabian sector. Some obliquity in the direction of attack is disclosed by the slight asymmetry of the epsilon, the earlier date of the folding in Western Europe, the crowded nature of the fold-belt there, the intense overfolding and overthrusting to the north-west and north in North America and Europe, but to the south in Asia, and the later date when Asia had its general movement reversed. A slight clockwise rotation and distortion of Laurasia is thus indicated, which is in agreement with the incipient spiral structure of Russia detected by Fujiwhara (34; Fig. 17).

Through dividing Laurasia into two portions, which reacted somewhat differently, the mobile Ural geosyncline by its sympathetic opening and closing played the part of a buffer zone and continued to do so until well into the Mesozoic. The Urals record in short the internal response of Laurasia to the ever-varying pressures on west, south and east, and the rôle so played by it was unquestionably a vital one.

POST-HERCYNIAN BUT PRE-ALPINE OROGENIES

While the later Permian saw the poleward retreat of Laurasia, the loss of momentum through its collisions slowed up Gondwana, if it did not temporarily arrest or even reverse its northward pursuit. Such consequent easing of the N.–S. pressure is disclosed by the deepening and lengthening of the Tethys, so that terrestrial life could nowhere cross it, and by the related opening of the Ural trough, whereby Asia became cut off from Europe.

Reduction in the speed of Gondwana would have produced compressive stresses within its mass, so that it ultimately gave way within the only weak zone available, namely the Samfrau geosyncline, wherein from Argentina to northern Queensland arose the Gondwanide foldings about the closing Permian and early Triassic with trend sensibly parallel to that of the Hercynian and curving away symmetrically at either end towards the north. Stratigraphical considerations indeed suggest a southerly retreat of the continent into this zone of collapse.

Gondwana came into touch, again, in the later Triassic with Laurasia, though mainly in the eastern sector. North America not only kept its distance, but drew away from Africa, the Appalachian region becoming the scene of extensive rifting—Palisade disturbance ; Central Europe was but slightly affected ; Eastern Europe developed minor folding—the Early Cimmerian orogeny ; the Urals became land and Europe and Asia were reunited ; the Tethys closed in a few spots, such as Persia, permitting an exchange of terrestrial life; while South-eastern Asia came into conflict with Australasia and developed strong folding (striking N.E.–S.W.) in itself and in Queensland, with some disturbances in Western Australia, New Caledonia and, perhaps, New Zealand.

In the Jurassic, because of the tendency of Gondwana to drift south again, the Tethys deepened enormously, particularly in the Mediterranean region, while the Ural trough reopened, and the consequent late Jurassic—Late Cimmerian—folding was relatively weak in North Africa (Tunis), Europe (Donetz, Persia, etc.) and Southern Asia. Meanwhile the fossa around Laurasia had been deepening, while that continent was drifting northwards displacing the Pacific and incidentally squeezing itself into a region of converging meridians. Strong marginal or intra-marginal plications were accordingly thrown up during late Jurassic times and, again, in mid or late Cretaceous times in advance of, and along the sides of, the drifting mass, that is to say, from California through Alaska, eastern Siberia, Korea and China, where, after crossing the old E.–W. Hercynian lines, they ultimately tended to follow the latter westwards into the interior of Asia.

Despite this ring of thickening foldings, an internal tension was developing through centrifugal action by the close of the Jurassic, confirmed by the fact that the Ural geosyncline remained throughout a sea-way.

On the contrary, the absence from Gondwana of active Jurassic diastrophism along both Pacific borders, but presence of inter-Cretaceous movements in Patagonia and along the Samfrau zone, bespeak the slower growth of the centrifugal stresses within the more rigid southern mass.

ALPINE OROGENY

By the close of the Cretaceous the lands as a whole still lay well to the *south of their present positions* with Africa and India close to Eurasia, but South America far from North America and Australasia some distance from Indo-China.

The detailed account given in Chapter IX of the two grand Fold-rings or Peripheral Mountain Systems (Figs. 19 and 20) has served to bring out their structural unity, the general synchronism of the several tectonic phases and the remarkable agreement of the latter with the consequences of such postulated centrifugal drift. In the case, however, of two such masses of enormous extent, not always at the same stage in their orogenic evolution, each breaking up and drifting not only radially but in groups across the face of the earth, and in places interfering with each other, dynamical generalisations are naturally not readily to be perceived—in other words, it becomes "hard to see the wood for the trees."

Nevertheless, right throughout the more active member is Gondwana, or, more precisely, the Afro-Arabian section—which had broken away from the remainder — reinforced, later, by India, and those masses repeatedly repelled all attempts by the Eurasian portion to return to its equatorial position and at the same time prevented their opponent from fracturing transversely, as it might have done, for instance, in the Ural-Persia region. Their clash gave rise to the inter-continental, double orogeny of the " twin zone " between Central America and the East Indies with general trend a little to the south of east. About this little need be said, since Staub [1] has well treated of the more important section between Spain and India. One can merely draw attention to the part of a " roller " played by Spain

[1] Staub (28), 32-55.

during the slight rotation of the impinging masses, weakened as they had been by the drifting-off of the Americas.

The problem of the breaking-up of Gondwana and its dispersal is by no means involved and has been fully dealt with.

The case of Laurasia is, on the contrary, somewhat intricate and calls for more explicit treatment, particularly because the explanations put forward by Taylor and Wegener have not always been clear or convincing and have therefore given rise to misunderstanding and acute criticism, some of it not undeserved.

Our interpretation is essentially as follows: With each advance of Afro-Arabia Laurasia was driven back, that is to say *northwards*, while the great crooked Atlantic-Arctic rift proceeded to open and Greenland to part from Europe. North America and Asia nevertheless remained intact and functioned as a double-lobed mass joined by the narrow and relatively weak link made by Alaska and the end of Siberia, just as the two Americas behave today.

In the beginning the North Pole lay out in the northern Pacific, and the great compound mass must be envisaged as creeping from the European side towards and over the pole, pushing up marginal and advance folds. When the front passed beyond the pole the flanks would necessarily have proceeded to *converge* and thus to overrun the centre, thereby leading to progressive bending in the Alaskan region. The closing-in upon the Pacific of America and Asia would furthermore have opposed the continued advance of Alaska into that ocean and might even have brought it to rest. Significantly, on both sides of Behring Strait the Pliocene was marked not as elsewhere by folding, but by simple uplift (Mertie). That region, furthermore, reacted as a beam to bending, developing tension on its northern side, that culminated in the rift of the Arctic Ocean and Behring Strait, and compression on the other, that is evinced in the beautifully symmetrical advance-fold of the Aleutian Islands 3000 km. in span, the bowed form being indicative of end-pressures to that arc. All this was convincingly set forth by Taylor [1] in word and picture. The fold and fracture patterns of the region show, moreover, all the indications of strong torsion.

Keith has asked why, if America and Asia had drifted towards one another, there is no transverse belt of compression in Alaska. It is clear that Alaska acted more as a hinge than as a strut, but folding is actually represented somewhat farther to the west, namely in Siberia in the Verkhoiansk region east of the Lena River, where the belt of deformation runs northwards into the Arctic Ocean, having been produced by late Mesozoic and perhaps even some early Tertiary squeezing between the Angara

[1] Taylor (10), 202.

and the Kolyma nuclei. On recognising this, the eastern corner of Siberia then becomes—as it actually is geologically and structurally—part of Alaska, the two having behaved as a unit thereafter with subsequent (circum-Pacific) deformation that followed the Okhotsk-Anadyr-Alaska arc.

In dealing with the great mass of Asia it becomes necessary to criticise or dissent from certain views expounded by Taylor, Argand and Staub. Taylor is seemingly incorrect when he regards the disjunctive Arctic basin as having opened radially around the *present* North Pole, for, according to Kreichgauer, Wegener and the author the pole must then have lain out in the Pacific. For that and other reasons the writer views the Arctic as essentially an extension of the zigzag Atlantic rift and as having opened originally on the European side of the present polar centre though ultimately reaching and occupying the latter position.

Argand, outdoing Suess, has given a spectacular though questionable picture of a highly " plastic " Asia pouring outwards during the Alpine diastrophism from the Angara platform in a succession of wrinkles with enormous crustal shortening in the N.–S. direction. Those following him have accordingly spoken of " Asia " as moving southwards *en masse* from the Cretaceous onwards. This the author cannot concede, holding that the several parts of that continent were considerably restricted in their freedom to expand and took widely divergent paths. Bowie also has doubted this general southerly drift.

From the evidence of the climatic girdles Asia moved as a whole *northwards* until well into the Tertiary. The north-western side remained firmly attached to the Russian block and was hence neutral, the south-western was invaded by the active mass of India and partially arrested, while the north-eastern, situated beyond the mobile Verkhoiansk zone, was hampered by its attachment to Alaska and the presence of the Kolyma nucleus. Since the rigid floor of the Pacific caused obstruction, the freer north-central and central portions of Asia were constrained to escape in the extreme south-west and south-east on either side of India into areas where rifts were actively opening. Incidentally, much of the folding in Indo-China currently supposed to be Alpine is more probably pre-Tertiary.

Strong compression took place between Angara and India in which were involved not only the Tertiary Tethys but a wide belt along its northern and a narrower one along its southern margin. Shear types of folding and vortices of opposite sign were developed at the two corners of the Indian block, the fold-system turning southwards through Burma and scarcely affecting Indo-China and south-west China. In Central Asia the

Permian and Jurassic ranges were broken and extensively re-modelled, though not with the freedom pictured by Argand, their original east-north-easterly courses being still preserved through long distances, as depicted on Lee's instructive tectonic map (Fig. 46). Great distortion of the continent must un-questionably have taken place, as is finely brought out on Argand's map, though much of the expansion in the southerly direction would seem to have been set off by simultaneous compression in the E.–W. one. Part of the currently inferred southward movement is dependent upon the common overturn-ing of folds towards the south, but such can also be interpreted as an underfolding due to a northward sub-crustal drag. Save in not accepting so much southerly drift for Asia, the author finds himself in close agreement with Staub's able synthesis. Such, furthermore, includes the recognition that the universal westerly creep of the Lands may in part have been responsible for the development and deepening of the various " seas "— China, Japan and Okhotsk—between Asia and its expanding exterior Island Arcs, as originally suggested by Wegener.

It is unnecessary to enlarge upon the problem of these arcs, which has been so well discussed by others, particularly by Tokuda, except to point out that the frequently-expressed homology between the East and West Indies is fallacious. In the former case *three* blocks were involved, arranged in a triangle of which India and Australia had moved so far apart that their weak connecting frontal folds were overwhelmed by the greater crustal waves bordering Indo-China. In the latter case, *two* nearly-opposed blocks were concerned with their frontal folds set nearly parallel, with the Caribbean Sea occupying the entire inter-fold region. The Bartlett Deep is, moreover, a rift: the Banda Deep, a crustal swirl, which Staub [1] has expressly compared with the pattern made by the Western Alps and Pyrenees (the north and south points being transposed).

It is well to note that outside the vast region mentioned above some isolated zones of Tertiary disturbance also occur, for instance that in west Spitzbergen, striking nearly N.–S. with strong overturning towards the east.

CONCLUSIONS

The above brief summary fully supports our contention that within the reassembled lands there has been from the palæozoic onwards a *regular clockwise shift* in the orientation of the dominant tectonic zones and of their conjugate branches, where such are present. This, which from the beginning of the Cambrian has amounted in all to somewhat less than 90 degrees,

[1] Staub (28), 90, Fig. 24.

will obviously represent an excessively slow shifting in direction
of the responsible forces, and therefore of crustal creeping.   It

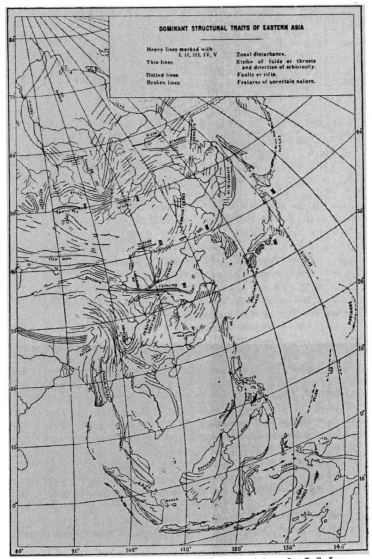

FIG. 46.—Dominant Structural Traits of Eastern Asia—after J. S. Lee.

furthermore accords well with Kreichgauer's hypothesis of the
dominant diastrophism being developed along or parallel to the

equator of the time.   This deduction, that *the lands as a whole have pursued a strongly curved path over the face of the globe,* with anti-clockwise rotation, is manifestly of high geophysical significance.

The large amount of compression and the great width involved in each of the major orogenies are universally admitted. When it is recollected that arcuate and sometimes spiral patterns are involved, that the Alpine " telescoping " was over wide regions superimposed upon the Hercynian, and that in turn upon the Caledonian, etc., a very considerable amount of crustal distortion must be conceded since the mid-palaeozoic at least. This outstanding fact greatly weakens, if it does not dispose of, the argument so frequently advanced, that the adherents of drift seem to mould their continents *ad lib.* Our contention, expressed earlier, that in such continental restorations appreciable allowance for distortion is justifiable for particular zones, is definitely upheld.

In conclusion, it will be observed that the above structural scheme for the Earth bears a distinct resemblance to that formulated by Lee (29), especially his Fig. 9, although the interpretation of the individual movements differs in many respects from his, owing mainly to the fact that he has not clearly differentiated between the effects of the several long-parted orogenies. He has nevertheless stressed the existence of appreciable movements of translation and rotation as affecting the continents, and his contribution is a notable one therefore.

Our scheme agrees closely with that of Staub for the Tertiary, and to a fair degree for those of the previous epochs, provided that the spaces between the lands, supposed by him to be now occupied by down-sunken portions of Laurasia and Gondwana, be regarded as due essentially to horizontal drift instead.

Over all we observe the one dominating action, namely, the centrifugal dispersals of the two major masses and the crowding in of the lands from different quarters upon the passive basin of the Pacific, which has thereby been forced to occupy an area about one-half of that of its former representative (Chapter X). Furthermore, from seismological evidence Gutenberg (36) has been able to show that the present boundaries of that basin are sharp—as could, indeed, be expected if the bordering ranges had deep roots—and that movements towards the centre are taking place today along faults that are dipping inwards beneath the encircling lands.

CHAPTER XVII

CAUSES OF CONTINENTAL DRIFT

INTRODUCTION

As pointed out in Chapter III, each Continental Block has on the whole *gained in area* as compared with its prototype of the palæozoic—not all round, but generally on its youthfully folded side or sides, with, however, a certain though usually unknown amount of loss along the remainder of the periphery.

Current theories of Diastrophism have incidentally restricted the cross-sections drawn through the crust to only short segments thereof, thereby obscuring the difficulties that would manifest themselves were such arcs to be continued right round the earth along great circles. One is faced, moreover, with the paradox that the crust must, after folding, be occupying more or less the same space as before, and that, too, after maybe more than one compression. Such authenticated telescoping demands either compensatory shrinkage of the earth's interior or else an equivalent amount of surface-stretching elsewhere in the shape of a rift system or disjunctive basin or a corresponding breadth of intrusive matter. The first of these would require an inordinate amount of shrinkage, far more than could be conceded, while the last is not to be detected in the field, wherefore Rifting alone remains to be considered and, with it by implication, Drifting.

The diastrophic scheme of current conception has furthermore the tendency to view the continents statically rather than dynamically, and hence leaves a sense of inertness of the lands, which is unreal.

Some, such as Keith (23) and Willis (32 ; 36), have welcomed igneous crystallisation as a source for tangential compression, which is hard to understand, since far less mechanical work would be needed were the magma simply to well up to the surface through fissures and so relieve the internal pressure.

FEASIBILITY OF SLIDING

Critics have repeatedly argued that no adequate force or combination of forces has yet been found which would be competent to cause continental sliding; indeed, this is probably the outstanding objection among those raised against the principle of drift. Any convincing explanation thereof would immediately convert a host of doubters, who, while recognising the unquestioned merits of the hypothesis, are confronted by this serious stumbling-block.

On the other hand, Rastall (29) has ingeniously pointed out that the problem forms a parallel to that of "continental glaciation," the fact that no full or generally accepted explanation has yet been given for past ice-ages not having prevented geologists from accepting the reality of such vast refrigerations even though the evidence therefor be essentially inductive.

The diastrophic record of the Earth is one long round of upheaval and sinking, of intense compression within restricted zones and of magmatic activity, all on a stupendous scale. Inter- and intra-continental movements of the order of several hundreds of kilometres at the least are universally recognised, with the admission that far larger distances might perhaps be involved. Current geology professes to explain the mechanism of such "revolutions," but fails lamentably to do so, partly because the potent influence of radioactivity is generally ignored, partly because of little-known or even unsuspected factors.

Opinion generally views the Sima as highly-heated and able with rise in temperature or fall in pressure to liquefy with consequent changes in volume. Resting upon it in approximate equilibrium is the lighter Sial in which heat is being generated by radioactivity at a rate which would generally appear to be *slightly greater* than can be conducted to the surface. That rate is discovered to be larger beneath the continents than the oceans. The lands are less dense and, therefore, stand higher than the oceanic basins, but the balance between them is perpetually being endangered by the processes of erosion and deposition. Furthermore, the Earth is a rapidly rotating mass that is subject to tidal and other forces. Unquestionably, all the elements making for *crustal instability* are present.

Nevertheless, most geologists regard Continental Drift as a physical impossibility and with such a conclusion most geophysicists agree, although, in view of our limited knowledge concerning the inaccessible sub-crust, it would perhaps be more proper for them to take up an inquiring or else a neutral rather than an *a priori* attitude. Mathematical analysis has furthermore been directed rather from the statical and dynamical than from the thermodynamical aspect. For instance, having demonstrated the variability of the accessible rind of the earth as regards temperature, density, mobility and so on, the geophysicist usually proceeds to divide the interior into a series of concentric shells. Since these are more or less equipotential surfaces, his mathematical analysis naturally fails to reveal forces competent to do anything improper. The immense depth [1] at which earthquakes can and do originate, and the varying rates at which such waves travel *via* different paths, should alone cast serious doubts upon any hypothesis involving a uniformly-layered earth beneath its admittedly heterogeneous skin.

The particular forces that have been invoked by Kreichgauer, Taylor, Wegener, Staub, Daly and others to produce drift in either an equatorial or meridional direction are subject to numerical calculation and prove unfortunately to be of a very low order, quite insufficient in themselves to cause movements of the magnitude postulated. On the other hand, they would be of great importance in *directing or guiding* the drifting blocks once the latter had been set in motion by other means. Hence perhaps the frequency of foldings trending roughly E.–W. and N.–S. in respect to the polar axis of the time. Baker's novel hypothesis, whereby continental disruption was effected by tidal action through a planetoid almost grazing the earth, although attractive and capable of explaining many difficulties, would seem at variance with the geological evidence, which demands, among other things, quite a lengthy period for the dispersion of the blocks.

After carefully analysing in the light of our interpretation the various opinions that have been expressed on the subject, everything would seem to point, as Holmes has made clear, to *internal rather than to external forces.*

IS THE EARTH CONTRACTING?

The view, that contraction of the Earth's interior, due to cooling or to molecular rearrangement—as under the older conception—or to closer packing—as under the Planetesimal Theory—could be responsible for crustal deformation, has the merit of simplicity and is therefore held by many persons, for example, Suess, Jeffreys, Stille, Kober and Bucher. The objections that have been voiced to such an hypothesis are nevertheless numerous and weighty. For instance: (1) the restriction of major compression to geosynclinal zones of deposition only after each had been filled with a great thickness of sediment; (2) the concentration of such compression within relatively brief periods of activity followed by long periods of quiescence,

[1] Gutenberg (36) has cited instances from 300 to as much as 700 km.

thereby implying discontinuity in shrinking ; (3) the absence of crowding of the lands by the oceans such as is contradicted by isostasy (Nevin) ; (4) the surprisingly intense deformation of the Tertiary at a stage when shrinkage should theoretically have been getting less ; (5) the absence of any fold-girdle across the Pacific ; and (6) the world-wide distribution of dykes and normal faults of different ages, which shows that interludes of stretching of the crust have been general.

One is the more inclined to reject the idea since Halm (35) has given some attractive reasons, based upon stellar evolution, for an *Expanding Earth,* and has indeed employed that hypothesis to explain the development of the continents from a former pangæa. While such expansion would very simply account for continental fracture, a difficulty arises through the varying amounts by which the blocks have become separated. Certain of Halm's conclusions are nevertheless striking and worthy of a close study by geophysicists.

ROTATIONAL FORCES

That centrifugal forces acting upon the higher-standing continents is tending to move those masses towards the equator has long been appreciated. This is the so-called " Polflucht " force favoured by Kreichgauer, Taylor and Staub as the explanation of equatorial folding. Arnold Heim (36) has made the pointed observation that the repeated reversals of movement in the orogenic zones scarcely seems to support the idea of the polflucht force as the original cause of compression. That " polar flight " was not the dominant factor is clearly shown by the deduced movement (Chapter XIII) of Gondwana *towards* the South Pole in the period preceding the Carboniferous and of Laurasia *towards* the North Pole after that epoch.

For middle latitudes Jeffreys finds this force to have a value of only about 5 grammes per square centimetre, which is manifestly insufficient to account for the intense plication observed. Judging from the general latitudinal trends of such major zones, it may nevertheless have helped to determine the initial deformation that was completed by other and more potent forces. Incidentally one should not overlook the possibility that crumpling-up along a particular line might be primarily due to distantly applied forces transmitted through a rigid crust. Under our hypothesis, for instance, the collision between one block and another could transfer part of the kinetic energy of the one to the other and even, perhaps, deform the remote side of the latter if a weaker zone occupied such a position. One can cite in analogy the impacts between the trucks of a long railway train, but with the movements slowed down enormously.

A more potent force would appear to be that due to changes in the amount of flattening of the earth in the polar regions, a factor favoured by Taylor. Arnold Heim (36) has made a valuable contribution by showing how small changes in the velocity of rotation—a fluctuation that receives astronomical support—would involve enormous amounts of energy and could therefore account for various tectonic movements. With such, again, one can associate the slipping of the crust or a part of the crust over the core, as deduced from the climatic evidence (Chapter XIII), with internal friction, shearing and loss of kinetic energy.

The displacement of continental masses would through altering the moment of inertia have produced unbalanced forces and thereby led to torsion in, and possible fracture of, the blocks, as visualised by Wegener. Disturbances would similarly be introduced through elevation or depression of parts of the crust, isostatic adjustment, radioactive heating, melting, crystallisation and so on.

Tidal forces have been invoked by Taylor as the chief cause of mountain building and continental displacement, while Wegener, Joly and Staub have all postulated some westerly drift of the sial and ascribed it to tidal friction, though Jeffreys and Lambert have pointed out the insignificant value possessed by the available force.

Among other factors to which appeal has been made are precession, changes in the obliquity of the Earth's axis and so forth. It is nevertheless seriously doubted whether any of these other factors that are based upon the Earth's rotation, either singly or in combination, would be adequate to account for orogenesis or drift.

CONVECTION CURRENTS

The hypothesis of *Sub-crustal Streaming*, long ago propounded by Ampferer and since supported by Joly, Jeffreys, Schwinner and others, and whole-heartedly by Holmes, has recently been challenged by Borchert (32), whose reasons do not, however, appear convincing. In similar fashion and on *a priori* grounds the existence of two or more superimposed currents in the atmosphere or the ocean could equally be denied. In view of the small depth of the circulatory system in both cases, namely the troposphere and the upper part of the ocean respectively, one could readily concede quite a small thickness to our postulated zone of fluid or plastic sima, nor need such sub-crustal fluidity have at any instant been universal and not local. Indeed, it would seem impossible to doubt that under differential heating a *system of magmatic streaming must develop* whenever the

Y

viscosity of the material becomes sufficiently lowered.   Support for such a view is to be got from the phenomena of the Earth's ever and rapidly varying magnetic field.

We can accordingly picture a feeble, more or less general " planetary " circulation upon which would be superimposed impermanent though stronger deviations due to unequal heating and melting—largely through the presence of cakes of sial, as Holmes has so skilfully worked out—as well as to tectonic causes.   The orogenic process itself must introduce modifications, for, where the material was already close to fusion-point, " strain-melting " could take place along shear - planes and thereby determine sliding.   While Gutenberg (36) has from the study of deep-seated earthquakes postulated a high viscosity even at depths of some hundreds of kilometres, it is important to note that such deep foci are confined to the Pacific basin, which is currently regarded as a " rigid " block.   We are therefore not prevented from assuming a greater mobility within the sub-crust for the rest of the globe.   That a drag upon the base of the overlying sial could thereby be produced by such currents can scarcely be doubted, more particularly in the case of downward protuberances into the sima, where, incidentally, heat would be generated by friction as Bull (21) has pointed out. C. L. Pekeris has calculated the shearing stress on the bottom of the crust through the convective circulation to be of the order of 10^7 dyne/cm^2.

Various schemes embodying the above ideas have been developed by Joly, Holmes, Groeber, Staub, Schwinner, Daly, von Seidlitz and others, which have one and all contributed materially to the important theory of Magmatic Currents as the cause of diastrophism.   The writer, nevertheless, feels that, in the form hitherto stated, *convection by itself is not wholly competent to account for continental drift in the full meaning of the term*, and that some modification of the theory is accordingly demanded.

In our interpretation every revolution has involved a certain amount of " drift " and such movement must hence be regarded as a normal consequence of the orogenic cycle.   The *Energy* therefor must be derived from some internal source that is itself cyclic, and of such the only outstanding one is that provided by the Earth's radioactive content.   *Drifting must indeed be viewed as a function of Radioactivity.*

BOUNDARY RELATIONS OF SIAL AND SIMA

There has been a good deal of misconception regarding the precise relations of these two shells during the process of drift, and much of the criticism in this connection has been singularly

inept. Through following Wegener too closely, the general
picture has usually been one of oceans floored by bare sima,
whereas the evidence points strongly to quite a thickness of
material of at least intermediate composition, if not of acid
character, intervening between the oceanic waters and the sima
below. Whether such floor yields by folding, plastic flow or
melting is actually of small moment in the problem, since it
is this lighter and more brittle covering which mainly provides
the tangible proof of the deformation going on beneath it.

That the sial block during its deformation can carry up
with it or in some way involve the upper part of the peridotite,
and occasionally the eclogite, layer, is signalled by the frequency
with which dense ultrabasic eruptive rocks have made their
appearance within the cores of anticlines and overfolds, with the
implication that such invasion must have been more general,
only erosion has not gone deep enough to expose such bodies.
While not absent from the older orogenic zones, they notably
mark out the Alpine foldings and penetrate strata so young
as Eocene or even later, as Benson's stimulating review (26)
brings out.

One can hence surmise that under the intense compression
shearing would largely take place along a "sole" slicing
through the upper part of the sima and base of the sial, thereby
enabling such ultrabasic matter to make its way into higher
levels. Through analogy rifting should follow planes extending
obliquely downwards through the sial into the top of the sima.
The drift of a block could moreover be viewed as prompted by
a chain of processes operating about the sial-sima boundary that
have involved fluxing at the advancing edge, transfer of the
melt to the rear and its recrystallisation there, rather in the
same way as "regelation" in ice. Various and probably
complex factors would have determined whether the resulting
advance became manifested in a series of short pulses or was
concentrated into one or more large pushes.

THE RÔLE OF THE FOSSA

Before applying any of the known or suggested causes of
horizontal drift to our particular scheme, it will be of advantage
to set down the course of the continental movements, as deduced
in Chapters XIII and XVI. Starting off in the mid-palæozoic
with a single pangæa that embraced the units of Laurasia and
Gondwana, together with their ever-varying inter-geosyncline,
we have been able to establish that (1) this compound body crept
away from the North Pole until the Permo-Carboniferous, but
reversed its motion thereafter ; (2) it developed an anti-clockwise rotation which lessened with time ; (3) the tropical sides

of both masses never trespassed so far beyond their equator as to take the respective centres of gravity of the bodies across that line; (4) the main, pre-Jurassic deformations were dominantly latitudinal and nearly equatorial; (5) those diastrophisms accord well with the hypothesis of alternate mutual approaching and retreating of the two masses; (6) the poles did *not* form the centres of dispersal for these masses; (7) on fracture, the fragments, wherever not restricted, drifted radially outwards, certain blocks in each hemisphere moving towards and even over the corresponding pole; (8) such dispersive motion was intermittent and presumably rhythmic.

All this strongly suggests that, while the centrifugal or polflucht force due to the Earth's rotation played a part, and almost certainly an important part, it was nevertheless, *neither before nor after the disruption of the lands, the prime cause of continental drift*. The dominant agent in crustal disruption was undoubtedly that simple yet remarkably endowed structure, the *Fossa* or *Rim Syncline*, which together with its partial form, the *Marginal Geosyncline*, can be regarded as *essentially responsible for continental sliding*.

The Fossa was a well-developed feature around not only the two main masses, but their component shields, such as that made by Laurentia *plus* Greenland—which incidentally resisted fracture until the Tertiary,—the Scandinavian-Russian platform, etc. Where long-established, this peripheral furrow became a source of weakness because of (*a*) the continued depression of its base; (*b*) the heating-up of its accumulating sediments by radioactive substances; (*c*) stresses set up by the excess pressures at the continental margin against the ocean floor—as brought out by Wegener and emphasised by van der Gracht, though commonly overlooked; (*d*) paramorphic expansion beneath the lightening block—since not all such volume-increase would take place in the vertical direction; and (*e*) progressive flexing at the continental margin. One can furthermore view the frequently associated depression known as the "foredeep" as a second, though interior, fossa endowed, nevertheless, with rather similar properties.

Beneath the fossa there would accordingly develop an inverted bulge projecting downwards into the sima to become accentuated with time and with marginal folding, thereby giving this ideal sial-cap a form like that of a saucepan lid. Such a shape would eminently favour a sub-crustal convective circulation of the kind depicted by Holmes, reproduced here (Fig. 47), with current rising beneath and about the centre, spreading out radially, striking the protruding "roots" or "saucepan edge" of the fossa, deflecting downwards and returning to the centre as a counter-current. Furthermore, as shown by Pekeris, the crust would be pushed upwards under the warmer (continental) regions and pulled downwards under the colder (oceanic) ones, thereby tending to dome-up the lands. Such may explain the " bulge " of Africa.

The cap would thereupon be placed under *radial tension*

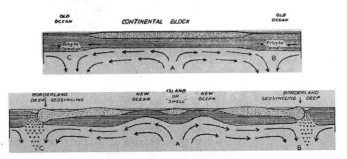

FIG. 47.—Two stages in the Fracturing of a Continental Block by sub-crustal currents—after A. Holmes.

and would tend to stretch—with the development of one or more interior depressions—and might even fracture—with the ascent of magma through the resulting fissures. Of such a scheme Holmes has given an enlightening account. Our arrangement, which is very similar, is pictured in Fig. 48. Even

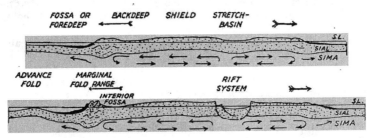

FIG. 48.—Hypothetical stages in the Fracturing of a Continental Block. Black, Orogenic Sediments ; large arrows, Direction of Stretching.

if the sial-cap failed to disrupt, it could *become enlarged through outward migration of the fossa*, a view favoured by the well-established observation that the continents have, on the whole, grown outwards with time. We thus arrive at a conception rather like that contained in Gutenberg's " Fliesstheorie." It is suggested that the vast floods of basalt that welled out from fissures at the close of the Triassic—more especially in Gondwana—marked a preliminary attempt at such disruption. Presumably, the escaping of such quantities of magma reduced for some time thereafter the internal energy to below the requisite minimum for partition.

No matter the length or direction of the path taken, each fragment of the cap has in its travel acquired the same general structural pattern from front to rear, which suggests that its drift was engineered by sub-crustal currents, was self-induced or was due to both these causes.

The " twin-zone " of the Earth can furthermore be pictured as the region of interference of the Northern and Southern Fossæ. On the contrary, if the marginal geosyncline be incomplete or have reached the critical state in places alone, then only such sections would be able to collapse, and an orogenic zone or zones of arcuate pattern would develop. Indeed, the intra-continental zones of deformation can be regarded as variants of the fossa through possessing a large or an infinite radius of curvature, with a correspondingly lesser freedom to migrate, for which reasons they have seemingly not prompted disruption. Our next step must therefore be to establish how the *outward drag on the fossa could have been maintained after disruption.*

Of additional importance is that adjunct of the Fossa, the *Advance Fold* or *Anticline*, thrown up from the ocean floor ahead of the drifting block, which by its migratory and undulating progress has given birth to the so-called " border lands." This structure will correspond closely with the " Moving Anticline " expounded so lucidly by Chubb (34) for the island chains of the Pacific.

During the Past its rôle is deemed to have been an equally vital one through (*a*) constituting the source off-shore of the " continental " sediments that has had to be so frequently deduced on geological grounds ; (*b*) forming the unstable highway for terrestrial life; and (*c*) functioning as a temporary deflector or barrier to the spreading of marine life. Many of the difficulties met with in palæogeography can be eased by the recognition of this type of structure, which, it is suspected, may perhaps prove to be a fairly reliable indicator of former drift.

AUTOGENOUS SLIDING

When denying the feasibility of sliding, critics tend to view the block as inert, as drifting horizontally and passively over the face of the globe and as remaining generally unaltered in shape, attitude, distribution of mass, etc. They do not visualise such transport as having been *determined by the block itself,* and are hence compelled to assume the existence of *external orces.* Regarding in a similar light a child seated upon a garden-swing, these same persons should from analogy deny any possible consequent motion for the combination since no outside forces would be available to produce swinging. Yet the child can and does swing himself *ad lib.*

They have overlooked the possibility that the crustal block could in a sense be a living and slowly-evolving unit, that it could hardly be inert and that various important characters possessed by it could not have remained unaltered during its drifting, which, according to our ideas, must have occupied a period of many millions of years. A pointer is indeed provided by the dynamical property of the *geosyncline to migrate laterally* —normally seawards in the case of marginal types—the slow undulation hinting at a self-propagating kind of movement. Stratigraphical geology is crowded with fine instances of major geosynclines that reveal such uni-directional progress.

If the influence of the Intermediate layer has been correctly gauged, the leading edge of the block should become depressed and the fractured rearward margin raised, thereby producing a slight forward tilt of the mass. With time and travel such tilting should become accentuated through erosion of the rising " stern " and weighting of the " prow " by coastal or marine sediments, etc., which effects would in turn be amplified by related changes in the Intermediate layer. Such progressive shifting of the centre of gravity—and also of the metacentric or flotation height—can be compared with the action of the child in proceeding to move the swing, to effect which he must naturally *perform a definite amount of work against gravity.*

In the case of the moving block at least four main sources of energy, apart from convection currents, would have been available : (*a*) that derived from the sun in producing erosion and deposition ; (*b*) thermal changes in the crust and sub-crust, of opposite sign normally beneath the leading and trailing edges ; (*c*) sub-crustal melting or crystallisation with changes in volume of the material; (*d*) radioactivity, for, as the mass creeps over fresh sub-crust, it would produce a blanketing effect while leaving behind it a region wherein different lithological and thermal conditions had developed or were developing.

Our problem becomes indeed not only a mechanical but a thermodynamical one.

It is suggested that, with melting taking place in the collapsing front, and magma ascending, differentiating and solidifying in the rift opening in the rear, a horizontal component of pressure might be induced great enough, when sub-crustal conditions became favourable, to overcome basal friction and frontal obstruction, and so propel the block forwards. In short, such movement is conceived as *self-determined or autogenous.*

This mechanism can be likened to that of a " heat engine " with the sima constituting the connecting medium between points at different temperatures, i.e. between the " heater " and the " cooler." Because of the slowness of the process such transfer is assumed to be isothermal rather than adiabatic and, therefore, following the " Carnot cycle." The rhythmic nature of the process on not only a mega- but a microscopic scale is held to play a vital part in the movement in analogy with the muscular spasms in the child when swinging himself. The asymmetrical arrangement of structures and forces would furthermore conduce towards creep in the one direction with little or no back-sliding during rest-periods. On the whole, therefore, or up to a point at least, *such motion would be irreversible.* The breaking-down of crustal rigidity would moreover be materially assisted by the earth-tides due to Sun and Moon. Shearing along the base of the block could hardly have taken place through the solid sial or sima, as supposed by Wegener, but only when and where the presumably high viscosity of the material had been sufficiently reduced through progressive softening or melting. Such softening would be more or less restricted to the *foundations of the block* because of the radioactive nature and blanketing effect of the mass itself, thereby creating a state of crustal instability beneath it. Sliding would naturally be attended by rifting on the one side, and by wrinkling on the other, in which latter would be involved not only the brittle crust but the progressively softening sub-crust. Horizontal movement would be governed essentially by the rate of yielding in the zone of collapse as the shells proceeded to fold and/or melt. Such movement would in time bring about further softening and yielding along the leading edge of the block and thereby determine another advance, and so on. The resulting creep would accordingly tend to be intermittent rather than continuous, which deduction is in agreement with the rhythmic nature of the orogenic cycle.

Over lengthy portions of each geological epoch surplus energy seems to have been dissipated in widely distributed movements, radial and tangential, of a minor and often conflicting character, and more particularly was this the case during the beginning and middle of each epoch. The concluding stage was, however, normally marked by a loosening of the bonds uniting the block to its foundation, resulting in a " revolution," after which the sima proceeded to solidify, and the two layers to become welded together again. If, however, the influence of convective currents be admitted, then the block must be pictured as tending to be carried forward by them, just as an ice-raft by a river.

Of the vast store of energy ceaselessly generated by radioactive substances only a limited proportion would seem to have escaped to the atmosphere or ocean through the medium of volcanic eruption, thereby leaving a surplus available for moving continents, compressing geosynclines, elevating the lands or in other endothermic reactions. *It is this stupendous transformation of internal energy that would appear to be primarily responsible for the resolidification following radioactive melting.*

Joly's illuminating theory of Thermal Cycles can now be applied, modified, however, through the recognition of (*a*) such periodical absorption of energy ; (*b*) a greater inequality in the depth and distribution of the earth's shells ; and (*c*) intermittent drifting of the sial blocks over the sima. This would effectively dispose of the approximately equal spacing between the revolutions and of the contemporaneity of mountain-building over the whole earth demanded by that particular theory.

It may be remarked that the mere loading or unloading of a section of a sial block could conceivably lead to movement of the latter in the horizontal direction, since secondary actions would be set agoing around as well as within and below the mass. The simple transfer of a weight from the bow to the stern of a boat would admittedly only tilt the latter, but, if that load were to be suitably linked to some mechanism, such as a paddle, the energy derived from gravity could now be induced to propel the vessel. Such a mechanism, it is submitted, is at hand in the *Paramorphic Layer*. In analogy the child shifts his position on the swing, and the latter starts its oscillations.

For such as well as for other reasons, the writer disagrees with those, such as Taylor [1] and Bucher,[2] who rigidly exclude isostatic adjustment from among the causes of drift. Isostasy can perhaps better be regarded as the trigger which releases a train of potential forces having a horizontal component. Viewed in this light isostasy not only modifies, it creates !

ECLECTIC HYPOTHESIS

The problem of Drift being so essentially a *hydrodynamical* one, a comparison with the equally complex atmospheric circulation will serve to bring out some useful analogies, a viewpoint already suggested by Holmes (28-29). We thereupon have three elements : (*a*) the upper atmosphere or stratosphere, increasing in density downwards, relatively undisturbed and at a more or less constant temperature ; (*b*) the lower atmosphere or troposphere beneath it, compressible, increasing in density downwards, heated and cooled both regularly and irregularly and hence subjected to continual minor and occasional major movements in both the vertical and horizontal plane, with fluctuation of temperature and other variations ; and (*c*) the incompressible

[1] Taylor (26), 65.   [2] Bucher (33), 61, Opinion 8.

floor made by sea and land with contrasted differences in level and temperature.

Under the influence of the Sun, coupled with the rotation of the Earth, the troposphere has developed a convection system which is nevertheless confined to the surprisingly small depth of about a dozen kilometres. As is well known, the *general* or *planetary circulation* has been determined by currents ascending at the equator and descending at the tropics, etc., which give rise to the so-called permanent belts of high and low pressure (trending approximately parallel to the thermal equator), to the polar anticyclones and so on.

Furthermore, and of yet higher importance, are the impermanent cyclonic disturbances that with rotative tendencies sweep with high velocity through the troposphere, spreading confusion and leaving new conditions in their wake.

Analogously the crust shows : (*a*) the thin, practically incompressible and brittle upper portion, passing downwards into (*b*) the warmer, " paramorphically compressible " lower portion, and that in turn into (*c*), the region of melting and sub-crustal shearing and streaming underlain by (*d*), the presumably "solid" core, these several shells being heated up by their radioactive contents as well as by conduction from the still deeper interior.

We can, therefore, presume that under the Earth's rotation a more or less stable or planetary circulation would be developed directed generally towards or away from the poles, which would indeed have been responsible for geosynclinal evolution in all its phases with roughly equatorial trend, and also for the general creeping of Laurasia and Gondwana over the face of the Earth.

Of greater moment, however, would have been the deviations primarily brought about by those thickened portions of the sial that constituted the continents, which not only were strongly radioactive in themselves, but blanketed the underlying heated interior and thereby generated " hot spots " above the sima. A convective circulation analogous to the cyclonic systems of the atmosphere would thereupon have been induced with radial and rotational tendencies that would have been superimposed upon the normal planetary movement. Such is regarded as having essentially determined the ultimate fracture of the sial-caps and the dispersal of the fragments with all their consequences.

It is significant that both in Laurasia and Gondwana such hypothetical cyclonic " centres " lay neither near the equator nor the poles of the time, that is to say the fracturing masses were not strictly co-polar, and by arriving at such conclusion the author differs fundamentally from Taylor and Wegener. Once disruption had been effected, " autogenous sliding " would have taken charge and the fragments would have proceeded on their way automatically, their motions governed by the various factors met with *en route* as well as those generated internally, just as in the case of a passing cyclonic depression of the atmosphere. Throughout an important part would obviously have been played by forces due to the earth's rotation, namely tidal action, precession, etc., working respectively with or against the dominant planetary circulation.

Though such distinction does not seem to have been clearly made in the past, it is considered that the diastrophisms, under which both Laurasia and Gondwana respectively (*a*) drifted *en masse* with mutual interference (down to the Mesozoic), and (*b*) broke up with sliding of the fragments (during the late Mesozoic and Tertiary), arose in different ways and were hence *fundamentally dissimilar*. The first was dominantly "planetary," the second "non-planetary."

To conclude, *Continental Sliding* is deemed to be essentially the outcome of cyclonic disturbances that have affected the more or less "permanent" convective circulation in the sub-crust. It is furthermore regarded not as a special process applying to a particular period or region, but as an inherent property of the crust and as having operated throughout geological time as well as over the entire globe, only for the epochs prior to about the mid-palæozoic the evidence is as yet too fragmentary for its proper decipherment.

By analogy the energy demanded for such sliding could be derived, first, from magmatic currents conforming to a cyclonic distribution, and, secondly, from sources situated within and below the moving blocks, the postulated fluxing of the leading edge and solidifying at the rear giving rise to a closed thermodynamical system of the isothermal type. Calculations will show that without some such consumption of thermal energy the Earth would today be maintaining a higher crustal temperature, so that a surplus must therefore have been available for promoting drift. It cannot, therefore, be maintained on physical grounds that crustal sliding is impossible.

Since such a principle receives the fullest of support from every other line of evidence, as set out in the preceding pages, it is insisted in conclusion that the Hypothesis of Continental Drift must now be regarded no longer as a mere speculation, but as a definitely established and fundamental truth that can brilliantly and effectively reveal the past history of our Earth.

CONCLUSION

In the foregoing pages an endeavour has been made to set forth in an elementary fashion certain individual ideas of

Earth Structure and Evolution. In no way, however, does such profess to be a general treatise on the vast subject of Drift. The incompleteness of the account as regards huge stretches of the Earth is fully recognised, as well as its imperfections from the geophysical aspect.

It is nevertheless put forward, first, as a provisional attempt to interpret the host of phenomena that do not appear to be adequately, if at all, explained under current theories of Geology, Biology or Geophysics, and, secondly, as an inducement to others to develop that important branch of the science that can appropriately be termed " Comparative Geology," that is to say, the study of continental " fragments."

Particularly to be desired are critical comparisons on somewhat similar lines to those developed here between such lands as (a) Nova Scotia, Newfoundland, British Isles and Iberia ; (b) Greenland, Labrador and the Canadian Archipelago—more especially in regard to " rifting " ; (c) Greenland, Scotland, Spitzbergen and Scandinavia ; (d) Alaska and north-eastern Siberia ; (e) West Africa and northern South America ; (f) New Guinea and North-eastern Australia. The complex tectonics of the Arctic and Antarctic and of the East and West Indies should also come in for revision from the hydrodynamical viewpoint. The " ironing out " of the various compression-zones should then enable the continental fragments to be reassembled on the artificial globe with far more precision than has been possible in this initial effort.

Only through some such methods as these, it is felt, shall we be able to arrive at a closer approximation towards the truth regarding our wonderful planet—the goal for which we are all earnestly striving.

The question remains, whether the Hypothesis is competent to reveal anything having an *economic bearing*. It can be answered that the doctrine of past Continental *rapprochement* will in certain instances demand analogies or even similarities in the mineralisations of such formerly confluent lands, as was stressed by the writer (29) in the case of Brazil and Africa. If our particular reconstruction be anywhere correct, the future discovery in the Enderby quadrant of Antarctica of rare minerals such as monazite or thorianite should occasion no surprise.

The Hypothesis of Drift will, however, find its essential application in the field of Petroleum Geology, wherein the orogenic control of sedimentation plays so vital and foretelling a part. That such special aspect is of some practical consequence can indeed be inferred from the fact that the illuminating *Symposium on Continental Drift*, owed its inception to the American Association of Petroleum Geologists, Tulsa, U.S.A.

GLOSSARY

Acid. Rocks rich in silica and poor in oxides of lime, iron or magnesia.

Amphibolite. A metamorphic rock consisting of amphibole (a variety of hornblende) with some lime-soda felspar, a little quartz and often some garnet.

Amygdaloid. A volcanic rock containing variously sized hollows—usually steam or gas cavities—now filled with secondary minerals.

Andesite. An igneous rock—usually a lava—of intermediate composition containing a lime-soda felspar and a dark mineral—pyroxene, hornblende or mica. It is the volcanic equivalent of the plutonic rock diorite.

Angara (from the Angara River). The ancient nucleus of northern Siberia.

Anticline. An arch-like upfold in stratified rocks, that dips away on both sides from its crest or axis.

Autogenous. Self-generated.

Backdeep. The depression developed on the opposite side of a fold-range to the Foredeep.

Basalt. A fine-grained, dark, heavy lava of basic character composed of lime-soda felspar and augite (pyroxene), with or without olivine. It is the volcanic equivalent of the plutonic rock gabbro.

Basic. Rock rich in oxides of lime, iron or magnesia and poor in silica.

Batholith. A plug- or dome-shaped body of plutonic rock, sometimes of great size—the lower part of which is inaccessible and is hence unknown—intruded from below into crustal rocks.

Bathyal. (*a*) Sediments of a fine grained nature and largely of a chemical origin deposited in the deeper parts of the ocean ; (*b*) life of that environment.

Charnockite. A hypersthene granite. The charnockite series is a varied group of related intrusive rocks all containing hypersthene.

Clastic. Sedimentary rock composed of fragments of minerals or rocks.

Conglomerate. A sedimentary rock composed of coarse rounded fragments set in a variable matrix.

Consequent Drainage. The pattern developed on a surface sloping nearly evenly in one direction—namely, a series of streams coursing regularly and in sub-parallel fashion down the dip.

Crystallines. Deep-seated plutonic rocks such as granite, gneiss, gabbro, amphibolite or crystalline schist.

Dacite. A more acid variety of andesite containing a little quartz.

Deeps. Oceanic areas of exceptional depth—conventionally exceeding about 5500 metres

Diastrophism (adj. diastrophic). The process of deformation of the earth's crust—by elevation, sinking, folding, faulting, etc.

Diorite. A plutonic rock of intermediate composition composed of lime-soda felspar and hornblende, brown mica or augite (pyroxene).

Disjunctive Basin. A sunk-land, sea or ocean produced essentially by crustal fracture and the drawing apart of the opposed sides.

Dolerite. An intrusive rock of basic composition and moderately coarse texture formed of lime-soda felspar, augite and sometimes olivine.

Dyke or *Dike*. A fissure extending more or less vertically through any formation and filled with an igneous rock introduced from below.

Eclogite. A heavy metamorphic rock of intermediate to basic composition formed essentially of garnet and pyroxene.

Epeirogeny or *Epeirogenesis* (*adj.* epeirogenic). The process under which extensive areas of land or sea are raised, lowered, warped or faulted, usually without appreciable folding.

Epicentre. The point on the earth's surface situated vertically above the centre of origin of an earthquake.

Eustatic (literally "well-standing") *movement*. A displacement of the strand line affecting all parts of the globe in the same sense almost simultaneously.

Facies (*adj.* facial). (*a*) The particular development of deposits of any period—littoral, neritic, lacustrine, æolian, etc. ; (*b*) The general faunal development of any period evolved within, and characteristic of, a particular region.

Fault. A break in the continuity of a bed or mass of rock or of the earth's crust with displacement along the line of fracture.

Felspar. A silicate of aluminium represented by a number of kinds in which the bases potash, soda, lime-soda or lime are respectively dominant.

Foliation. The arrangement of the minerals of a crystalline rock in roughly parallel layers.

Foredeep. In the current sense, the depression adjoining a line of fold-ranges due essentially to loading of the crust by the weight of the rising and sometimes overriding mountain mass, and hence situated between the latter and the rest of the continent.

Foreland. The little disturbed continent towards which crustal folding or over-folding has been directed.

Fossa. A geosynclinal depression developing all around a continental mass.

Gabbro. A basic plutonic rock of coarse grain composed of lime-soda felspar and augite, hornblende or olivine.

Geanticline (*adj.* geanticlinal). An up-bowed or up-arched deformation—sometimes compound—of the earth's crust along a lengthy belt.

Geosyncline (*adj.* geosynclinal). A trough-shaped depression in the earth's surface—often compound—which becomes filled by transgressional waters and for ages collects detrital sediments.

Geotherm or *Isogeotherm*. An imaginary shell beneath the earth's surface having the same temperature at every point thereon.

Gneiss. A foliated rock of the same mineral composition as granite.

Gondwana (from the Gond territory of India). The ancient southern continent of which South America, Africa, Madagascar, India, Australasia and Antarctica formed parts.

Graben (plural, Gräben). A rift valley.

Grain. The general trend of the foliation, strike of beds or direction of fold-ranges in any region—in analogy with the grain in wood.

Granite. An acid plutonic rock composed of quartz, potash felspar and mica—sometimes hornblende.

Granulite. A metamorphic rock of granular texture composed of quartz, felspar, pyroxene and garnet in varying proportions.

Gravity Anomaly. The difference between the observed and calculated value of the intensity of gravity at any point.

Horst. A raised crustal block bounded by faults or flexures.

Hypsometric Curve. The graphic representation of the proportion, to the total surface of the earth, of the areas standing at equal heights above or depths below sea-level.

Intermediate Rocks. Kinds intermediate in their chemical composition between the acid and basic types.

Isostasy (adj. isostatic). The principle under which any section of the earth's crust adjusts itself to variations in loading or unloading.

Kimberlite. Ultrabasic volcanic rock rich in magnesia and containing olivine, pyroxene, mica, ilmenite, garnet and sometimes diamond, set in a serpentinous ground-mass.

Laccolith. A lens-shaped body of plutonic rock introduced along a plane of bedding in sediments.

Laterite. A residual deposit formed through rock-weathering in a hot moist climate and generally containing some free hydrated oxides of iron and aluminium.

Laurasia. The ancient northern continent made by North America, Greenland, Europe and Asia (excluding India).

Linkage. The abrupt joining up of one line of folding with another.

Littoral. (*a*) Sediments laid down along a shore-line or the inner part of the continental shelf, and including coarse to medium grained materials; (*b*) the life of that environment.

Magma. The primitive molten matter within or below the crust, from which the various eruptive rocks have been derived.

Metamorphic. A rock in which the mineral and often the chemical composition has been altered by temperature or pressure.

Nappe. A crustal over-fold of so extreme a type that both limbs lie at low angles, one upon the other, often with the disappearance of the lower limb through stretching or sliding.

Negative Movement. The retreat of the strand line consequent upon the rising of the land.

Neritic. (*a*) Sediments of the moderately deep seas, generally of somewhat fine grained nature deposited off the continental shelf; (*b*) the life of that environment.

Orogeny or *Orogenesis (adj.* orogenic). The process of development of zones of strong deformation in the earth's crust, as contrasted with Epeirogeny.

Over-fold. Term used when folded beds have been tilted to beyond the vertical and thereby inverted. If they now dip southward the over-folding is said to be to the *north.*

Over-thrust. Term used when a part of a bed or mass of rock or part of the earth's crust has been pushed over the remainder along a fault of low angle to the horizontal.

Paralic. Sedimentary deposits partly of marine, partly of fresh-water character, and commonly alternating.

Paramorphism (adj. paramorphic). A change in mineral, without alteration in chemical, composition.

Peneplain. A surface departing but little from a plain, evolved through prolonged erosion.

Peridotite. An ultrabasic plutonic rock rich in magnesia and iron, composed of olivine and other ferro-magnesian silicates such as pyroxene, hornblende, garnet, etc.

Phase. (*a*) The equivalent of Facies; (*b*) a single large, or else a series of lesser orogenic movements closely spaced in time.

Plutonic. An igneous rock, medium to coarse in grain that has solidified with slow cooling deep below the earth's surface.

Positive Movement. The advance of the strand line upon the land consequent upon the sinking of the latter.

Pyroxene. A silicate composed mainly of iron, lime and magnesia characteristic of the intermediate and basic rocks, of which augite is the most abundant kind.

Quartz. Rock crystal—the crystalline form of silica—and one of the commonest rock-forming minerals.

Quartzite. A cemented, compact, altered sandstone.

Rhyolite. An acid lava consisting largely of glass or altered glass carrying small crystals of quartz and potash felspar.

Ria. A much-indented type of coast produced essentially by subsidence of a folded topography.

Rift Valley or *Rift.* A trough-like feature usually with flattish floor formed by the subsidence of a strip of the crust, normally between bounding faults.

Schist. A metamorphic rock that breaks into leaves or slabs owing to the abundance of flaky minerals such as mica or fibrous minerals such as hornblende.

Serpentine. An ultrabasic rock in which the constituent silicates have become changed by hydration to a greenish material.

Shield. A wide extensively worn-down platform of ancient rocks within a continent that has experienced little subsequent deformation.

Sial or *Sal* (from *Si*lica +*Al*umina). The outer, lighter and more siliceous shell of the earth.

Sima (from *Si*lica +*Ma*gnesia). The deeper, heavier and more basic matter supporting the outer shell of the earth.

Syncline (*adj.* synclinal). A trough-like fold in stratified rocks (an anticline inverted) dipping towards its lowest line or axis.

Syntaxis. The crowding together with or without merging of several lines of folding.

Taphrogeny or *Taphrogenesis.* The process through which large-scale rift valleys or gräben are formed.

Tectonic. A descriptive term applied to features in the crust produced by the forces of deformation, e.g. fold-ranges, rift valleys, etc.

Tillite. A compacted till or glacial boulder-clay belonging to some former epoch.

Virgation. The converging and ultimate merging of two or more lines of folding.

Welt. An extensive, elevated portion of the earth's crust, above or below the sea, elongated in shape.

BIBLIOGRAPHY

ALLAN, R. S. (35). The Fauna of the Reefton Beds (Devonian), New Zealand. *Geol. Surv. N.Z.*, Pal. Bull. No. 14, 1–72.

ANDREWS, E. C. (16). The Geological History of the Australian Flowering Plants. *Am. Jl. Sci.* xlii, 171–232.

(25). Structural Unity of the Pacific: Evidence of the Ore Deposits. *Ec. Geol.* 20, 707.

ANTEVS, E. (29). Maps of the Pleistocene Glaciations. *Bull. Geol. Soc. Amer.* 40, 631–720.

ARGAND, E. (24). La Tectonique de l'Asie. *Cong. Géol. Inter.* XIII, 1, 171–372.

ARLDT, T. (17). Handbuch der Paläogeographie. 1–495, *Leipzig.*

BACKLUND, H. G. (35). Zur tektonischen Gliederung Asiens. *Geograf. Ann. Sven Hedin.* 242–54.

BAILEY, E. B. (29). The Palæozoic Mountain Systems of Europe and America. Pres. Add. Sect. C, *Brit. Assoc. A.S. Glasgow* (1928), 57–76.

BAKER, H. B. (11). The Origin of the Moon. *Detroit Free Press*, April 23.

(12). The Origin of Continental Forms, II. *Mich. Acad. Sci.*, Ann. Rep., 1912, 116–41.

(13a). The Origin of Continental Forms, III. *Mich. Acad. Sci.*, Ann. Rep., 1913, 107–13.

(13b). The Origin of Continental Forms, IV. *Ibid.* 26–32.

(14). The Origin of Continental Forms, V. *Mich. Acad. Sci.*, Ann. Rep., 1914, 99–103.

(32). The Atlantic Rift and its Meaning. 1–305, *Detroit.*

(36). Structural Features crossing the North Atlantic. *Mich. Acad. Sci.* March 20. Also published as " Structural Features Crossing Atlantic Ocean." *Pan-Amer. Geol.* lxvi, 1–11.

BARACCHI, P. (14). Astronomy and Geodesy in Australia. *Handb. Brit. Assoc. A.S.*, 1914, 326–90.

BARTON, D. C., and HICKEY, M. (33). The Continental Margin at Texas-Louisiana Gulf Coast. Nat. Res. Council. *Trans. Am. Geoph. Union*, 16–20.

BEETZ, P. F. W. (34). Geology of South-West Angola, between Cunene and Lunda Axis. *Trans. Geol. Soc. S.Af.* xxxvi, 137–76. Map.

BENSON, W. N. (23). Palæozoic and Mesozoic Seas in Australasia. *Trans. N.Z. Inst.* 54, 1–62.

(24). The Structural Features of the Margin of Australasia. *Trans. N.Z. Inst.* 55, 99–137.

(26). The Tectonic Conditions accompanying the Intrusion of Basic and Ultrabasic Igneous Rocks. *Nat. Acad. Sci. Wash.* xix, Mem. No. 1, 1–90.

BERRY, E. W. (33). Carboniferous Plants interbedded in the Marine Section of Bolivia. *Am. Jl. Sci.* xxv, 49–54.

BESAIRIE, H. (30). Recherches géologiques à Madagascar. *Toulouse*, 1–272.

BILLINGS, M. (29). Structural Geology of the Eastern Part of the Boston Basin. *Am. Jl. Sci.* xviii, 97–137.

BLACK, Davidson (31). Palæogeography and Polar Shift. *Bull. Geol. Soc. China*, x, 106–57.

BÖGGILD, O. B. (17). Grönland. *Handb. Region. Geol.* iv, 2a, 1–38, Heidelberg.

BOGOLEPOW, M. (30). Die Dehnung der Lithosphäre. *Zeits. Deut. Geol. Ges.* 82, 206–28, Fig. 1–8.

BORCHERT, H. (32). Über den Werdegang der subpazifischen Schicht und verwandte Probleme. *Zeits. Deut. Geol. Ges.* 84, 761–78.

BORN, A. (33). Über Werden und Zerfall von Kontinentalschollen. *Fort. d. Geol. u. Pal.* x, 32, 347–422.

BOWIE, W. (35). Significance of Gravity Anomalies at Stations in the West Indies. *Bull. Geol. Soc. Am.* 46, 869–78.

BROOKS, C. E. P. (26). Climate through the Ages. 1–439, *London*.

BROUWER, H. A. (20). On the Crustal Movements in the Region of the Curving Rows of Islands in the Eastern Part of the East Indian Archipelago. *Kon. Akad. Wetens. Amsterdam* (1916), Proc. xxii, 772–82.

— (21). De alkaligesteenten . . . van Brasilië en Zuid-Afrika. *Kon Akad. Wetens. Amsterdam*, xxix, 1005–20.

BRYAN, W. H. (25). Earth Movements in Queensland. *Proc. Roy. Soc. Qld.* xxxvii, 2–82.

— (26). Earlier Palæogeography of Queensland. *Proc. Roy. Soc. Qld.* xxxviii, 79–102.

BRYAN, W. H., and WHITEHOUSE, F. W. (27). Later Palæogeography of Queensland. *Proc. Roy. Soc. Qld.* xxxviii, 103.

BUBNOFF, S. von (26). Geologie von Europa, I. 1–322, *Berlin*.

BUCHER, W. H. (33). The Deformation of the Earth's Crust. 1–518, *Princeton*.

BULL, A. J. (21). A Hypothesis of Mountain Building. *Geol. Mag.* 364–67.

CASE, E. C. (19). The Environment of Vertebrate Life in the Late Palæozoic in North America. Publ. No. 283, *Carnegie Inst. Wash.*

— (26). Environment of Tetrapod Life in the Late Palæozoic of Regions other than North America. Publ. 375, *Carnegie. Inst. Wash.*

CHAMBERLIN, R. T. (28). Some of the Objections to Wegener's Theory. Theory of Continental Drift : a Symposium. 83–7, *Tulsa*.

CHUBB, L. C. (34). The Structure of the Pacific Basin. *Geol. Mag.* 289–302.

COLEMAN, A. P. (26). Ice Ages : Recent and Ancient. 1–296, *London*.

COLLET, L. W. (27) ; 2nd ed. (35). The Structure of the Alps. 1–304, *London*.

CONDIT, D. D. *et al.* (36). Geology of Northwest Basin, Western Australia. *Bull. Amer. Assoc. Petr. Geol.* 20, 1028–70.

CRICKMAY, G. W. (32). Evidence of Taconic Orogeny in Matapedia Valley, Quebec. *Am. Jl. Sci.* xxiv, 368–86.

DALY, R. A. (23). The Earth's Crust and its Stability. *Am. Jl. Soc.* v. 349.

— (25). Relation of Mountain-building to Igneous Action. *Proc. Amer. Phil. Soc.* 64, 283–307.

— (26). Our Mobile Earth. 1–342, *New York*.

— (28). The Outer Shell of the Earth. *Am. Jl. Sci.* xv, 108–35.

— (30). Nature of Certain Discontinuities in the Earth. *Bull. Seism. Soc. Amer.* 20, 41–52.

— (33). Igneous Rocks and the Depths of the Earth. 1–598, *New York*.

— (34). The Changing World of the Ice Age. 1–271, *New Haven*.

DAVID, T. W. E. (14). The Geology of the Commonwealth [of Australia] *Federal Handbook, Brit. Assoc. A.S.*, 1914, 241–325.

— (32). Explanatory Notes to accompany a New Geological Map of the Commonwealth of Australia. 1–177, *Sydney*.

DAVID, T. W. E., and SÜSSMILCH, C. A. (19). Sequence, Glaciation and Correlation of the Carboniferous of the Hunter River District. *Proc. Roy. Soc. N.S.W.* 53, 246–338.

(31). The Upper Palæozoic Glaciations of Australia. *Bull. Geol. Soc. Amer.* 42, 481–522.

(36). The Carboniferous and Permian Periods in Australia. *Cong. Géol. Inter.* XVI, 1, 629–43.

DE LURY, J. S. (35). Geologic Deductions from Earthquakes of Deep Focus. *Jl. Geol.* xliii, 759–64.

DIXEY, F. (30). The Karroo of the Lower Shiré-Zambezi Area. *Cong. Géol. Inter.* XV, 2, 120–42.

(30a). A Provisional Correlation of the Karroo north of the Zambezi. *Ibid.* 143–60.

(35). The Transgression of the Upper Karroo and its Counterpart in Gondwanaland. *Trans. Geol. Soc. S.Af.* xxxviii, 73–89.

DOUGLAS, G. V. (34). On the Theory of Continental Drift. *Canadian Inst. Min. Metal.* 1–7.

DOUGLAS, J. A. (20). Geological Section through the Andes of Peru and Bolivia, II. *Quart. Jl. Geol. Soc.* lxxvi, 1–59.

ELLSWORTH, E. Lincoln (36). My Flight across Antarctica. *Nat. Geogr. Mag.* July, 1–35.

EVANS, J. A. (23). The Wegener Hypothesis of Continental Drift. *Nature,* No. 2786, 393–4.

(24). Introduction to Wegener's " The Origin of Continents and Oceans." vii–xii, *London.*

(25). Regions of Tension. *Quart. Jl. Geol. Soc.* lxxxi, lxxx–cxxii.

(26). Regions of Compression. *Ibid.* lxxii, lx–cii.

FERMOR, L. L. (13). Preliminary Note on Garnet as a Geological Barometer and on an Infra-plutonic Zone in the Earth's Crust. *Rec. Geol. Surv. India,* xliii, i, 41–7.

(14). The Relationship of Isostasy, Earthquakes and Vulcanicity to the Earth's Infra-plutonic Shell. *Geol. Mag.* 65–7.

FLEURY, E. (24). Les Plissements hercyniens au Portugal. *Cong. Géol. Inter.* XIII, i, 489–506.

FORREST, H. E. (35). The Atlantean Continent. 2nd. ed. 1–352, *London.*

FOURMARIER, P. (28). Les Traits directrices de l'évolution géologique du continent africain. *Cong. Géol. Inter.* XIV, 3, 839–85.

FOX, C. S. (31). The Gondwana System and Related Formations. *Mem. Geol. Surv. India,* lviii, 1–241.

(34). The Lower Gondwana Coalfields of India. *Ibid.* lix, 1–386.

FREBOLD, H. (35). Geologie von Spitzbergen, der Bäreninsel, des König Karl- und Franz-Joseph-Landes. 1–195, *Berlin.*

FROMAGET, J. (29). Note préliminaire sur la stratigraphie des formations secondaires et sur l'âge des mouvements majeurs en Indochine. *Bull. Serv. Géol. Indochine.* xviii, 5, 1–33.

FUJIWHARA, S., TSUJIMURA, T., and KUSAMITSU, S. (34). On the Earth-Vortex, Echelon Faults and Allied Phenomena. *Ergb. Kosm. Phys.* ii, 303–360, *Leipzig.*

FULLER, R. E. (31). The Geomorphology and Volcanic Sequence of Steens Mountain in Southeastern Oregon. *Univ. Washington,* 3, i, 1–130.

FULLER, R. E., and WATERS, A. C. (29). The Nature of the Origin of the Horst and Graben Structure of Southern Oregon. *Jl. Geol.* 37, 204–39.

GERTH, H. (32). Geologie Südamerikas, vol. i, 1–199, *Berlin.*

(35). *Ibid.* vol. ii, 200–389, *Berlin.*

GILLIGAN, A. (31). A Contribution to the Geological History of the North Atlantic Region. *Proc. Yorks. Geol. Soc.* xxi, iv. 301–21.

GLENNIE, E. A. (36). Gravity Anomalies in the United States. *Jl. Geol.*
    xliv, 765–82.
GORANSON, R. W. (28). The Density of the Island of Hawaii and Density
    Distribution in the Earth's Crust. *Am. Jl. Sci.* xvi, 89–120.
GOULD, L. M. (33). Some Geographical Results of the Byrd Antarctic
    Expedition. *Ann. Rept. Smith. Inst.* for 1932, 235–50.
    (35). Structure of the Queen Maud Mountains, Antarctica. *Bull. Geol.
    Soc. Amer.* 46, 973–84.
GRABAU, A. W. (23–4). Stratigraphy of China. Pt. 1, *Geol. Surv. China*,
    1–528.
    (24). Migration of Geosynclines. *Bull. Geol. Surv. China*, 2, 207–349.
    (28). Stratigraphy of China. Pt. 2, *Geol. Surv. China*, 1–774.
GREGORY, J. W. (13). The Nature and Origin of Fiords. 1–542, *London*.
    (21). The Rift Valleys and Geology of East Africa. 1–479, *London*.
    (28). Wegener's Hypothesis. Theory of Continental Drift: a Sym-
    posium, 93–6, *Tulsa*.
    (29). The Geological History of the Atlantic Ocean. *Quart. Jl. Geol.
    Soc.* lxxxv, lxviii–cxxii.
    (30). The Geological History of the Pacific Ocean. *Quart. Jl. Geol. Soc.*
    lxxxvi, lxxii–cxxxvi.
    (31). The Earthquake off the Newfoundland Banks of 18th November,
    1929. *Geogr. Jl.* lxxvii, 123–39.
GREGORY, J. W., *et al.* (29). The Structure of Asia. 1–227, *London*.
GROEBER, P. (27). Ensayo sobre Tectónica Teórica y Provincias mag-
    máticas. *Bol. Acad. Nac. Cien. Cordoba*, xxx, 177–229.
GUTENBERG, B. (27). Die Veränderung der Erdkrusts durch Fliessbewe-
    gungen der Kontinentalscholle. *Gerlands. Beitr. z. Geophysik.* 16,
    239–47 ; 18, 281–91.
    (33). Tilting due to Glacial Melting. *Jl. Geol.* 41, 449–67.
    (36). Structure of the Earth's Crust and the Spreading of the Continents.
    Bull. Geol. Soc. Amer. 47, 1587–1610.
HALM, J. K. E. (35). An Astronomical Aspect of the Evolution of the
    Earth. *Jl. Astron. Soc. S.Af.* iv, 1–28.
HARKER, A. (09). The Natural History of Igneous Rocks, 1–384, *London*.
    (32). Metamorphism, 1–360, *London*.
HARRINGTON, H. (34). Sobre la Presencia de Restos de la Flora de
    " Glossopteris " en las Sierras australes de Buenos Aires. *Revis. Mus.
    La Plata.* xxxiv, 303–38.
HAUGHTON, S. H. (30). The Origin and Age of the Karroo Reptilia.
    Cong. Geol. Inter. XV, ii, 252–62.
HESS, H. H. (32). Interpretation of Gravity Anomalies and Sounding
    Profiles in the West Indies. *Trans. Amer. Geoph. Union*, 26–33.
    (33). Interpretation of Geological and Geophysical Observations. The
    Navy-Princeton Gravity Expedition to the West Indies in 1932. *U.S.
    Hydrographic Office*, 27–54.
HEWITT, J. (23). Remarks on the Distribution of Animals in South Africa.
    S.Af. Jl. Sci. xx, 96–123.
    (25). Facts and Theories on the Distribution of Scorpions in South Africa.
    Trans. Roy. Soc. S.Af. xii, 249–76.
HEYL, G.R. (36). Princeton University Contribution to the Geology of
    Newfoundland. Bull. 3, *St John's*.
HINDS, N. E. A. (34). The Jurassic Age of the Last Granitoid Intrusives in
    the Klamath Mountains and Sierra Nevada, California. *Am. Jl. Sci.*
    xxvii, 182–92.
HOBBS, W. H. (23). The Asiatic Arcs. *Bull. Geol. Soc. Amer.* 34, 243–52.
HÖGBOM, A. G. (13). Fennoskandia. *Handb. reg. Geol.* IV. 3, 1–197,
    Heidelberg.

HOLLAND, T. H. (15). Presidential Address. *Rep. Brit. Assn. A.S.*, 1914, 344–58.
(33). The Geological Age of the Glacial Horizon at the Base of the Gondwana System. *Quart. Jl. Geol. Soc.* lxxxix, lxiv-lxxxvi.
HOLMES, A. (26). Contributions to the Theory of Magmatic Cycles. *Geol. Mag.* 306–29.
(29). A Review of the Continental Drift Hypothesis. *Min. Mag.* 40, 205, 286 and 340.
(30). Petrographic Methods and Calculations. 2nd ed. 1–515, *London.*
(31). Radioactivity and Earth Movements. *Trans. Geol. Soc. Glasgow,* xviii, 559–606.
(33). The Thermal History of the Earth. *Jl. Wash. Acad. Sci.* 23, 4, 169–95.
HOLMES, A., and PANETH, F. A. (36). Helium-Ratios of Rocks and Minerals from the Diamond Pipes of South Africa. *Proc. Roy. Soc. Lond.* Ser. A, 154, 385–413.
HOLTEDAHL, O. (20). Paleogeography and Diastrophism in the Atlantic-Arctic Region during Paleozoic Time. *Am. Jl. Sci.* xlix, 1–25.
(21). The Scandinavian " Mountain Problem," *Quart. Jl. Geol. Soc.* lxxvi, 387–402.
(25). Some Points of structural Resemblance between Spitsbergen and Great Britain and between Europe and North America. Videns. Akad. 4, 1–20, *Oslo.*
IRMSCHER, E. (22). Pflanzenverbreitung und Entwicklung der Kontinente. *Mitt. Inst. allgem. Botan.* 5, 15–235, *Hamburg.*
JEFFREYS, H. (29). The Earth: its Origin, History and Physical Constitution. 2nd ed. 1–346, *Cambridge.*
JESSEN, O. (36). Reisen und Forschungen in Angola. 1–397, *Berlin.*
JOLY, J. (23). The Movements of the Earth's Surface Crust. *Phil. Mag.* Ser. 6, xlv, 270 ; xlvi, 271.
(24). Radioactivity and the Surface History of the Earth. 1–40, *Oxford.*
(25). The Surface History of the Earth. 1–192, *Oxford.*
(28). Continental Movement. Theory of Continental Drift. 88–9, *Tulsa.*
KEIDEL, (16). La Geológia de las Sierras de la Provincia de Buenos Aires y sus relaciones con la montanas de Sud África y los Andes. *An. Minist. Agric. Argent.* XI, num. 3.
KEITH, A. (23). Outlines of Appalachian Structure. *Bull. Geog. Soc. Amer.* 34, 309–80.
(28). Structural Symmetry of North America. *Ibid.* 39, 321–85.
KERFORNE, F. (24). Contribution a l'étude de la tectonique du massif armoricain. *Cong. Géol. Inter.* XIII, 1, 557–63.
KILIAN, C. (25). Essai de synthèse de la géologie du Sahara sud constantinois et du Sahara central. *Cong. Géol. Inter.* XIII, 2, 887–944.
KING, P. B. (32). An Outline of the Structural Geology of the United States. Guidebook 28, *Cong. Géol. Inter.* XVI.
KING, W. W. (34). The Downtonian and Dittonian Strata of Great Britain and North-Western Europe. *Quart. Jl. Geol. Soc.* xc, 526–67.
KOBER, L. (21) ; 2nd ed. (28). Der Bau der Erde, 1–500, *Berlin.*
(23). Bau und Entstehung der Alpen. 1–283, *Berlin.*
(33). Der Orogentheorie. 1–184, *Berlin.*
KOCH, L. (26). A new Fault Zone in Northwest Greenland. *Am. Jl. Sci.* xii, 301–10.
(35). Geologie von Grönland. 1–159, *Berlin.*
KÖPPEN, W., and WEGENER, A. (24). Die Klimate der geologischen Vorzeit. 1–255, *Berlin.*
KREICHGAUER, D. (02) ; 2nd ed. (26). Die Äquatorfrage. 1–394, *Steyl.*

KRENKEL, E. (22). Die Bruchzonen Ostafrikas. *Berlin.*
  (25). Geologie Afrikas, I. 1–461, *Berlin.*
  (28). *Ibid.* II. 463–1000, *Berlin.*
KRIGE, L. J. (26). On Mountain Building and Continental Sliding. *S.Af.*
  Jl. Sci. xxiii, 206–15.
  (30). Magmatic Cycles, Continental Drift and Ice Ages, *Proc. Geol. Soc.*
  S.Af. 32, xxi-xl.
KUENEN, P. H. (35). Geological Results. The Snellius-Expedition in the
  Eastern Part of the Netherlands East Indies. 1929–1930, v, i. 1–124.
  (Reviewed by Schuchert).
LAKE, P. (22). Wegener's Displacement Theory. *Geol. Mag.* 59, 338–46.
  (23). Wegener's Hypothesis of Continental Drift. *Nature,* Feb., 226–28.
  (33). Gutenberg's Fliesstheorie : a Theory of Continental Spreading.
  Geol. Mag. 70, 116–21.
LEE, J. S. (29). Some Characteristic Structural Types in Eastern Asia and
  their Bearing upon the Problem of Continental Movements. *Geol.*
  Mag. 76, 358–75 ; 457–73 ; 501–22.
LEE, W. T. (23). Building of the Southern Rocky Mountains. *Bull. Geol.*
  Soc. Amer. 34, 285–308.
LEES, G. M. (28). The Geology and Tectonics of Oman and of Parts of
  South-eastern Arabia. *Quart. Jl. Geol. Soc.* lxxxiv, 585–670.
LEME, A. B. P. (29). État des connaissances géologiques sur le Brésil. *Bull.*
  Soc. Géol. France. xxix, 1–2, 35–87.
LEMOINE, P. (13). Afrique occidentale. *Handb. region. Geol.* VII, 6A.
  1–80, *Heidelberg.*
  (16). Madagascar. *Ibid.* VII, 4, *Heidelberg.*
LEUCHS, K. (16). Zentralasien. *Ibid.* V, 7, 1–138, *Heidelberg.*
LONGWELL, C. R. (23). Kober's Theory of Orogeny. *Bull. Geol. Soc.*
  Amer. 34, 231–242.
  (28). Some Physical Tests of the Displacement Hypothesis. Theory of
  Continental Drift. 145–57, *Tulsa.*
  (35). Is the " Roots of Mountains " Concept Dead ? *Am. Jl. Sci.*
  xxix, 81–92.
LYDEKKER, R. (11). " Zoological Distribution." *Encyl. Brit.* 11th ed.
  1002–18.
MAACK, R. (34). Gondwanaschichten in Süd-Brasilien und ihre Bezie-
  hungen zur Kaokoformation Südwestafrikas. *Zeits. Ges. Erdk. Berlin,*
  194–222.
MATLEY, C. A. (26). The Geology of the Cayman Islands. *Quart. Jl.*
  Geol. Soc. lxxxii, 352–86.
MAWSON, D. (34). The Kerguelen Archipelago. *Geogr. Jl.* lxxxiii,
  18–27.
MEINESZ, F. A. V. (31). Gravity Anomalies in the East Indian Archipelago.
  Geogr. Jl. lxxvii, 323–32.
  (33). Ergebnisse der Schwerkraft: Beobachtungen auf dem Meere in den
  Jahren 1923–1932. *Gerlands. Beitr. Erg. Kosm. Phys.* supp. 2, ii.
MEISNER, O. (18). Isostasie und Küstentypus. *Peter. Mitt.* 64, 221.
MELTON, F. (30). Review of " Theory of Continental Drift." *Jl. Geol.* 38,
  284–6.
MERTIE, J. B., Jun. (30). Mountain Building in Alaska. *Am. Jl. Sci.* xx,
  101–24.
MEYRICK, E. (25). Wegener's Hypothesis and the distribution of the Micro-
  Lepidoptera. *Nature,* 95, 834–5.
MICHAELSEN, W. (22). Die Verbreitung der Oligochäten im Lichte der
  Wegenerschen Theorie. *Verh. natur. Ver. Hamburg.* 1–37, 1921.
MOUTA, F., and O'DONNELL, H. (33). Notice explicative carte géologique
  de l'Angola. *Minist, Colonias. Lisboa,* 1–87, Map.

MUSHKETOV, D. J. (36). Modern Conceptions of the tectonics of Central Asia. *Cong. Géol. Inter.* XVI, 2, 885–94.

NEVIN, C. M. (31). Principles of Structural Geology. 1–303, *New York.*

NISSEN, H. (34 ?). The Origin of the Moon. 1–156, *Minneapolis.*

NOPSCA, F. (34). The Influence òf Geological and Climatological Factors on the Distribution of non-marine Fossil Reptiles and Stegocephalia. *Quart. Jl. Geol. Soc.* xc, 76–140.

NORIN, E. (30). An Occurrence of late palæozoic Tillite in the Kuruk-Tagh Mountains, Central Asia. *Cong. Géol. Inter.* XV, ii, 74–6.

OBST, E. (13). Der östliche Abschnitt der grossen östafrikanischen Störungzone. *Mitt. geogr. Ges. Hamburg*, 27, 187.

ODELL, N. E. (33). The Mountains of Northern Labrador, (part 2). *Geogr. Jl.* lxxxii, 315–24.

OPPENHEIM, V. (34). Rochas Gondwanicas è Geologia do Petroleo do Brasil Meridional. Bol. N. 5. *Dep. Nac. Produc. Miner.* 1–129, Rio.

PARSONS, E. (29). The Origin of the Great Rift Valleys as Evidenced by the Geology of Coastal Kenya. *Trans. Geol. Soc. S.Af.* xxxi, 63–96.

PEACOCK, M. A. (35). Fiord-Land of British Columbia. *Bull. Geol. Soc. Amer.* 46, 633–96.

PENCK, A. (21). Wegeners Hypothese der kontinentalen Verschiebungen. *Zeits. Ges. Erdk. Berlin.* 1921, 110–20.

PEKERIS, C. L. (35). Monthly notices. *Roy. Astrom. Soc.* Geoph. Suppl. iii. 343–67.

PICKERING, W. H. (97). The·Place of Origin of the Moon. *Jl. Geol.* 15, 23–38.

(24). The Separation of the Continents by Fission. *Geol. Mag.* 61, 31–4.

PJETURSS, H. (10). Island. *Handb. region. Geol.* IV., i, 1–22, *Heidelberg.*

PRATJE, O. (28). Beitrag zur Bodengestaltung des Südatlantischen Ozeans. *Cent. f. Min. G. u. Pal.* Abt. B. 3, 129–52.

RASTALL, R. H. (29). On Continental Drift and Cognate Subjects. *Geol. Mag.* 66, 447–56.

REED, F. R. C. (21). The Geology of the British Empire. 1–480, *London.*

RICHARZ, S. (28). Problems of the Equator. *Pan.-Amer. Geol.* xlix, 21–34.

SAGUI, C. K. (32). La Teoria di Wegener; sulla deriva dei Continenti. *Boll. Soc. Geol. Ital.* li, 248–52.

SAHNI, B. (35). Permo-Carboniferous Life Provinces, with Special Reference to India. *Current Science*, iv, 385–90.

(36). Wegener's Theory of Continental Drift in the Light of Palæobotanical Evidence. *Jl. Indian. Bot. Soc.* xv, 319–32.

SALOMON-CALVI, W. (33). Die permokarbonischen Eiszeiten. 1–156, *Leipzig.*

SANDFORD, K. S. (26). The Geology of North-East Land (Spitzbergen). *Quart. Jl. Geol. Soc.* lxxxii, 615–65.

SAYLES, R. W. (14). The Squantum Tillite. Bull. 66, *Mus. Comp. Zool. Harvard*, 141–75.

SCRIVENOR, J. B. (31). The Geology of·Malaya. 1–217, *London.*

SCHUCHERT, C. (23). Sites and Nature of the North American Geosynclines. *Bull. Geol. Soc. Amer.* 34, 151–230.

(26). Stille's Analysis and Synthesis of the Mountain Structures of the Earth. *Am. Jl. Sci.* xii, 277–92.

(28). The Hypothesis of Continental Displacement. Theory of Continental Drift, 104–44, *Tulsa.*

(29). Geological History of the Antillian Region. *Bull. Geol. Soc. Amer.* 40, 337–59.

(30). Cretaceous and Cenozoic Continental Connections according to von
Huene. *Am. Jl. Sci.* xix, 55–66.
(32). Gondwana Land Bridges. *Bull. Geol. Soc. Amer.* 43, 875–916.
(36). Geologic Interpretation of the Bathymetry of the East Indian Archipelago. *Am. Jl. Sci.* xxxii, 292–7.
(36a). Historical Geology of the Antillean-Caribbean Region. 1–811,
New York.

SCHUCHERT, C., and DUNBAR, C. O. (34). Stratigraphy of Western
Newfoundland. Mem. 1, *Geol. Soc. Amer.* 1–123.

SEWARD, A. C. (24). The later Records of Plant Life. *Quart. Jl. Geol. Soc.*
lxxxi, lxi–xcvii.
(30). Botanical Records of the Rocks. *Rept. Brit. Assn.,* 1929. 199–216.
(32). Carboniferous Plants from Sinai. *Quart. Jl. Geol. Soc.* lxxxviii,
350–57.

SEWELL, R. B. S. (34). The John Murray Expedition to the Arabian Sea.
Nature, 133, 669–72 ; 134, 685–8.

SHERLOCK, R. L. (34). Notes on the Amazon. *Geol. Mag.* 70, 112–16.

SMIT SIBINGA, G. L. (27). Wegener's Theorie en het ontstaan van den
oostelijken O. J. Archipel. *Tijds. Kon. Ned. Aardrijks. Gen.* xliv.
pt. 5, 581–98.

SÖRGEL, W. (16). Die atlantische Spalte : kritische Bemerkungen zu
A. Wegeners Theorie der Kontinentalverschiebung. *Zeits. Deuts.
Geol. Gesell.* Monatsb. 68, 200–39.

STAUB, R. (28). Der Bewegungsmechanismus der Erde. 1–270, *Berlin.*
(28a). Gedanken zum Strukturbild Spaniens. *Cong. Géol. Inter.* XIV,
3, 949–96.

STEERS, J. A. (32). The Unstable Earth. 1–341, *London.*

STILLE, H. (24). Grundfragen der vergleichenden Tektonik. 1–443,
Berlin.

STOCKLEY, G. M. (36). A Further Contribution on the Karroo Rocks of
Tanganyika Territory. *Quart. Jl. Geol. Soc.* xcii, 1–31.

SUESS, E. (85 ; 88 ; 01). Das Antlitz der Erde. *Wien.* Translated as
" The Face of the Earth," Parts I–V (04 ; 06 ; 08 ; 09 ; 24), *Oxford.*

SUESS, F. E. (36). Europäische und nordamerikanische Gebirgszusam-
menhänge. *Cong. Géol. Inter.* XVI, 2, 815–28.

SÜSSMILCH, C. A. (25). The Carboniferous Period in Eastern Australia.
Rept. Aust. and N.Z. Assn. Adv. Sci. 1935, 83–118.

TABER, S. (22). The Great Fault Troughs of the Antilles. *Jl. Geol.,* 30,
89–114.
(25). The Active Fault Zones of the Greater Antilles. *Cong. Géol. Inter.*
XIII, ii, 731–6.
(27). Fault Troughs. *Jl. Geol.,* 35, 577–606.

TAMS, E. (21). Über die Fortpflanzungsgeschwindigkeit der seismischen
Oberflächenwellen längs kontinentaler und ozeanischer Wege. *Centr. f.
Min. G. u. Pal.,* 1921, 44–52 ; 75–83.

TAYLOR, F. B. (10). Bearing of the Tertiary Mountain Belt on the Origin
of the Earth's Plan. *Bull. Geol. Soc. Amer.* 21, 179–226.
(23). The Lateral Migration of Land Masses. *Proc. Wash. Acad. Sci.*
13, 445–47.
(25). Movement of Continental Masses under the Action of Tidal Forces.
Pan-Amer. Geol. xliii, 15–50.
(26). Greater Asia and Isostasy. *Amer. Jl. Sci:* xii, 47–67.
(28). Sliding Continents and Tidal and Rotational Forces. Theory of
Continental Drift. 158–77, *Tulsa.*
(28a). North America and Asia ; a Comparison in Tertiary Diastrophism. *Bull. Geol. Soc. Amer.* 39, 985–1000.

TERMIER, P. (25). The Drifting of the Continents. *Ann. Rept. Smiths. Inst. for* 1924, 219–36.

TERRA, H. DE. (36). Himalayan and Alpine Orogenies. *Cong. Géol. Inter.* XVI, 2, 859–71.

TILLEY, C. E. (36). Enderbite. *Geol. Mag.* 73, 312–16.

TING YING, H. (36). On the Devonian Equator located by the Growth Rate of Tetracorals. *Jl. Geol. Soc. Japan*, 43, 353–56; *Trans. Proc. Pal. Soc. Japan*, 3, 43–9.

TOIT, A. L. DU (21). Land Connections between the other Continents and South Africa in the Past. *S.Af. Jl. Sci.* xviii, 120–40.

(21a). The Carboniferous Glaciation of South Africa. *Trans. Geol. Soc. S.Af.* xxiv, 188–217.

(27). A Geological Comparison of South America with South Africa. *Carnegie Inst. Wash.*, Publ. 381, 1–157.

(29). Some Reflections upon a Geological Comparison of South Africa with South America. *Proc. Geol. Soc. S.Af.* xxxi, xix-xxxviii.

(29a). The Volcanic Belt of the Lebombo. *Trans. Roy. Soc. S.Af.* xviii, 189–217.

TOLLNER, H. (34). Astronomische Ortsbestimmungen auf Jan Mayen. *Sitzb. Akad. Wiss. Wien.*, Math-Nat. Kl., IIa, 143, 304.

TRASK, P. D. (36). Review of Schuchert's "Historical Geology of the Antillean-Caribbean Region. . . ." *Ec. Geol.* xxxi, 770–73.

TYRRELL, G. W. (15). The Petrology of South Georgia. *Trans. Roy. Soc. Edin.* 50, iv, 823–36.

(31). The South Sandwich Is. *Discovery Reports*, iii, 191–98, Cambridge.

VEATCH, A. C. (35). Evolution of the Congo Basin. Mem. No. 3, *Geol. Soc. Amer.* 1–183.

VERSLUYS, J. (34). The Distribution of Marine Animals and the History of the Continents. *Rept. Brit. Assn. A.S.*, for 1934, 317.

VOITESTI, I. P. (28). Contributions à la connaissance de l'extension des Nummulites de grande taille, dans les régions carpathiques. *Cong. Géol. Inter.* XIV, 3, 1143–72.

VON HUENE (25). Die südafrikanische Karroo-Formation als geologisches und faunistisches Lebensbild. *Forts. Geol. Pal.* 12, 1–124.

(29). Los Saurisquios y Ornithisquios del Cretaceo argentino. *An. Mus. La Plata*, III, 2a, 1–196.

VON IHERING, H. (27). Geschichte des Atlantischen Ozeans. *Jena.*

(31). Land-Bridges across the Atlantic and Pacific Oceans during the Kainozoic Era. *Quart. J. Geol. Soc.* lxxxvii, 376–91.

WADE, A. (14). Madagascar and Its Oil Lands. *Jl. Inst. Petr. Tech.* 15, 2–29.

WADIA, D. N. (19). 2nd ed. (26). Geology of India for Students, 1–400, *London.*

(31). The Syntaxis of the North-West Himalaya: its Rocks, Tectonics and Orogeny. *Rec. Geol. Surv. Ind.* lxv, 190–220.

WAGER, L. R. (33). The Form and Age of the Greenland Ice Cap. *Geol. Mag.* 70, 145–56.

WAGNER, P. A. (28). The Evidence of the Kimberlite Pipes on the Constitution of the Outer part of the Earth. *S.Af. Jl. Sci.* 25, 127–48.

WARD, L. K. (25). Notes on the Geological Structure of Central Australia. *Trans. Roy. Soc. S.Aus.* xlix, 61–84.

WARRING, C. B. (87). The Evolution of Continents. *Trans. Vassar. Bros. Inst. New York.* IV, ii, 256–71. See *Jl. Geol.* xlii, 223, 1934.

WASHINGTON, H. S. (23). Comagmatic Regions and the Wegener Hypothesis. *Jl. Wash. Acad. Sci.* 13, 339–47.

WATERSCHOOT VAN DER GRACHT, W. A. J. M. (28). The Problem of Continental Drift. The Theory of Continental Drift. 1–75 ; 197–226, *Tulsa.*

(31). The Permo-Carboniferous Orogeny in the South-Central United States. *Verh. Kon. Akad. Wet. Amster.* II, xxvii, 3, 1–170.

(33). De Laat-Palæozoische Plooïingsphase in Noord-Amerika. *Tijds. Kon. Ned. Aardr. Gen.* 1, 6, 903–29.

WATTS, W. W. (35). Form, Drift, and Rhythm of the Continents. *Rept. Brit. Assn. A.S.,* 1935, 1–21.

WAYLAND, E. J. (30). Rift Valleys and Lake Victoria. *Cong. Géol. Inter.* XV, ii, 323–53.

(30a). The African Bulge. *Geogr. Jl.,* 75, 381–3.

WEGMANN, C. E. (35). Preliminary Report on the Caledonian Orogeny in Christian X's Land. *Meddel. om Gronland.* 103, 3, 1–59.

WEGENER, A. (12). Die Entstehung der Kontinente. *Peterm. Mitt.,* 185–195 ; 253–56 ; 305–9.

(12a). Die Entstehung der Kontinente. *Geol. Rundsch.* 3, iv, 276–92.

(15). Die Entstehung der Kontinente und Ozeane. 1–94, *Braunschweig*; 2nd ed. (30), 1–135 ; 3rd ed. (22), 1–144 ; 4th ed. (29), 1–231.

(24). The Origin of Continents and Oceans. (English translation by J. G. A. Skerl.) 1–212, *London.*

(28). Two Notes concerning my Theory of Continental Drift. Theory of Continental Drift. 97–103, *Tulsa.*

WEST, W. D. (34). Some Recent Advances in Indian Geology (2), *Current Science,* iii, 185–8 ; 286–91.

(37). Earthquakes in India. *Pres. Add. Sec. C. Twenty-fourth Indian Sci. Congr.* 1937.

WHITE, D. (24). Gravity Observations from the Standpoint of the Local Geology. *Bull. Geol. Soc. Amer.* 35, 207–78.

WILLARD, B. (35). Devonian Ice in Pennsylvania. *Jl. Geol.* xliii, 214–19.

WILLIAMS, A. F. (32). The Genesis of the Diamond, I, 1–352 ; II, 353–636, *London.*

WILLIS, B. (29). Continental Genesis. *Bull. Geol. Soc. Amer.* 40, 281–336.

(32). Isthmian Links. *Ibid.* 43, 917–52.

(32a). Radioactivity and Theorizing. *Am. Jl. Sci.* xxiii, 193–226.

(36). East African Plateaus and Rift Valleys. Pub. No. 470, *Carnegie Inst. Wash.* 1–358.

WINDHAUSEN, A. (31). Geología Argentina (Segunda Parte), 1–646, *Buenos Aires.*

WING EASTON, N. (21). Het onstaan van den maleischen Archipel, bezien in het licht van Wegener's hypothesen. *Tijds. Kon. Neder. Aard. Gen.* 38, iv, 484–512.

(21a). On some Extensions of Wegener's Hypothesis and their Bearing upon the Meaning of the Terms Geosynclines and Isostasy. *Ver. Geol.- Mijn. Gen. Neder.,* Geol. Ser. v, 113–33.

WOODRING, W. P. (24). Tertiary History of the North Atlantic Ocean. *Bull. Geol. Soc. Amer.* 35, 425–35.

WOODWORTH, J. B. (23). Cross Sections of the Appalachians in Southern New England. *Bull. Geol. Soc. Amer.* 34, 253–62.

GENERAL INDEX

NOTE.—*v.* = *vide*

INDEX OF AUTHORS

CPSIA information can be obtained
at www.ICGtesting.com
Printed in the USA
LVHW081124131221
706047LV00002B/28

9 781014 139290